Philosophy of Engineering and Technology

Volume 30

The Philosophy of Engineering and Technology book series provides the multifaceted and rapidly growing discipline of philosophy of technology with a central overarching and integrative platform. Specifically it publishes edited volumes and monographs in: the phenomenology, anthropology and socio-politics of technology and engineering the emergent fields of the ontology and epistemology of artifacts, design, knowledge bases, and instrumentation engineering ethics and the ethics of specific technologies ranging from nuclear technologies to the converging nano-, bio-, information and cognitive technologies written from philosophical and practitioners perspectives and authored by philosophers and practitioners. The series also welcomes proposals that bring these fields together or advance philosophy of engineering and technology in other integrative ways. Proposals should include: A short synopsis of the work or the introduction chapter. The proposed Table of Contents The CV of the lead author(s). If available: one sample chapter. We aim to make a first decision within 1 month of submission. In case of a positive first decision the work will be provisionally contracted: the final decision about publication will depend upon the result of the anonymous peer review of the complete manuscript. We aim to have the completework peer-reviewed within 3 months of submission. The series discourages the submission of manuscripts that contain reprints of previous published material and/or manuscripts that are below 150 pages / 75,000 words. For inquiries and submission of proposals authors can contact the editor-in-chief Pieter Vermaas via: p.e.vermaas@tudelft.nl, or contact one of the associate editors.

More information about this series at http://www.springer.com/series/8657

Sven Ove Hansson
Editor

Technology and Mathematics

Philosophical and Historical Investigations

Springer

Editor
Sven Ove Hansson
Division of Philosophy
Royal Institute of Technology (KTH)
Stockholm, Sweden

ISSN 1879-7202 ISSN 1879-7210 (electronic)
Philosophy of Engineering and Technology
ISBN 978-3-030-06722-9 ISBN 978-3-319-93779-3 (eBook)
https://doi.org/10.1007/978-3-319-93779-3

Softcover re-print of the Hardcover 1st edition 2018

This Springer imprint is published by the registered company Springer Nature Switzerland AG
The registered company address is: Gewerbestrasse 11, 6330 Cham, Switzerland

Preface

The relationship between mathematics and the natural sciences has been subject to much discussion and investigation from both historical and philosophical points of view. Since I work at a technological university whose major educational task is to teach and train future engineers, I have plenty of opportunities to observe and reflect on the equally interesting relationship between mathematics and technology. But when searching the literature, I found very little on the topic.

It does not take much reflection to realise that this gap in the literature needs to be filled. Already in pre-literate times, craftspeople depended on their mathematical acumen. Early makers of bronze and other mixtures must have understood the notion of proportions. The weaving of fabrics with beautiful symmetrical patterns required considerable mastery of geometry, and so did many of the ancient and medieval building projects. In modern times, the role of mathematics in technology has been further strengthened. Since the nineteenth century, engineering relies heavily on mechanics, electrodynamics, and other mathematics-based physical theories. Conversely, mathematics depends increasingly on electronic computing. There have been substantial philosophical discussions on computer-mediated proofs and, of course, on the notion of computability, but the technological implications seem to have gone largely unnoticed in these deliberations.

In this book, investigations of a wide range of aspects on the technology–mathematics relationship have been brought together for the first time. Hopefully, this can inspire further studies. There is much more to be done in this area!

I would like to thank the publisher and the series editor Pieter Vermaas for their strong support and the contributing authors for their dedication and all the work they have put into this project.

Stockholm, Sweden
February 9, 2018

Sven Ove Hansson

Contents

About the Authors

Selmer Bringsjord is professor of cognitive science, computer science, logic and philosophy, and management and technology, at Rensselaer Polytechnic Institute (RPI), where he is director of the Rensselaer AI and Reasoning Laboratory. He obtained the PhD in philosophy from Brown University and BA in Philosophy at the University of Pennsylvania. He specialises in the logico-mathematical and philosophical foundations of artificial intelligence (AI) and cognitive science, and in collaboratively engineering AI systems with human-level cognition, where the engineering is based on novel computational formal logics inspired in large part by Leibniz. (His chief collaborator is co-author-here Naveen Sundar Govindarajulu.) Bringsjord is the author of *What Robots Can & Can't Be*, *Artificial Intelligence and Literary Creativity: Inside the Mind of Brutus, A Storytelling Machine* (with David Ferrucci), *Superminds: People Harness Hypercomputation*, and numerous additional books and papers, most of which, alas, are unpublished.

Michael E. Cuffaro is a postdoctoral research fellow of the Rotman Institute of Philosophy at the University of Western Ontario, and an external member of the Munich Center for Mathematical Philosophy at LMU Munich. His research focuses on the philosophy of physics, the philosophy of computing, as well as on the history of and more general questions within the philosophy of science and mathematics. He is co-editor of the interdisciplinary volume: *Physical Perspectives on Computation, Computational Perspectives on Physics* (Cambridge University Press, 2018) and co-author of the *Stanford Encyclopedia of Philosophy* entry on "Quantum Computation". He has published in such journals as: *The British Journal for the Philosophy of Science*, *Philosophy of Science*, *Studies in History and Philosophy of Modern Physics*, *History of Philosophy Quarterly*, and *Politics Philosophy and Economics*.

Mauro Dorato teaches philosophy of science in the Department of Philosophy, Communication and Media Studies at the University of Rome "Roma Tre". He is member of the Academy of Europe and of the Académie Internationale de Philosophie des Sciences. He has served in the steering committee of the European

Philosophy of Science Association, was president of the Italian Society for History and Philosophy of Science, and is currently co-editor of the series *College Publication*. He is the author of over 100 publications in international and Italian journals and has published three books in Italian and two books in English, *Time and Reality* (Clueb, Bologna, 1995) and *The Software of the Universe* (Ashgate, 2005). In addition to the philosophy of time, he has worked on scientific realism, the metaphysics of quantum mechanics, and laws of nature.

Laura Felline is a postdoc in philosophy of science in the Department of Philosophy, Communication and Media Studies at the University of Rome "Roma Tre". She works on mechanistic explanation, structural explanation and the foundations of physics.

Naveen Sundar Govindarajulu is senior research scientist at Rensselaer Polytechnic Institute (RPI) and assistant director of the Rensselaer AI and Reasoning Laboratory. He obtained his PhD in computer science from RPI and MSc in physics and BE in electrical and electronics engineering from the Birla Institute of Technology and Science, Pilani (BITS-Pilani). He has performed fundamental research in a wide range of research labs. His experience includes natural-language processing research done at Yahoo Research, image recognition at HP Labs, prototyping fingerprint-recognition systems for the Indian Space Research Organization, and modelling fundamental physics problems at the Tata Institute for Fundamental Research (one of India's premier physics research institutes). His research at RPI has focused on logic-based AI; achievements include a game for crowd-sourcing the solving of computationally hard (specifically NP-complete and $\boldsymbol{\Sigma}_1$) problems through easy-to-play games, and building real-time sophisticated logic-based reasoning systems for robotic systems, including ethical reasoning systems.

Sven Ove Hansson is professor of philosophy in the Department of Philosophy and History at the Royal Institute of Technology, Stockholm. He is member of the Royal Swedish Academy of Engineering Sciences (IVA) and was President of the Society for Philosophy and Technology in 2011–2013. He is editor-in-chief of *Theoria* and of the two book series *Outstanding Contributions to Logic* and *Philosophy, Technology and Society*. He is the author of more than 350 refereed articles and book chapters in logic, epistemology, philosophy of science and technology, decision theory, the philosophy of risk, and moral and political philosophy. His recent books include *The Ethics of Risk* (2013), *The Role of Technology in Science. Philosophical Perspectives* (edited, 2015), *The Argumentative Turn in Policy Analysis. Reasoning About Uncertainty* (edited with Gertrude Hirsch Hadorn, 2016), *The Ethics of Technology. Methods and Approaches* (edited, 2017), and *Descriptor Revision. Belief Change Through Direct Choice* (2017).

Ronald Kline is Bovay professor of history and ethics of engineering at Cornell University, where he holds a joint appointment between the Science and Technology

Studies Department and the School of Electrical Engineering. He is past president of the Society for the History of Technology, which awarded him Leonardo da Vinci Medal in 2016, and past president of the IEEE Society for Social Implications of Technology. He has published numerous articles on the intellectual, cultural, and social history of technology, and three books: *Steinmetz: Engineer and Socialist* (1992), *Consumers in the Country: Technology and Social Change in Rural America* (2000), and *The Cybernetics Moment: Or Why We Call Our Age the Information Age* (2015), all with Johns Hopkins University Press.

Wolfgang Lenzen was professor of philosophy at the University of Osnabrück, Germany, from 1981 until his retirement in 2011. He has published books on philosophy of science (*Theorien der Bestätigung*, 1972), philosophical logic (*Recent Work in Epistemic Logic*, 1978; *Glauben, Wissen und Wahrscheinlichkeit*, 1980), Leibniz (*Das System der Leibnizschen Logik*, 1990; *Calculus Universalis*, 2004), and applied ethics (*Liebe, Leben, Tod*, 1999; *Sex, Leben, Tod und Gewalt* 2013). His current fields of research are philosophy of mind and history of logic. In addition to the academic works, he has published a collection of reports about his personal experiences in long-distance running, bicycling, and mountaineering (*Magische Ziele*, 2007) as well as a book about the adventures of bicycling around the world (*Das letzte magische Ziel*, 2016).

Mark Priestley is an independent researcher in the history and philosophy of computing. He has worked as a computer programmer and was a principal lecturer at the University of Westminster, where he was head of the Department of Software Engineering for a number of years. He is currently on the editorial board of the *IEEE Annals of the History of Computing*. He has written a number of papers on the history of computing, and especially programming, including *When Technology Became Language* (with David Nofre and Gerard Alberts, 2014), which was awarded the inaugural SIGCIS Mahoney Prize in 2015 for an outstanding article in the history of computing and information technology. His recent books include *A Science of Operations* (2011), which was awarded a Special Commendation in the 2013 Fernando Gil International Prize in Philosophy of Science, and *ENIAC in Action* (with Tom Haigh and Cripsin Rope, 2016).

Tor Sandqvist is an associate professor of philosophy in the Department of Philosophy and History at the Royal Institute of Technology, Stockholm. His research publications include papers on proof theory, semantic justifications of classical and intuitionistic logic, belief revision, and the analysis of counterfactuals. He teaches logic, philosophy of mathematics, and philosophy of science, and also takes an interest in meta-ethics and computability theory.

Doron Swade (PhD, C.Eng, FBCS, CITP, MBE) is an engineer, historian, and museum professional. He is a leading authority on the life and work of Charles Babbage and was responsible for the successful construction of the first complete Babbage calculating engine built to the original nineteenth-century designs. He was curator of computing at the Science Museum, London, and later assistant director and head of collections. He studied physics, electronics, engineering, philosophy of science, machine intelligence, and history at various universities including Cambridge University and University College London. He has published three books (one co-authored) and many articles on the history of computing, curatorship, and museology. He is currently researching Babbage's mechanical notation at Royal Holloway University of London. He was awarded an MBE for services to the history of computing in 2009.

Sara L. Uckelman is lecturer in logic and philosophy of language at Durham University. She did her PhD in logic at the University of Amsterdam, writing on *Modalities in Medieval Logic*. Her logical research is focused on modern modal and dynamic logics and the ways in which these can be used to inform our understanding of developments in logic in the eleventh-fourteenth centuries. More philosophically, she is interested in questions of semantics and metaphysics arising from fictional discourse, especially the study of fictional languages. By night, she is a writer of speculative fiction, and her short stories have been published by Hic Dragones and *Pilcrow & Dagger*. When not pursuing any of these activities, she can often be found doing medieval re-enactment with her husband and daughter, and serving as the Managing Editor of the *Dictionary of Medieval Names from European Sources*.

Phil Wilson is an applied mathematician and senior lecturer in the School of Mathematics and Statistics, University of Canterbury, Aotearoa New Zealand. He has a PhD in mathematics from University College London, where he worked on theoretical fluid dynamics. Phil held two postdoctoral research positions at the University of Tokyo, working on mathematical modelling of red blood cells. He has been interested in the philosophical implications of applied mathematics for his entire career, and has combined this interest with his passion for popularising science by exploring philosophical issues in his published popular science writing. His recent research includes studies of the flow of wind in cities, the interaction between neuron metabolism and blood flow in the brain, the lift off and transport of dust on Mars, the clogging of diesel generators following volcanic eruptions, and the theory of detecting vortices in fluid flow.

Sandy Zabell is a professor of mathematics and statistics at Northwestern University. One of his research interests is the history and philosophical foundations of probability and statistics. In recent years he has been particularly interested in the history of cryptography during World War II, and has written several papers about Alan Turing and the use of statistics at Bletchley Park.

Part I
Introductory

Chapter 1
Introduction

Sven Ove Hansson

Abstract This is a brief introduction to a multi-author book that provides both historical and philosophical perspectives on the relationship between technology and mathematics. It consists mainly in summaries of the chapters that follow. The books has three main parts: The Historical Connection, Technology in Mathematics, and Mathematics in Technology.

Mathematics and technology are closely knit together in several ways. Most obviously, modern technology would be unthinkable without mathematics. Engineers receive a much more thorough mathematical education than most other professions, and that is because they need it. Present-day technology is largely based on scientific theories such as solid and fluid mechanics, electrodynamics, thermodynamics, and quantum mechanics, all of which require considerable mathematical training. Engineers often also need additional mathematical tools, for instance for simulation, optimization, and statistical analysis.

The relationship between technology and mathematics is a reciprocal one. Technology needs mathematics, but mathematics also needs technology. When computing power has increased, so has the mathematical use of computers. Mathematicians use them not only for calculations, but also for numerous other tasks, including the search for proofs and validations. Furthermore, the very notion of computability has a central role at the foundations of mathematics. What we can compute is in important respects a technological question. Therefore, issues from the philosophy of technology come to light in studies of the foundations of mathematics.

But in spite of all these connections, very few studies have focused on how the two disciplines are related. This book is the first broad investigation of their interrelations. The chapters that follow will show how mathematics and technology

S. O. Hansson (✉)
Division of Philosophy, Royal Institute of Technology (KTH), Stockholm, Sweden
e-mail: soh@kth.se

S. O. Hansson (ed.), *Technology and Mathematics*, Philosophy of Engineering and Technology 30, https://doi.org/10.1007/978-3-319-93779-3_1

have influenced each other throughout human history, and continue to do so today. They will also show that the technology-mathematics connection gives rise to a multitude of philosophical issues in need of further investigations.

1.1 The Historical Connection

A series of six chapters puts focus on various aspects of the historical connections between mathematics and technology. Chapter 2, *Mathematics and Technology Before the Modern Era*, reaches back to the Palaeolithic age, featuring a tally stick about 11,000 years old that testifies to an early practice of the art of counting. In many preliterate societies, the most advanced mathematical activities were carried out by weavers, who were predominantly female. Cloths with intricate geometrical patterns are known from indigenous cultures around the world. To produce them, number series had to be constructed on the basis of geometrical insights. Considerable geometrical knowledge was also involved in the great building projects of ancient and medieval times. The complex geometrical patterns on the walls and ceilings of medieval Islamic buildings bear witness to a high level of mathematical proficiency, as do the rose windows of Gothic cathedrals from the same period. In both cases, ruler-and-compass constructions were used. We do not know if craftspeople picked up this technique from learned geometers, or if it was the other way around. A few contacts between mathematicians and mathematically-minded craftspeople have been documented, but the extent and contents of such contacts cannot be inferred from the available sources.

In Chap. 3, *Computation in Medieval Western Europe*, Sara Uckelman introduces the history of computation from the seventh century to the beginning Renaissance, focusing on three major intellectual developments. The first of these is the calendar calculations in the seventh to ninth centuries that we know from Irish and English sources. In order to solve practical and ecclesiastical problems, such as ensuring that Easter was celebrated at the right time, careful calculations were necessary, and they had to be based on as precise astronomical observations as possible. New developments in Arab astronomy were essential for the accuracy of these calculations. In the thirteenth to fifteenth centuries, a new calculatory tradition was developed by scholars studying physics in an Aristotelian tradition. Contrary to the calendric calculations, these studies were not based in monasteries but in the more secular environment of the universities. Scholars at Oxford (the Oxford Calculators) took the lead. They developed precise notions of velocity, acceleration, and other important concepts in mechanics, and showed how these concepts could be used in mathematical accounts of natural phenomena. The third development described in this chapter is Ramon Llull's (c.1232–c.1315) use of mathematical principles for drawing logical conclusions from a set of premises. His basic ideas were combinatorial, and he used templates with movable parts to perform his derivations. His constructions may seem simplistic to a modern reader, but they were far from trivial to his contemporaries, and they were highly influential in the Renaissance and the early modern period.

Gottfried Leibniz's (1646–1716) impressive contributions to several aspects of computation are the subject of Chap. 4 by Wolfgang Lenzen, *Leibniz and the Calculus Ratiocinator*. Leibniz invented a "four species" calculating machine, i.e. a machine capable of all the four basic arithmetic operations: addition, subtraction, multiplication and division. He also realized the potential of the binary system of numbers, and invented two types of calculating machines for binary numbers. Calculations had a central role in his philosophy. He believed that it would be possible in principle to calculate infallibly the truth-value of any proposition. This would require a universal language (characteristica universalis) in which all concepts were expressed in a way that mirrored their logical interrelations. Leibniz seems to have believed this to be possible; for instance he proposed that God could have created the world by creating numbers that correspond to various properties of the actual world. But in practice he came to focus on how logical validity can be determined by calculative methods. Lenzen walks us through some of the logical writings by Leibniz that precursed ideas to be developed in the centuries that followed, including modal logic, quantifiers, and a rudimentary set theory. Many of the ideas that have shaped modern computing are foreshadowed in various places in his publications and manuscripts.

Doron Swade's Chap. 5, *Mathematics and Mechanical Computation*, begins with a brief history of mechanical calculation, including the technical problems that made it difficult even in the late nineteenth century to construct and manufacture reliable calculating machines. His focus is on the pioneering work of Charles Babbage (1791–1871), who invented two general-purpose computational machines, the Difference Engine and the programmable Analytical Engine. Neither of these impressive constructions was built until a Difference Engine was completed in 1991 under Swade's direction for the Science Museum in London. Babbage promoted his machines as means to produce reliable mathematical tables, a task of considerable practical importance at the time. However, he also outlined how computing machines could be used to solve equations for which no analytical solution was available. Babbage foresaw that computation by machine would lead to the development of new forms of mathematical analysis. Many of the major principles of modern computer programming can be found in his work and in that of his friend and ally Ada Lovelace (1815–1852). Their achievements illustrate what Swade calls a "two-way relationship between mathematics and machine": On the one hand, the machine was based on mathematical principles that had been developed previously to organize the work of human computists. On the other hand, the technological principles inherent in the machine inspired new mathematical ideas.

Modern computers are general-purpose machines. We usually take them to be constructed as such, but that has not always been the case. In Chap. 6, *The Mathematical Origins of Modern Computing*, Mark Priestley investigates the construction of two key machines in the pioneering period of electronic computing, the ENIAC and the EDVAC. They were both developed in the USA in the 1940s. The ENIAC was made for calculating missile trajectories and the EDVAC for processing wind tunnel data. Both tasks require the solution of large systems of differential equations. This involves multiple repetitions of small sequences of mathematical operations, each of which employs numerical results from its predecessors. Priestley shows how these historical contingencies "deeply affected the ways in which computers could be deployed in areas outside of mathematics". In the design of hardware, swift retrieval of stored intermediate results was more important than fast input or output operations. Both the hardware and the software were constructed to facilitate the use of techniques that had been established in the management of large-scale manual calculation tasks, namely the division of complex tasks into a large number of simple subtasks. (Charles Babbage had used the same strategy.) In the 1950s, when computers started to be used for other tasks than mathematical calculations, new programming methods had to be introduced for these new purposes.

Sandy Zabell begins Chap. 7, *Cryptology, Mathematics, and Technology*, by noting that cryptology provides "an ideal case study of the synergy between mathematics and technology". He divides the history of cryptology into four major phases. In the first of these, which lasted until the end of World War I, the vast majority of cryptographic systems were based on manual encrypting and decrypting. Only a very limited mathematical or technological competence was usually needed for either constructing or cracking a cipher or code. The second period was the era of encoding and decoding machines that were based on mechanical or electromechanical principles. The most famous of these was the German *Enigma* that was deciphered during the Second World War by ingenious Polish and British cryptanalysts, among the latter Alan Turing. In this period, cryptography became thoroughly mathematized. The third era, starting in the early 1970s, was marked by the introduction of digital computers for encryption and decryption. They made it possible to employ more advanced codes and to change cryptographic systems without having to replace physical equipment. In the same decade, cryptography shifted into the fourth and still on-going era, that of public key systems. These are cryptographic systems, based on number theory, that do not require prior exchange of a secret key over a secure means of communication. This is the mathematics-based technology that is used today on a massive scale for financial transactions and secure messaging over the Internet.

1.2 Technology in Mathematics

Four chapters discuss the impact of computer technology on current and future mathematics. The first of them is devoted to computer-mediated proofs and the other three to various aspects of computations and computability.

According to the four colour theorem, you never need more than four colours to colour the regions of a map on a Euclidean plane so that no two regions with a common border (other than a corner) have the same colour. This was proved in 1976 by letting a computer check through a large number of cases. The proof was too long for a human to verify all its details. It triggered an extensive and still on-going philosophical discussion on whether we can rely on such a proof in the same way that we rely on a proof that is short enough for a human to go through in detail. In Chap. 8, *The Epistemology of Computer-Mediated Proofs*, Selmer Bringsjord and Naveen Sundar Govindarajulu generalize this discussion, asking what level of belief a human is justified in having in a conclusion based on some argument, if her access to the conclusion and the argument is mediated by a computer. In this more generalized form, the question is accessible to a detailed philosophical analysis that distinguishes between different types of proofs and other arguments, different types of computers and computer mediation, and different types of belief and knowledge. This results in a framework that makes it possible to answer questions about the epistemic status of computer-mediated proofs in a more nuanced way than previously.

The relationship between technological computations and the mathematical concept of computability provides one of the best avenues to studies of the technology–mathematics relationship. Chapter 9 by Sven Ove Hansson, *Mathematical and Technological Computability*, begins by showing how modern studies of computability connect with a long tradition of attempts to convert all forms of mathematical reasoning into routine manipulation of symbols. In the early twentieth century, many mathematicians believed that such a conversion had been achieved through recent advances in the rigorization and formalization of mathematical proofs. In 1936 Alan Turing proposed a simple procedure – now called the Turing machine – which he claimed would be able to perform all symbol manipulations (computations) that human beings can perform by strictly following a set of unambiguous instructions. The chapter explains in some detail why this bold claim is a highly plausible one. It also discusses some of the sketches that have been made of technological devices with a computational capacity surpassing that of a Turing machine. Many of these proposals refer to physical events that would not normally be counted as computations. It is argued that computations are technological processes into which an intelligent being enters an input, and receives an output. This would exclude many of the schemes for computing devices that are said to surpass the capacity of a Turing machine.

Quantum computation is based on information theoretical accounts of quantum mechanics, and in order to understand the former we need to understand the latter. In Chap. 10, *On Explaining Non-Dynamically the Quantum Correlations via Quantum Information Theory: What It Takes*, Mauro Dorato and Laura Felline introduce some of the major philosophical issues involved in quantum information. They do this from the perspective of an influential information-theoretical axiomatization of quantum theory that was proposed by Clifton, Bub, and Halvorson in 2003. This approach describes the physical world in terms of how information is transferred and transformed. The authors put focus on the concepts of an explanation and a "structural explanation". For instance, Einstein's postulation of a curved space-time makes gravity a part of the structure of the universe and therefore not subject to truly causal explanations. Does quantum information theory make non-locality and entanglement structural in the same sense? This is still an open question, but it is clarified in important respects in this chapter.

In spite of significant progress, quantum computing is still far from practical use, but it has given rise to extensive philosophical discussions. Previous speculations that quantum computation could transcend the limits of Turing computability have not been substantiated in a more detailed analysis. Instead, discussions have increasingly turned to issues of computational complexity, i.e. (to put it simply) how fast the computational resources required to compute $f(n)$ for a given function f and a natural number n increase with n. In Chap. 11, *Universality, Invariance and the Foundations of Computational Complexity in the Light of the Quantum Computer*, Michael Cuffaro discusses the possibility of a "quantum speed-up", i.e. that quantum computers may outperform classical computers (technically: that they may perform better in solving certain mathematically and technologically significant problems). One of the implications of this would be that computational complexity theory would have to pay more attention to machine-specific issues. Current discussions of computational complexity usually refer to a level of abstraction that makes all computational models equivalent, since each of them can efficiently simulate each of the others. Investigations of quantum computation may lead to an increased focus on questions concerning certain classes of computers, rather than all computers. However, in Cuffaro's view this is not as radical a break with current computational complexity theory as some might think. As he sees it, complexity theory is "at its core, a practical science" that applies idealized mathematical concepts to improve our understanding of actual operations performed on real-world computers. The analysis of quantum computing serves to remind us of the actual purpose of this "conceptual bridge between the study of mathematics and the study of technology".

1.3 Mathematics in Technology

The last section of the book consists of four chapters on the role of mathematics in technology. The first of them highlights the differences between mathematical modelling in technology and in the social sciences by investigating a historical example of transdisciplinary transfer of modelling techniques. It is followed by a chapter that describes a conflict in the late nineteenth century over the extent and nature of mathematics teaching in the education of engineers. The last two chapters discuss the "unreasonable effectiveness" of mathematics in empirical applications.

Mathematical control theory and its engineering applications in servomechanisms have been essential for the control of steam and combustion engines, airplanes, turbines, and many other technologies. In Chap. 12, *Mathematical Models of Technological and Social Complexity*, Ronald Kline investigates the attempts made in the decades following World War II to extend this engineering approach to complex social phenomena. Herbert Simon (1916–2001) applied servomechanism theory to the optimization and control of production in a manufacturing unit.

Others applied these ideas in economics, political science, sociology, anthropology, and psychology. The American engineer Jay Forrester (1918–2016) constructed large models of complex social phenomena, using techniques from engineering. He used a multitude of numerical variables, connected non-linearly with multiple loops, to describe the workings of a social system, such as a company or a city. The resulting equation systems were way too complex for analytical treatment, but with the new tools for computerized approximation, predictions could be made about the behaviour of these systems. The most famous application was the controversial Club of Rome report *Limits to Growth* in 1972. Forrester has received much criticism for oversimplifying social phenomena and not taking results and models from the social sciences into account. As Kline himself notes, the chapter combines three approaches to the interconnectedness of mathematics and technology: "the technological origins of mathematical modelling in cybernetics and System Dynamics; the use of digital computers to create models in System Dynamics; and the conception of scientific models, themselves, as technologies".

Mathematics has been a core discipline in engineering education since its beginnings in the late eighteenth century. The introduction and early history of mathematics teaching for engineers is the starting point of Chap. 13 by Sven Ove Hansson, *The Rise and Fall of the Anti-Mathematical Movement*. However, its main focus is on a little known counter-reaction to modern mathematics among German professors in the engineering disciplines in the 1890s. This was a short-lived movement that hardly survived into the twentieth century, but it managed to achieve reductions in the mathematical curricula of several German technological colleges (now technological universities). Some members of this movement agitated for the dismissal of all mathematicians from the engineering schools. Instead, the (reduced) courses in mathematics would be taught by engineers. The movement denounced the use of abstract and rigorous methods in mathematics, preferring traditional methods that were considered to be more intuitive. Such resistance to precise methods reappeared in the 1920s and 1930s in the more ominous context of the Nazi movement for "German mathematics". Its adherents pushed for allegedly more intuitive methods in mathematics which they contrasted with the rigorous "Jewish" mathematics that dominated in academia.

In a famous speech in 1959, Eugene Wigner voiced his bafflement over the "unreasonable effectiveness of mathematics in the natural sciences". Again and again, theories from pure mathematics have turned out to be eminently useful in both science and technology. In Chap. 14, *Remarks on the Empirical Applicability of Mathematics*, Tor Sandqvist attempts to demystify the empirical effectiveness of mathematics. He focuses on what is arguably its most astonishing aspect, namely the role of mathematics in successful predictions of future events. Sandqvist treats this as a version of the philosophical problem of induction. It is amazing and possibly inexplicable, he says, that the universe exhibits regularities that allow us to predict the future on the basis of the past. However, the fact that we can use mathematics to describe these regularities does not necessarily add to the amazement. It can be explained by the observation that "the development of mathematics always takes place under the influence of simplicity considerations similar to those guiding human concept formation and inductive projections in general".

In Chap. 15, *What the Applicability of Mathematics Says About Its Philosophy*, Phillip Wilson approaches the same issue from another angle. He turns the question around and asks: What does the existence of applied mathematics teach us about the philosophy of mathematics? To answer that question he explores the four dominant traditions on the nature of mathematics: Platonism, logicism, formalism, and intuitionism. They have all mostly been discussed in relation to pure mathematics. In their modern forms, they are concerned with much the same key issues, such as the nature of numbers and sets, the status of infinite structures, and what constitutes a valid mathematical proof. Approaching these four standpoints from the perspective of applied mathematics puts them in an uncustomary context, in terms of both their ontological and their epistemological implications. Wilson concludes that although the lens of applied mathematics cannot adjudicate between these four major standpoints, it helps us to bring into focus the questions that have to be addressed when formulating and defending philosophical standpoints about mathematics.

Part II
The Historical Connection

Chapter 2
Mathematics and Technology Before the Modern Era

Sven Ove Hansson

Abstract The use of technology to support mathematics goes back to ancient tally sticks, khipus, counting boards, and abacuses. The reciprocal relationship, the use of mathematics to support technology, also has a long history. Preliterate weavers, most of them women, combined geometrical and arithmetical thinking to construct number series that give rise to intricate symmetrical patterns on the cloth. Egyptian scribes performed the technical calculations needed for large building projects. Islamic master builders covered walls and ceilings with complex geometric patterns, constructed with advanced ruler-and-compass methods. In Europe, medieval masons used the same tools to construct intricate geometrical patterns for instance in rose windows. These masters lacked formal mathematical schooling, but they developed advanced skills in constructive geometry. Even today, the practical mathematics of the crafts is often based on traditions that differ from school mathematics.

Some human cultures appear to have very little mathematics. For instance, a few indigenous communities do not have the practice of counting. Just like us, they can easily distinguish between 1, 2, 3, 4 and 5 objects by direct visual impression, and just like us they see directly that 20 objects are more in number than 12 objects. However, they do not know the process of counting, and therefore their languages do not contain numbers higher than those needed to report direct visual impressions of number (Pica et al. 2004). This does not prevent them from having an otherwise advanced culture, and studies in one such community show that its members can deal with differences in numerosity in other ways than counting (Dehaene et al. 2008). The ability to count is not something we are born with. It had to be invented, and now it has to be passed on from generation to generation.

S. O. Hansson (✉)
Division of Philosophy, Royal Institute of Technology (KTH), Stockholm, Sweden
e-mail: soh@kth.se

S. O. Hansson (ed.), *Technology and Mathematics*, Philosophy of Engineering and Technology 30, https://doi.org/10.1007/978-3-319-93779-3_2

2.1 Technologies for Counting and Arithmetic

But the art of counting is known in the vast majority of human communities. We often do it with the help of one-to-one correspondences with sets of small objects such as stones, twigs, or pieces of wood. For instance, inhabitants of the Nggela Islands (part of Solomon Islands) keep track the number of guests at a feast by collecting a small item from each of them as they arrive. In many places, for instance in Borneo, Melanesia and the Philippines, knots on a string are used for counting and for keeping a record of numbers (Sizer 2000). The Incas used khipus, sets of connected knotted strings, for book-keeping and the levy of taxation (Urton and Brezine 2005; Gilsdorf 2010) (Fig. 2.1).

An even safer way to keep records of numbers is to make notches on durable objects such as bones or pieces of wood. This method is known from many parts of the world (Sizer 2000, p. 260), and it has a long history. A small bone the size of a pencil that was excavated in Congo has three columns with in total 167 tally marks (Fig. 2.2). It is about 11,000 years old, and bears witness to our ancestors' ability to write down numbers long before they could write words (Huylebrouck 1996). Other, much older, bones with notches have also been found, but their interpretation as tally marks is controversial (Vogelsang et al. 2010, p. 197; d'Errico et al. 2012, pp. 13216 and 13219; Cain 2006). In modern societies, more advanced tally sticks using a positional system for higher numbers have been used to document debts. Such tallies were still used in both England and France at the beginning of the twentieth century (Stone 1975). The use of cuts on the body to record numbers has also been reported (Lagercrantz 1973).

In the traditional Basque system for counting sheep, two of these technical means for counting were combined in a most efficient way:

> Counting invariably involves two men; one does the actual counting and one records the hundreds. The counter carries 5 small stones (or nails, or some other small item that can be easily held in the hand) and counts either silently or aloud up to 20. When he reaches 20, he transfers a stone from one hand to another, and after transferring the 5th stone, he shouts ehun! [which means 'hundred'] and the recorder makes a mark by notching a stick or piece of wood. After the last rock has been transferred to the opposite hand, the counter begins again and shifts the rocks back to his original hand, not losing count of the moving sheep. When the last sheep has passed through the passage-way, he shouts the number aloud and counts the rocks in his hand. The number said aloud is one between 1–20 and the rocks in his hand represent the multiples of 20. Thus, by combining these with the number of notches made by the other person, the total number of sheep in the band is obtained. (Araujo 1975, pp. 142–143)

These different means to record numbers – stones, knots and notches – have been reported from indigenous cultures all around the world. Similar technologies for simple arithmetic, adding and subtracting, are also widespread. Already in preliterate societies, these operations were usually performed with small objects such as stones or twigs that were moved around to represent the operation (Sizer 1991, p. 54). In many cultures, special counting-boards for arithmetic were constructed. For instance, the Incas used counting boards for their calculations (and khipus to

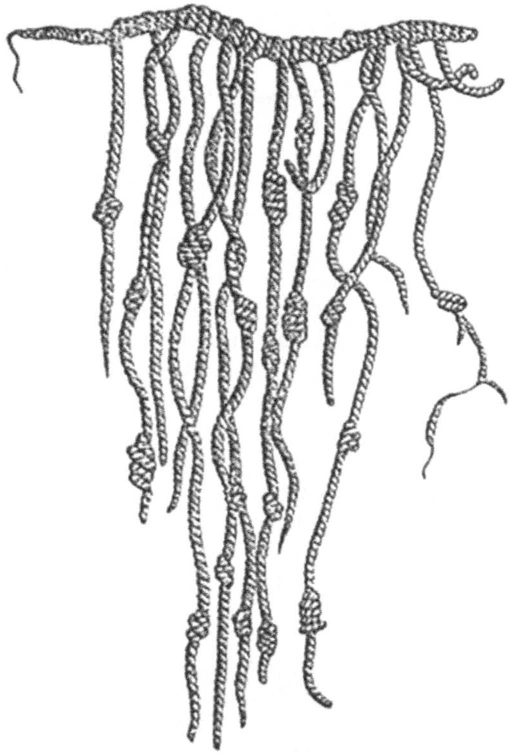

Fig. 2.1 An Inka khipu. (From Meyers Konversationslexikon, 1888)

Fig. 2.2 The Ishango bone, a Stonge Age tally stick found in Congo

record the outcome, when that was needed) (Gilsdorf 2010). In medieval Europe, before cheap paper became available, calculations were performed on an abacus or a counting-board, or in a sand tray (Acker 1994; Periton 2015). As late as the early twentieth century, writing slates were used in schools instead of paper for economic reasons (Davies 2005).

Thus, the use of tools to support arithmetic has a long history. The same is true of the reciprocal relation, the use of mathematics to support technology.

2.2 The Mathematics of Weaving

One of the foremost early uses of mathematics belongs to a traditionally female occupation, namely weaving. Textiles from about 10,000 BCE have been found in the Guitarrero Cave in northern Peru (Jolie et al. 2011), and imprints of woven material have been found in even older archaeological sites. We do not know much about Stone Age weavers, but we can see from present-day hand weaving that the craft of weaving provides excellent opportunities for developing mathematical thinking. Indigenous women all around the world have woven elaborate geometrical patterns with intricate symmetries. In order to do this, they have to combine geometric and arithmetical thinking to construct the number series and numerical relationships that give rise to the desired patterns on the fabric (Karlslake 1987, p. 394). In addition, weavers often have to calculate beforehand how much material they need for a particular piece of fabric (Figs. 2.3 and 2.4).

In traditional cultures in Central and Southern Africa, cloths with complex geometrical designs are highly valued. The women who weave them perform the most advanced mathematical activities in these societies (Gerdes 2000; Harris 1987). Similarly, textiles with symmetrical patterns, both geometrical and figurative, were much esteemed by the Incas. The construction of such patterns must have

Fig. 2.3 A Navaho weaver. (Photograph by Roland W. Reed)

Fig. 2.4 Cloth from the Shoowa people in what is now the Democratic Republic of Congo. (Courtesy of the Brooklyn Museum)

been one of the most advanced mathematical activities in their culture as well. The tradition is still alive in some Andean communities:

> [M]aster weavers called *Mamas* (a Quechua word, not the word for mother) . . . are women who most likely started weaving when they were girls and reached a high level of expertise. They are generally treated with special respect within their community. The *Mamas*' abilities in counting and understanding patterns of symmetry and in geometry are part of that expertise. The ethnomathematical aspect of this situation is this: if we asked one of these women to explain geometric or symmetry properties in terms of lines, rotations, polygons, and so forth, they probably would not explain them in such textbook-like terms. Yet, they clearly understand these mathematical concepts. The difference is that their understanding comes from the perspective of a weaver who must create a cultural product and who wants to include certain patterns. (Gilsdorf 2014, p. 9)

Mathematics is also involved in other textile-related activities such as braiding, beadwork, basketry, and the traditionally male activity of rope-making (Chahine 2013; Albanese 2015; Albanese et al. 2014; Albanese and Perales 2014; Hirsch-Dubin 2009).

With larger societies came additional mathematical activities. Clay tablets from ancient Iraq testify to extensive accounting. Mesopotamian surveyors were tasked with calculating the areas of fields with different geometric shapes (Robson 2000). From ancient Egypt, several mathematical texts have been preserved. They are actually textbooks for scribes, who seem to have received a considerable dose of mathematics as part of their education (Ritter 2000). In addition to accounting they had to perform the calculations needed in surveying and construction. Surveying was much in need due to the yearly flooding of the Nile. Each year, agricultural fields had to be reconstructed when the Nile receded. Since the area of arable land often changed after the inundation, it was often necessary to redistribute land, and then the areas of differently shaped fields had to be calculated. These calculations were also important for taxation (Barnard 2014).

Scribes were required to calculate the amount of stones and other building material that was required in the pharaoh's big construction projects. They were trained to calculate the height of a pyramid, based on its edge and how much the side slanted. These and other calculations were probably used to guide the actual construction activities. Remaining marks on some Egyptian buildings indicate that the horizontal displacement of a sloped object was used as a form of angular measurement (Imhausen 2006, p. 21). The use of such measurements must have required some understanding of geometry. In addition, calculations relating to the workforce, such as the required quantities of food and beer, had to be performed.

Most technological operations in pre-modern societies were performed by craftspeople from whom we have no written evidence. In some cases, their mathematical abilities can be inferred from the archaeological evidence. For instance, the notion of proportionality is needed to produce alloys such as bronze with reliable quality, something that was achieved in several ancient civilizations (Malina 1983). Archaeological evidence from Raqqa in eastern Syria shows that glassmakers in the early Islamic period used a chemical dilution line to optimize the properties of glass (Henderson et al. 2004). However, we do not know how they performed the calculations behind these remarkable experiments.

2.3 Geometric Wonders of the Islamic World

Fortunately, there is one group of ancient craftspeople about whom we know more than about the others, namely those engaged in building construction. This is because many of their most advanced building projects, such as the great churches and mosques, are still available for our study.

Geometrical knowledge has probably been used since preliterate times in the construction of buildings. For instance, builders in several indigenous cultures have known how to make a small house rectangular (Each pair of opposite side beams should have the same length, and then the layout should be adjusted so that the diagonals have equal length.) (Sizer 1991, p. 56). But buildings from the High and Late Middle Ages in Europe, Northern Africa, and the Middle East reveal that their builders had access to a rich tradition of much more advanced geometrical knowledge. This is perhaps most obvious from the elaborate geometrical patterns displayed on the walls and ceilings of Islamic buildings. Many of these patterns exhibit mathematically advanced symmetries. The traditional way to construct them was by ruler and compass, an art that was passed over from master to apprentice (Hankin 1925; Thalal et al. 2011) (Fig. 2.5).

Fig. 2.5 God as a master mason, using a compass when creating the world. (From Codex Vindobonensis 2554, written in France around 1250, now in the Österreichische Nationalbibliothek)

Ruler-and-compass construction is well known from Euclid (fl. 300 BCE) and other Greek geometers. It may have been a Greek invention. At any rate, the Egyptians do not seem to have known the compass (Shelby 1965). The origin of its use in the learned tradition is obscure. Plutarch claims that Plato (c.425–c.348 BCE) sharply criticized mathematicians who tried to show the truth of geometrical statements with "mechanical arrangements" that were "patent to the senses" rather than relying on pure thought (Plutarch 1917, p. 471). This has been interpreted as reprobation of constructions by means of other tools than ruler and compass (Evans and Carman 2014, pp. 151–152).

Fig. 2.6 Geometric patterns on a house in Pyrgi, made by contemporary masons using ancient ruler-and-compass methods

Was ruler-and-compass construction an invention by learned geometers, who handed it over to craftsmen needing it for practical purposes? Or was it originally a practical work method, discovered and developed by craftsmen, which men of letters transformed from a practical way to use tools to a theoretical restriction on abstract mathematical reasoning? We will probably never know which of these hypotheses is true.[1] What we do know, however, is that the method serves both purposes remarkably well. Contemporary Moroccan carpenters still construct complex geometrical patterns with the same ruler-and-compass methods that their predecessors used a millennium ago (Aboufadil et al. 2013). And in the Greek village Pyrgi, house façades are decorated with geometrical patterns made by traditional craftsmen who have learned the ruler-and-compass methods by apprenticeship (Stathopoulou 2006) (Fig. 2.6).

One of the best proofs of the mathematical proficiency of the Islamic master builders can be found in the shrine of Darbi Imam in Isfahan, Iran, which was constructed in 1453 (Fig. 2.7). It exhibits advanced tilings, which were not understood mathematically until five centuries later. Like Penrose patterns, these patterns are quasi-crystalline, which means that they fill the plane perfectly, but

[1] According to Plato, at least one Athenian stone mason, namely Socrates, was versed in the learned geometry of his time. See McLarty (2005) for a useful discussion of the geometric ideas of the Platonic Socrates.

Fig. 2.7 Tesselations in the Darbi Imam shrine

do not repeat themselves regularly like the more common types of tiling[2] (Lu and Steinhardt 2007). No documentation of the mathematical thinking behind this remarkable achievement seems to have been preserved.

2.4 Medieval Master Builders in Europe

The compass was as highly valued by Christian masons as by their Middle East colleagues. European masons were often portrayed holding a compass. They commonly used a large compass of the type that would now be called a pair of dividers, with legs ending in needle points. Contrary to the compasses used in latter-day technical drawing, it was not intended for drawing on paper. The master mason made marks directly on the building site. The compass was a useful instrument for that purpose since the layout of large buildings such as churches was based on geometrical principles (Bucher 1972). Marks were also made on the raw material for structural components, such as pieces of timber to be sawn or stones to be cut (Shelby 1965). In a few cases, setting-out marks made with a compass on a stone are still preserved and visible in the building (Branner 1960). When several similar stones or pieces of timber had to be prepared, the mason made his marks on a thin plank from which a template was cut. Large building sites such as a cathedral had a special place, a "tracing house", where these templates were kept (Shelby 1971).

[2]More precisely: They lack translational symmetry.

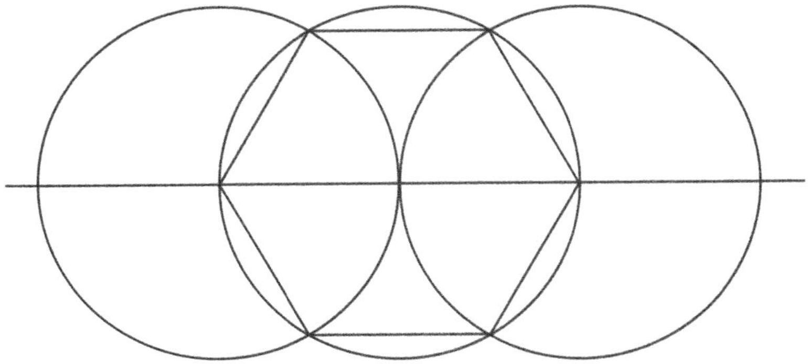

Fig. 2.8 The classical Euclidean construction of a regular hexagon. The construction begins with the horizontal line, followed by the circle in the middle and then the other two circles. This is an exact construction, i.e. any errors will depend on the physical execution, not the mathematical principles

In the early Middle Ages, master masons were usually illiterate, but beginning in the thirteenth century at least some of them learned how to read and write. However, they had no formal mathematical schooling. Their geometrical skills were transferred orally from masters to apprentices. Much of the most advanced knowledge in their craft seems to have been lost with the end of Gothic building, but a couple of master masons wrote small books in which parts of it have been preserved. These books make it clear that in their own view, geometry had a fundamental role in their craft (Shelby 1970, 1972). They explained how to construct a right angle, an equilateral triangle, a square, a pentagon, a hexagon or an octagon (Fig. 2.8). These geometrical procedures were components of the constructions used to set out marks on stones and other structures destined for various functions in a building. For instance, the construction of voussoirs (wedge-shaped stones in a vault) was particularly important, and close attention had to be paid to their geometrical proportions.

Most of these constructions were exact (in the Euclidean sense), but some were approximations. One example of the latter can be found in the book *Geometria deutsch* that was published by the German master builder Matthäus Roritzer, (c.1435–c.1495) in the late 1480s. One of his constructions was a method to draw a line equally long as the circumference of a circle (Fig. 2.9). At the time, doing this exactly was an intriguing, unsolved mathematical problem. (400 years later Ferdinand von Lindemann proved a theorem from which it follows that no such construction is possible with ruler and compass.) The construction consists essentially in marking the diameter of the circle three times in a row on a line, and then adding a seventh of the diameter (which is easily constructible). This amounts to approximating π as 22/7. Roritzer paid no attention to the small error (Shelby 1972). In fact, he had a good reason not to do so, namely that the error must have

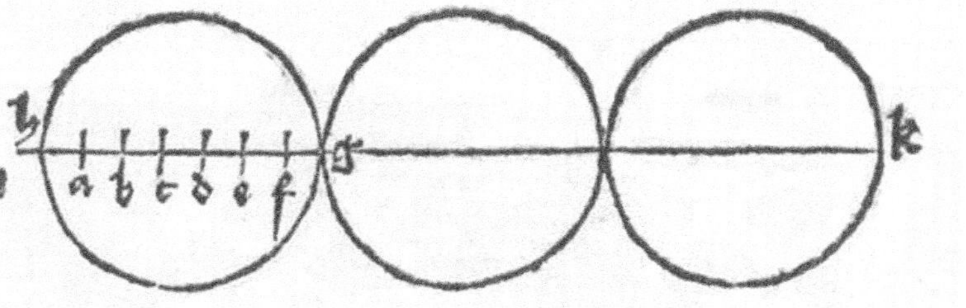

Fig. 2.9 Matthäus Roritzer's (approximate) construction of a line with the same length as the circumference of a given circle. (From his Geometria deutsch.) Presumably, his readers knew how to divide a given line, such as the diameter of a circle, into seven equal parts with a ruler and a compass

been negligible in practical applications on a building site.[3] For instance, if the task was to cut a strip of some material to be fitted around a circular shape with a diameter of one meter, then the error caused by this approximation would make the strip about 1.3 mm too long, which would almost certainly be negligible in comparison with the other uncertainties involved in such a work process.

The Gothic cathedrals had large rose windows, i.e. round windows with symmetrically arranged rib-work of stone. They were constructed with ruler and compass, and some of them had quite advanced geometrical patterns. The cathedral in Orvieto in central Italy has a large rose window in the form of a regular 22-sided polygon (icosikaidigon) on its façade (Fig. 2.10). The window was constructed in the fourteenth century. The vast majority of Gothic rose windows were based on a regular polygon that the mason could construct exactly with a compass and a straightedge, but no such construction of a 22-sided polygon was known by them (and we now know that no such construction is possible). Detailed measurements of the window indicate that it may have been constructed with a fairly advanced approximate ruler-and-compass method (Ginovart et al. 2016).

2.5 Contacts with Mathematicians?

Euclidean geometry was part of the medieval learned tradition. The way in which Euclid deduced theorems from a few basic axioms was a model not only for mathematicians but also for scholars working in other disciplines. We do not know to what extent learned geometers communicated with the craftspeople who put geometry to practical use, but a few such contacts have been documented. In his autobiography,

[3]This is an early example of the difference between technological and more theoretical ideals of precision (Hansson 2007).

Fig. 2.10 Rose window from the Orvieto cathedral

the Syriac mathematician Ibrahim ibn Sinan (908–946) recounted that he once told a technically clever craftsman how to construct a sundial (Saliba 1999, pp. 641–642). The Persian mathematician and astronomer Abu al-Wafa' Buzjani (940–c.998), who lived in Baghdad, wrote a book on the geometrical constructions that craftsmen had use for. Ruler-and-compass constructions of regular polygons were prominently featured in the book (Raynaud 2012). However, it is not know to what extent it actually reached its intended audience.

The Iranian polymath Al-Biruni (973–1048) commented on the difference between the arithmetical solutions to mathematical problems that scholars preferred and the (presumably geometrical) methods used by most craftsmen. Interestingly, he mentioned that some artisans, in particular instrument makers, preferred the arithmetical methods to those favoured by other craftspeople. If this was a common pattern, then such a minority of mathematically inclined artisans may have formed important links between learned and applied mathematics in this period. After developing a fairly complicated method for calculating the *qibla* (direction of prayer), Al-Biruni described an approximate method that should be good enough for people in the building trades who were not versed in mathematics (Saliba 1999, p. 642).

In the next century, the Saxon philosopher Hugh of Saint Victor (c.1096–1141) wrote a short treatise, *Practica Geometriae*, in which he introduced a division of geometry into two parts, called "practical" and "theoretical".

> The entire discipline of geometry is either theoretical, that is, speculative, or practical, that is, active. The theoretical is that which investigates spaces and distances of rational dimensions only by speculative reasoning; the practical is that which is done by means of certain instruments, and which makes judgments by proportionally joining together one thing with another. (Hugh of Saint Victor, quoted in Shelby 1972, p. 401).

In his discussion of practical geometry, Hugh referred to the application of geometry to surveying. At the time, the trade of surveying seems to have been less mathematically advanced than that of building construction. It was, at least predominantly, based on straight lines and right angles (Price 1955).

The Spanish scholar Dominicus Gundissalinus (c.1115–c.1190) wrote a treatise on the classification of knowledge, in which he broadened Hugh's description of practical geometry. In his treatment, it covered two categories of practitioners, namely surveyors and craftsmen:

> Craftsmen are those who exert themselves by working in the constructive or mechanical arts – such as the carpenter in wood, the smith in iron, the mason in clay and stones, and likewise every artificer of the mechanical arts – according to practical geometry. Each indeed forms lines, surfaces, squares, circles, etc., in material bodies in the manner appropriate to his art... The office of practical geometry is, in the matter of surveying, to determine the particular dimensions by height, depth, and breadth; in the matter of fabricating, it is to set the prescribed lines, surfaces, figures, and magnitudes according to which that type of work is determined. (Dominicus Gundissalinus, quoted in Shelby 1972, p. 403)

Other writers on practical geometry followed Hugh of Saint Victor in limiting their attention to surveying. However, in at least one case, craftsmen in the European Late Middle Ages received some form of mathematical education. The teacher in question was none less than the Florentine polyhistor Filippo Brunelleschi (1377–1446), who is today best known as the discoverer of the linear perspective. When overseeing the construction of the Florence cathedral, he reportedly taught masons and carpenters how to interpret construction drawings. This was new to them, since they were used to wooden models. Seemingly, the instruction included training in the mathematical principles on which these drawings were based (Knobloch 2004, p. 4).

In 1486, the above-mentioned master mason Matthäus Roritzer published a book on the geometry of his trade. He dedicated it to the bishop Wilhelm von Reichenau (1426–1496). In the preface he described the bishop as a friend and patron of "the free art of geometry", and related that the two had discussed this topic on many occasions (Roriczer 1845, p. 13).

2.6 Mathematics on the Offensive

In the sixteenth century, the use of mathematics increased in several sectors of European societies. Perhaps most importantly, sea journeys to other continents required improvements in navigation that could only be achieved by mathematical means. Already in 1508, the Spanish *Casa de Contratación*, which oversaw overseas trade, introduced exams to make sure that navigators were proficient in the mathematical art of navigation (Keller 1985, p. 357). In countries where Church property was confiscated there was also an increased need of surveying. The introduction of triangulation made it possible to draw more accurate maps, but it also raised the demands on the mathematical skills of surveyors. (ibid, p. 358) In addition, several attempts were made to solve technical problems with the help of mathematics. For instance, new fortifications were increasingly based on geometrical design principles (Knobloch 2004).

To meet the increased demand for mathematics, a new group of professionals presented themselves in the early Renaissance: the mathematical practitioners. They were men with a university education and training in mathematics, who offered their services in all areas where mathematics was needed, including navigation, surveying, and fortification (Cormack 2006). Many of them wrote vernacular textbooks in arithmetic and geometry, at least in part intended for craftspeople and other members of what we would today call technological occupations. In the prefaces of such textbooks, as well as other publication venues, the usefulness of mathematics was proclaimed much more emphatically than what had been common previously.

In 1543, the Italian mathematician Niccolò Fontana Tartaglia (c.1499–1557) published the first translation of Euclid into Italian. In the preface he offered a list of the applications of geometry, including building construction, surveying and geography, painting, and the construction of war machines and fortifications (Keller 1985, p. 350). Eight years later, Robert Recorde (c.1512–1558), who was one of the first mathematical practitioners in England, wrote a poem in praise of practical geometry, which he included in the preface of his textbook in the subject (Fig. 2.11):

> The Shippes on the sea with Saile and with Ore,
> were firste founde, and styll made, by Geometries lore
> Their Compas, their Carde their Pulleis, their Ankers,
> were founde by the skill of witty Geometers.
> To sette forth the Capstocke, and eche other parte,
> wold make a greate showe of Geometries arte.
> Carpenters, Caruers, Joiners and Masons,
> Painters and Limners with such occupations,
> Broderers, Goldesmithes, if they be cunning,
> Must yelde to Geometrye thanks for their learning (Stedall 2012, p. 65).

Fig. 2.11 Wall tablet commemorating Robert Recorde in St. Mary's Church in Tenby, Wales, close to where he was born. In addition to being an eminent mathematics educator, he was the inventor of the equals sign (=) and a prominent physician (Courtesy to Richard Hagen, Brisbane)

In an Italian treatise on geometry, published by Giovanni Peverone in 1558, a similar list was offered of crafts employing mathematics. Peverone emphasized in particular that without geometry, people would not be able to solve conflicts about the division of lands. (Keller 1985, p. 350) Writing in 1567, the French humanist and logician Petrus Ramus (1515–1572) put much emphasis on the importance of mathematics in mining. He did not explain the nature of its importance, but he probably referred to the use of machines such as levers, pulleys, and screw pumps, which operate on mathematical principles. The first English edition of Euclid was published in 1571 with a preface by the mathematician John Dee (1527–c.1608), who emphasized the usefulness of geometry in all kinds of trades:

> Besides this, how many a Common Artificer, is there, in these Realmes of England and Ireland, that dealeth with Numbers, Rule, & Cumpasse: Who, with their owne Skill and experience, already had, will be hable (by these good helpes and informations) to finde out, and deuise, new workes, straunge Engines, and Instrumentes: for sundry purposes in the Common Wealth? or for priuate pleasure? and for the better maintayning of their owne estate?. (Rampling 2011, p. 138)

Dee was anxious to point out that not only geometry, but also arithmetic, was useful for practical applications. Mint masters and goldsmiths could use it when mixing metals, physicians when making compound medicines, officers when ordering the troops, and lawyers when dividing property among heirs or between divorcing spouses (Rampling 2011, p. 141).

One reason for this emphasis on the mundane practical uses of mathematics was that at this time, mathematics was often associated with occult ideas and various forms of black magic. Many mathematicians – not least John Dee – contributed to this association by being deeply involved in astrological calculations. In addition, the strange symbols and diagrams that mathematicians relished could easily be interpreted as incantations of diabolic forces. In the 1550s, zealous officials in England took mathematical books for occult treatises and consequently committed them to the flames. Drawing attention to the practical usefulness of the mathematical arts was a "rhetoric of utility", employed by advocates of mathematical education who wanted to rid the subject of its sorcerous reputation (Neal 1999. Cf. Zetterberg 1980).

When reading these panegyrics of practical mathematics, it is important to remember their rhetorical purpose. They do not necessarily convey the actual usage of mathematics in the various crafts. We should also keep in mind that these texts were written long before the introduction of universal education. Most members of the labouring classes were still illiterate, and few of them had received any formal schooling in arithmetic or other mathematical skills. The basic education in mathematics that the mathematical practitioners pleaded for is now – at least to a considerable extent – realized in most countries of the world through compulsory education.

2.7 Epilogue

But even today, in spite of school mathematics, the practical mathematics of the crafts sometimes seems to live a life of its own. It does not necessarily coincide with school mathematics, and sometimes its mathematical nature is not even realized. Masons, carpet layers, and carpenters all use geometry in their daily work, often employing methods and ideas that differ from school geometry (Moreira and Pardal 2012; Masingila 1994; Millroy 1991). The work of tailors and dressmakers is one of the best examples of this. Their craft requires mastery of concepts such as angles, parallel lines, symmetry, and proportion. Body measures are often transferred to the cloth with the help of what is essentially a coordinate system (Hancock 1996). However, the mathematical nature of these skills is seldom fully recognized. Let us give the last word to the mathematics educator Munir Fasheh. He once made an interesting comparison between himself and his mother who was a seamstress:

> While I was using math to help empower other people, it was not empowering for me. It was, however, for my mother, whose theoretical awareness of math was completely undeveloped. Math was necessary for her in a much more profound and real sense than it was for me. My illiterate mother routinely took rectangles of fabric and, with few measurements and no patterns, cut them and turned them into beautiful, perfectly fitted clothing for people. In 1976 it struck me that the math she was using was beyond my comprehension; moreover, while math for me was a subject matter I studied and taught, for her it was basic to the operations of her understanding. In addition, mistakes in her work entailed practical consequences completely different from mistakes in my math . . . She never wanted any of her children to learn her profession; instead, she and my father worked very hard to see that we were educated and did not work with our hands. In face of this, it was a shock to me to realize the complexity and richness of my mother's relationship to mathematics. Mathematics was integrated into her world as it never was into mine. (Fasheh 1989)

References

Aboufadil, Y., Thalal, A., & Raghni, M. A. E. I. (2013). Symmetry groups of Moroccan geometric woodwork patterns. *Journal of Applied Crystallography, 46*(6), 1834–1841.

Acker, P. (1994). The emergence of an arithmetical mentality in Middle English literature. *Chaucer Review, 28*, 293–302.

Albanese, V. (2015). Etnomatemática de una Artesanía Argentina: identificando etnomodelos de trenzado. *Bolema, 29*, 493–507.

Albanese, V., & Perales, F. J. (2014). Pensar matemáticamente: una visión etnomatemática de la práctica artesanal soguera. *Revista Latinoamericana de Investigación en Matemática Educativa, 17*(3), 261–288.

Albanese, V., Oliveras, M. L., & Perales, F. J. (2014). Etnomatemáticas en artesanías de trenzado: aplicación de un modelo metodológico elaborado. *Bolema, 28*, 1–20.

Araujo, F. P. (1975). Counting sheep in Basque. *Anthropological Linguistics, 17*(4), 139–145.

Barnard, H. (2014). Surveying in Ancient Egypt. In *Encyclopaedia of the history of science, technology, and medicine in non-western cultures*. Dordrecht: Springer (online).

Branner, R. (1960). Villard de Honnecourt, Archimedes, and Chartres. *Journal of the Society of Architectural Historians, 19*(3), 91–96.

Bucher, F. (1972). Medieval architectural design methods, 800–1560. *Gesta, 11*(2), 37–51.

Cain, C. R. (2006). Implications of the marked artifacts of the Middle Stone Age of Africa. *Current Anthropology, 47*(4), 675–681.

Chahine, I. (2013). Juxtaposing form, function, and social symbolism: An ethnomathematical analysis of indigenous technologies in the Zulu culture. *Journal of Mathematics & Culture, 5*(1), 1–30.

Cormack, L. B. (2006). The commerce of utility: Teaching mathematical geography in Early Modern England. *Science & Education, 15*(2–4), 305–322.

d'Errico, F., Backwell, L., Villa, P., Degano, I., Lucejko, J. J., Bamford, M. K., Higham, T. F. G., Colombini, M. P., & Beaumont, P. B. (2012). Early evidence of San material culture represented by organic artifacts from Border Cave, South Africa. *PNAS, 109*(33), 13208–13213.

Davies, P. (2005). Writing slates and schooling. *Australasian Historical Archaeology, 23*, 63–69.

Dehaene, S., Izard, V., Spelke, E., & Pica, P. (2008). Log or linear? Distinct intuitions of the number scale in Western and Amazonian indigene cultures. *Science, 320*(5880), 1217–1220.

Evans, J., & Carman, C. C. (2014). Mechanical astronomy: A route to the ancient discovery of epicycles and eccentrics. In N. Sidoli & G. Van Brummelen (Eds.), *From Alexandria, through Baghdad: Surveys and studies in the Ancient Greek and Medieval Islamic Mathematical Sciences in Honor of J.L. Berggren* (pp. 145–174). Heidelberg: Springer.

Fasheh, M. (1989). Mathematics in a social context: Math within education as praxis versus within education as hegemony. In C. Keitel, P. Damerow, A. Bishop, & P. Gerdes (Eds.), *Mathematics, education, and society: reports and papers presented in the fifth day special programme on 'Mathematics, Education and Society' at the 6th International Congress on Mathematical Education, Budapest, 27 July – 3 August, 1988* (pp. 84–86). Paris: UNESCO.

Gerdes, P. (2000). On mathematical ideas in cultural traditions of central and Southern Africa. In H. Selin (Ed.), *Mathematics across cultures: The history of non-western mathematics* (pp. 313–343). Dordrecht: Kluwer.

Gilsdorf, T. (2010). Inca Mathematics. In H. Selin (Ed.), *Mathematics across cultures: The history of non-western mathematics* (pp. 189–203). Dordrecht: Kluwer.

Gilsdorf, Thomas E. (2014) Ethnomathematics of the Inkas. In *Encyclopaedia of the history of science, technology, and medicine in non-western cultures*. Springer. (online).

Ginovart, J. L. i, Samper, A., Herrera, B., Costa, A., & Coll, S. (2016). Geometry of the Icosikaidigon in Orvieto Cathedral. *Nexus Network Journal, 18*(2), 419–438.

Hancock, S. J. C. (1996). The mathematics and mathematical thinking of seamstresses. Paper presented at the Annual Meeting of the American Educational Research Association (New York, N.Y., 8 Apr 1996). http://files.eric.ed.gov/fulltext/ED399176.pdf.

Hankin, E. H. (1925) *The drawing of geometric patterns in saracenic art*. Memoirs of the Archaeological Survey of India, vol. 15.

Hansson, S. O. (2007). What is Technological Science? *Studies in History and Philosophy of Science, 38*, 523–527.

Harris, M. (1987). An example of traditional women's work as a mathematics resource. *For the Learning of Mathematics, 7*(3), 26–28.

Henderson, J., McLoughlin, S. D., & McPhail, D. S. (2004). Radical changes in Islamic glass technology: evidence for conservatism and experimentation with new glass recipes from early and middle Islamic Raqqa, Syria. *Archaeometry, 46*(3), 439–468.

Hirsch-Dubin, F. P. (2009). Mayan Elders, Mayan mathematics, and the weaving of resistance in Maguey Bag Production. *Journal of Mathematics and Culture, 4*(1), 63–83.

Huylebrouck, D. (1996). The bone that began the space odyssey. *Mathematical Intelligence, 18*(4), 56–60.

Imhausen, A. (2006). Ancient Egyptian mathematics: New perspectives on old sources. *Mathematical Intelligencer, 28*(1), 19–27.

Jolie, E. A., Lynch, T. F., Geib, P. R., & Adovasio, J. M. (2011). Cordage, textiles, and the late Pleistocene peopling of the Andes. *Current Anthropology, 52*(2), 285–296.

Karlslake, C. (1987). The language of woven images among the Tzotzil. *Canadian Journal of Native Studies, 7*(2), 385–397.

Keller, A. (1985). Mathematics, mechanics and the origins of the culture of mechanical invention. *Minerva, 23*, 348–361.

Knobloch, E. (2004). Mathematical methods in preindustrial technology and machines. In A. M. Gasca, M. Lucertini, & F. Nicolò (Eds.), *Technological concepts and mathematical models in the evolution of modern engineering system* (pp. 3–20). Basel: Birkhäuser.

Lagercrantz, S. (1973). Counting by means of tally sticks or cuts on the body in Africa. *Anthropos, 68*(3–4), 569–588.

Lu, P. J., & Steinhardt, P. J. (2007). Decagonal and quasi-crystalline tilings in medieval Islamic architecture. *Science, 315*(5815), 1106–1110.

Malina, J. (1983). Archaeology and experiment. *Norwegian Archaeological Review, 16*(2), 69–78.

Masingila, J. O. (1994). Mathematics practice in carpet laying. *Anthropology & Education Quarterly, 25*(4), 430–462.

McLarty, C. (2005). 'Mathematical Platonism' versus gathering the dead: What Socrates teaches Glaucon. *Philosophia Mathematica, 13*(2), 115–134.

Millroy, W. L. (1991). An ethnographic study of the mathematical ideas of a group of carpenters. *Learning and Individual Differences, 3*(1), 1–25.

Moreira, D., & Pardal, E. (2012). Mathematics in Masons' Workplace. *Adults Learning Mathematics: An International Journal, 7*(1), 31–47.

Neal, K. (1999). The rhetoric of utility: Avoiding occult associations for mathematics through profitability and pleasure. *History of Science, 37*(2), 151–178.

Periton, C. (2015). The medieval counting table revisited: a brief introduction and description of its use during the early modern period. *BSHM Bulletin: Journal of the British Society for the History of Mathematics, 30*(1), 35–49.

Pica, P., Lemer, C., Izard, V., & Dehaene, S. (2004). Exact and approximate arithmetic in an Amazonian indigene group. *Science, 306*(5695), 499–503.

Plutarch. (1917). Lives. *Loeb Classical Library* (vol. 87) (trans: Perrin, B.)

Price, D. J. (1955). Medieval land surveying and topographical maps. *Geographical Journal, 121*(1), 1–7.

Rampling, J. M. (2011). The Elizabethan mathematics of everything: John Dee's 'Mathematicall praeface' to Euclid's elements. *BSHM Bulletin: Journal of the British Society for the History of Mathematics, 26*(3), 135–146.

Raynaud, D. (2012). Abu al-Wafa' Latinus? A study of method. *Historia Mathematica, 39*, 34–83.

Ritter, J. (2000). Egyptian mathematics. In H. Selin (Ed.), *Mathematics across cultures: The history of non-western mathematics* (pp. 115–136). Dordrecht: Kluwer.

Robson, E. (2000). The uses of mathematics in ancient Iraq, 6000–600 BC. In H. Selin (Ed.), *Mathematics across cultures: The history of non-western mathematics* (pp. 93–113). Dordrecht: Kluwer.

Roriczer, M. ([1486] 1845). In A. Reichensperger (Ed.) *Das Büchlein der fialen Gerechtigkeit. Nach einem alten Drucke aus dem Jahre 1486 in die heutige Mundart übertragen und durch Anmerkungen erläutert*. Trier: Lintz.

Saliba, G. (1999). Review: Artisans and mathematicians in Medieval Islam. *Journal of the American Oriental Society, 119*(4), 637–645.

Shelby, L. R. (1965). Medieval masons' tools. II. Compass and square. *Technology and Culture, 6*(2), 236–248.

Shelby, L. R. (1970). The education of medieval master masons. *Mediaeval Studies, 32*, 1–26.

Shelby, L. R. (1971). Mediaeval masons' templates. *Journal of the Society of Architectural Historians, 30*(2), 140–154.

Shelby, L. R. (1972). The geometrical knowledge of mediaeval master masons. *Speculum, 47*(3), 395–421.

Sizer, W. S. (1991). Mathematical notions in preliterate societies. *Mathematical Intelligence, 13*(4), 53–60.

Sizer, W. S. (2000). Traditional mathematics in Pacific cultures. In H. Selin (Ed.), *Mathematics across cultures: The history of non-western mathematics* (pp. 253–287). Dordrecht: Kluwer.

Stathopoulou, C. (2006). Exploring informal mathematics of craftsmen in the designing tradition of 'Xysta' at Pyrgi of Chios. *For the Learning of Mathematics, 26*(3), 9–14.

Stedall, J. (2012). The pathway to knowledge and the English Euclidean tradition. In G. Roberts & F. Smith (Eds.), *Robert Recorde. The life and times of a Tudor mathematician* (pp. 57–72). Cardiff: Wales University Press.

Stone, W. E. (1975). The tally: An ancient accounting instrument. *Abacus, 11*(1), 49–57.

Thalal, A., Benatia, M. J., Jali, A., Aboufadil, Y., & Elidrissi Raghni, M. A. (2011). Islamic geometric patterns constructed by craftsmen working on wood. *Symmetry: Culture and Science, 22*, 103–130.

Urton, G., & Brezine, C. J. (2005). Khipu accounting in ancient Peru. *Science, 309*(5737), 1065–1067.

Vogelsang, R., Richter, J., Jacobs, Z., Eichhorn, B., Linseele, V., & Roberts, R. G. (2010). New excavations of Middle Stone Age deposits at Apollo 11 Rockshelter, Namibia: Stratigraphy, archaeology, chronology and past environments. *Journal of African Archaeology, 8*(2), 185–218.

Zetterberg, J. P. (1980). The mistaking of 'the Mathematicks' for magic in Tudor and Stuart England. *Sixteenth Century Journal, 11*(1), 83–97.

Chapter 3
Computation in Medieval Western Europe

Sara L. Uckelman

Abstract Practices that fall under the broad umbrella of 'computation' in the western European Middle Ages tend to be goal-oriented and directed at specific purposes, such as the computation of the date of Easter, the calculation of velocities, and the combinatorics of syllogisms and other logical arguments. In spite of this practical bent, disparate computational practices were increasingly built upon theoretical foundations. In this chapter, we discuss the theoretical principles underlying three areas of computation: computistics and the algorithms employed in computistics, as well as algorithms more generally; arithmetic and mathematical calculation, including the calculation of physical facts and theorems; and (possible) physical implementations of computing mechanisms.

3.1 Introduction

One cannot begin a discussion of the history of computation in the Middle Ages without first settling some definitions. What is 'computation'? What are 'the Middle Ages'? (We could also ask "What is 'history of'?", but we will forego that in the present context!) Typically, when one speaks of 'computation', one refers to the activity of a computer, i.e., mechanical and impersonal activity: A computer is a machine, and machines are (contra the hopes and dreams of some researchers in AI) unthinking (at least currently). On such a narrow view, one should immediately argue that there can be no history of computation prior to the invention of the computer, the machine which does the computation on your behalf. That would make for a very short chapter, so clearly we cannot accept this narrow view.

Instead, we do not take computation to be merely the activity of an unthinking machine but rather to cover a broader range of activities and processes which are united in their connections with calculation and reckoning. On such a view, a

S. L. Uckelman (✉)
Department of Philosophy, Durham University, Durham, UK
e-mail: s.l.uckelman@durham.ac.uk

S. O. Hansson (ed.), *Technology and Mathematics*, Philosophy of Engineering and Technology 30, https://doi.org/10.1007/978-3-319-93779-3_3

'computer' is not merely any*thing* that computes, but indeed any*one* who computes. Indeed, the word 'computer' was originally used in English to refer to people, as opposed to machines. This usage is found as early as the early seventeenth century (OED 2017b, s.v. computer). Earlier, the word for a 'person who computes' was Middle English 'compotiste' or 'compotister', found as early as the fourteenth century and deriving from Medieval Latin *compotista* (MED 2001–2014, s.v. compotiste). The Latin word *compotista* was used generally to describe any person who was a computer or calculator, as well as to pick out people doing a specific type of computation or calculation, namely the computation of the calendar. This discipline—calendar computation—was a branch of its own, known as computistics, and was of crucial importance to a society dominated by a church that needed to know when its movable feasts were to occur.

So much for computation; how about the Middle Ages? As with 'computation', we can either take a narrow or a wide view of our temporal scope. Ultimately, we do not wish to put any *termini* on our period of inquiry. Instead, we will pick out specific developments and aspects that are the most interesting for understanding the history of computation, and trace these facets rather than attempt to give a complete overview of the entire Middle Ages. Nevertheless, our temporal spread is great: Our earliest references will be to the Anglo-Irish computistic tradition and the Venerable Bede in the seventh to eighth century. We will spend extra time in the late 13th and early fourteenth century acquainting ourselves with Ramon Llull and the Merton Calculators, and then we will reach our terminus ad quem in algebraic algorithms developed in the Renaissance. By taking a concept- and procedure-oriented approach, we need not commit ourselves to a precise or exclusionary definition of the 'Middle Ages'.

Medieval computation tended to be goal-oriented, directed at specific purposes, such as the computation of the date of Easter, the calculation of velocities, and the combinatorics of syllogisms and other logical arguments. There are, of course, many reasons why one would prefer some sort of mindless method/mechanism/procedure/algorithm for such pragmatic ends: These mechanisms are both easier to retain and remember, and they reduce the possibility of error. It will come as no surprise, then, that many of our examples of 'computation' derive from contexts of educational reform.

But in spite of this practical bent that disparate medieval developments in computation shared, our interest throughout this chapter is primarily theoretical. We are interested in the principles underlying computation, rather than in the practical outcomes of computation or the tools used for performing them. As a result, we will omit from our scope geometrical constructions; practical engineering; and methods of reckoning and account—there will, alas, be rather a dearth of abacuses in this chapter. Instead, our energies will be concentrated primarily on three facets of computation: computistics and the algorithms employed in computistics, as well as algorithms more generally (3.2); arithmetic and mathematical calculation, including the calculation of physical facts and theorems (3.3); and (possible) physical implementations of computing mechanisms (3.4), with an account of one of the most important people in the history of computation prior to the invention of the computing machine—Ramon Llull.

3.2 Algorithms and Computistics

The concept or procedure most straightforwardly associated with computation is the algorithm. The word 'algorithm' is derived, via Latin *algorismus* + Greek αριϑμός 'number', from 'al-Khwārizmī', the name of a ninth-century Persian algebraist, who was responsible for many of the algorithms for solving algebraic equations that we know of today, such as the following algorithm for solving the equation $x^2 + 21 = 10x$:

> A square and 21 units equal 10 roots... The solution of this type of problem is obtained in the following manner. You take first one-half of the roots, giving in this instance 5, which multiplied by itself gives 25. From 25 subtract the 21 units to which we have just referred in connection with the squares. This gives 4, of which you extract the square root, which is 2. From the half of the roots, or 5, you take 2 away, and 3 remains, constituting one root of this square which itself is, of course, 9 (Tabak 2014, pp. 61–62).

However, 'algorithm' wasn't used to pick out the computational concept until the nineteenth century (OED 2017a, s.v. algorithm). Earlier, an 'algorithm' was simply the practice of using Arabic numerals. Johannes de Sacrobosco's *Liber ysagogarum Alchorismi*, an introduction to al-Khwārizmī's algebras and one of the earliest known Latin texts that used Hindu-Arabic numerals (Philipp and Nothaft 2014, p. 36), transformed the nature of calculation in western Europe in the Middle Ages and Renaissance. The text was written in the early part of the thirteenth century, and became part of the standard quadrivial curriculum in the universities of England, France, and northern Europe (Philipp and Nothaft 2013, p. 351).

Even though the word 'algorithm' didn't mean 'algorithm' until quite recently, medieval and Renaissance mathematicians still employed algorithms in their numerical computations. For example, Jordanus de Nemore, "one of the most important writers on mechanics and mathematics in the Latin West" (Folkerts and Lorch 2007, p. 2), wrote several *algorismus* treatises in the thirteenth century containing basic arithmetic operations as well as a procedure for the extraction of square roots using the Arabic number system (Folkerts and Lorch 2007, p. 5), although his treatises lacked the generality and sophistication of Sacrobosco's. But despite the widespread incorporation of algorithms into mathematical practice, in both the Middle Ages and the Renaissance, "the algorithms developed ... were also difficult and sometimes even counterintuitive. A lack of insight into effective notation, poor mathematical technique, and an inadequate understanding of what a number is sometimes made recognizing that they had found a solution difficult for them" (Tabak 2014, p. 60).

There is a tension between the practical or applied aspects of algorithms—algorithms are generally developed for a purpose—and their difficulty and counterintuitiveness (which was by no means restricted to the Renaissance algorithms!). We can see this tension clearly in the discipline of computistics, or the calculation of the calendar. The *computi* genre, outlining methods of computing the date of Easter, originated in Ireland in the seventh century (Philipp and Nothaft 2013, p. 348), stemming from controversies between the early Irish and English churches over how to calculate the date (Hawk 2012, pp. 44–45). But the dating of Easter was not merely an Anglo-Irish concern. As Nothaft puts it:

> For most of the Middle Ages up to the Gregorian reform of the calendar of 1582, the feasts and calendrical rhythms of Western Europe were governed by a single unified system of ecclesiastical time reckoning, which took account of the courses of both the Sun and the Moon... During the early Middle Ages, the practical necessity of instructing Christian monks and clerics in the use of these reckoning tools led to the development of a specific genre of learned text, the *computus*, which incorporated modules of knowledge from a wide variety of fields, most importantly arithmetic and astronomy, but also theology, history, etymology, medicine, and natural philosophy (Philipp and Nothaft 2014, p. 35).

Early calculations of Easter were based on the model of a 19-year lunar cycle developed in the third to fourth century and adapted for the Julian calendar (Costa 2012, p. 300; Philipp and Nothaft 2014, pp. 35–36). This model was transmitted to the West in the sixth century by Dionysius Exiguus (Philipp and Nothaft 2014, pp. 35–36), and was refined as mathematics and astronomy improved. Eventually, the calendrical calculations were overhauled in the twelfth and thirteenth centuries with the integration of Jewish and Arabic calendrical sources that developed in Iberia independently of the Christian tradition (Costa 2012, p. 301; Philipp and Nothaft 2015).

According to many modern commentators, the genre reached its apex with the Venerable Bede's *De temporum ratione* of 725 (itself an enlargement of an earlier treatise, *De temporibus*, from 703). The book included chapters on both practical topics, such as the conversion between Greek and Latin numerals,[1] as well as on more theoretical ideas, such as Bede's distinction between "the immutable cycles of natural time" and the linear time of human events (Costa 2012, p. 300). The linear time of human events requires accurate calendars founded upon astronomical observation, and thus this text can be seen as one of the first which displays computistics to be a science, including calculation as a central component. For many years Bede's text was "the undisputed milestone of Western computistics" (Philipp and Nothaft 2012, p. 14), and it significantly influenced later texts, such as Rabanus Maurus's *Liber de computo* (Hawk 2012, p. 37). But this view of Bede's texts, as the first real contribution to the field, has recently been challenged by the study of early Irish computists active in the era between Isidore and Bede (Graff 2010, p. 327). One such treatise is the Munich Computus (Warntjes 2010). This text was composed in 718–19, but was based "a substantial substratum" from 689 (Palmer 2010, p. 129). The Munich Computus is, like other texts of its type,

> principally arranged by the early medieval divisions of time, moving from the smallest units (the atom) to the passing of cycles, and ends with a brief chronicle of sorts framing the whole of human history in 532-year Easter cycles (Palmer 2010, p. 130),

and also includes material on the calculation of the leap-year day. It also compares the relative merits of different ways of calculating lunar and solar calendars, eventually favoring the Greek methods of Dionysius Exiguus over the Roman

[1] While Bede's treatise is the earliest known text to include such a conversion, cf. Hawk (2012, pp. 35, 37), it was by no means the only computistic text to incorporate such material (Cróiní 1982, pp. 283–285).

and Irish tables exemplified by Victorius of Aquitaine.[2] Another such treatise is a recently-discovered Irish computus which possibly pre-dates the Munich treatise and contains a similar comparison between the Victorian and the Dionysiac methods of reckoning (Warntjes 2005, p. 63).

From this we can see the range of application of computistics. Beyond the calculation of the date of Easter, computistics also incorporated specialized algorithms designed for computing other important aspects of the calendar, such as intervals between events. With these algorithms, not only could a skilled computist take data specific to a day and then "correctly locate a record in a long sequence of years [he] could also compute how many years had elapsed between two similarly dated events" (McCarthy 1994, p. 76). The algorithms the computist used were "mechanical but abstruse" and "well-suited to ensuring that the understanding and managing of historical records would remain the preserve of the privileged few who had been trained in the necessary computistic techniques" (McCarthy 1994, p. 76). Nothaft argues that "the Easter *computus* in which primitive algorithms (*argumenta*), memorised to perform various calendrical and chronological calculations, came to play a central role" was "the only major form of 'applied mathematics' known to early medieval scholars" (Philipp and Nothaft 2013, p. 348).

The *computus* treatises were also a means by which new developments in Arabic astronomy were translated to the West; for example, two mid-twelfth-century *computus* treatises written in southeastern Germany employ the Arabic lunar calendar (Philipp and Nothaft 2014, p. 36). The importation of the new Arabic material was necessary to rectify the defects of the 19-year cycle, which caused the standard calendar to no longer be in sync with the actual lunar phases by the end of the eleventh century, and leading "the church to celebrate Easter on the technically wrong date" (Philipp and Nothaft 2014, pp. 36–37). The introduction of the new Arabic lunar calendar allowed for more accurate calendrical computations on the basis of more precise data (Costa 2012, p. 301).

What of our claim that new aspects of computation flourished in the context of educational development and reform? Palmer notes that even to historians, computistics "can seem rather obscure and otherworldly" (Palmer 2010, p. 31). But in spite of this seeming obscurity, there is clear evidence that these treatises were spread throughout Europe. We have noted above how computistic manuscripts show a rich exchange of ideas between Ireland and England, reflecting the contemporary learning culture. On the continent, the transmission of computistic texts is associated with the desire of ninth-century Carolingians to "supplement Latin sources with attempts at Greek learning" (Hawk 2012, pp. 29, 30), and Palmer argues that "alongside the other works of the Regensburg library of the period, [computistics] is properly revealed as an integral part of early medieval learning in general" (Palmer 2010, p. 31). Hawk argues that we can see evidence for the Carolingian educational reforms in the context of contemporary glosses on Bede's *De temporum ratione*

[2]For further information on the significance of computistic texts on the development of early medieval science, see Borst (1993, 2006).

(Hawk 2012, p. 44), and this is corroborated by the wide spread of computistical manuscripts across Europe (Philipp and Nothaft 2012; Warntjes 2010). This spread is not only geographical but also temporal, with collections of computistic treatises being newly copied as late as the tenth century (Bisagni 2013–2014, p. 116).

3.3 Calculation

In this section we attend to the view of 'computation' that involves calculation. Calculation itself is manifest in many different ways. On the one hand, it can cover specific calculatory acts which result in a determinate outcome, for example, that process by which we calculate that $2 + 2 = 4$. On the other hand, it can cover a general methodological approach towards the solving of certain types of problems, whether arithmetic, philosophical, physical, or astronomical.[3]

Our discussion here jumps forward a few hundred years from the computistical texts of the previous section. The developments we cover are rooted in the foundational bedrock of Aristotelian mechanics, which entered the Latin West in the twelfth century. The new Aristotelian translations were read, disseminated, and, eventually, criticised and modified, over the course of the thirteenth to fifteenth centuries. Two trends in the study of mechanics in this period can be identified: What Murdoch calls the dynamic, Pseudo-Aristotelian approach which was "basically philosophical in character" and "dynamical in approach", but was "lacking a mathematical procedure of proof", and the Archimedean approach, which was "rigorously mathematical", but non-dynamical (Murdoch 1962, p. 122). Though the Islamic philosophers and mathematicians had already been melding these two approaches, it was not until the thirteenth century that such a mingling happened in the Latin West. This mingling resulted in the combination of the (Pseudo-)Aristotelian dynamical methods with the rigor of mathematics exhibited by the Archimedean approach. An example of this is Jordanus de Nemore's mid-thirteenth century treatise on statics, *Elementa super demonstrationem ponderum*, "in which the dynamical approach of Aristotelian physics is combined with the abstract mathematical physics of Archimedes" (Folkerts and Lorch 2007, p. 4). Texts such as Jordanus's provide the foundation of the general application of calculatory methods for problem solving.

The locus of this transformation of Aristotelian logic and natural philosophy was the universities, which were the primary site of the reception and dissemination of the new Aristotelian translations. The thirteenth century saw the rise of doctrinal conflicts between Aristotle's views and orthodox catholic doctrine, with the result that by the end of that century, the study of Aristotelian natural philosophy was concentrated within the Arts masters, with the discussion of any question theological in nature restricted to the theologians. (Thus, the secular nature of the computistic developments in the Aristotelian tradition can be distinguished

[3]For computational aspects of astronomy, see Chabas and Goldstein (2014) and McCluskey (1998).

with the ecclesiastical embedding of computistics.) By the fourteenth century, the secular study of Aristotle was well-embedded, and was clearly reflected in the works of a group of philosophers, logicians, and mathematicians working at the University of Oxford, known as the 'Oxford Calculators'.[4] As a group, the works of the Calculators are marked by an approach to problems of velocity, infinity, continuity, proportion, movement, etc., that combines calculatory methods with logic. Their achievements include "exact definitions of uniform motion and uniform acceleration [and] a proper grasp of the notion of instantaneous velocity" (Murdoch 1962, p. 123). Among the people who were either members of the Calculators or associated with them are Richard Kilvington (c. 1302–1361),[5] Thomas Bradwardine (c. 1295–1349), William Heytesbury (before 1313–1372/3), John Dumbleton (†c. 1349), Richard Swyneshed (†1355), Richard Billingham (fl. 1340s–1350s), Thomas Buckingham (†1349), and Roger Swyneshed (c.1335–c.1365).

We do not at present have the opportunity to survey all of the relevant works and results produced by these men, and so will content ourselves with highlighting some of their specific contributions to the mathematicization of physics and natural philosophy. In 1328, Thomas Bradwardine wrote a treatise *De proportionibus velocitatum in motibus*, which was later printed at Paris in 1495 and at Venice in 1505. In this treatise, he "devised a mathematical formula to establish the relationship between the force applied to an object, the resistance to its motion, and the velocity that results" and he also "speculated that in a vacuum, objects of different weights would fall at the same speed" (Wagner and Briggs 2016, p. 173). The same law appears, in more than 50 different mathematical versions, in Richard Swyneshed's 1350 book on calculation, helpfully entitled *Liber Calculationum* (printed at Padua in 1477 and at Venice in 1520). However, neither Bradwardine nor Swyneshed determined exactly the correct form of the law; this was left to William Heytesbury, who first correctly articulated the 'mean speed theorem' or the 'Merton rule of uniform acceleration' in 1335:

> A moving body will travel in an equal period of time a distance exactly equal to that which it would travel if it were moving continuously at its mean speed (Hannam 2010, p. 180).

An arithmetic proof of this theorem was given by John Dumbleton (Freely 2013, p. 159).

Interestingly, Heytesbury's statement of the mean speed theorem occurs not in a treatise on physics or mathematics, but of philosophy, in his *Regule solvendi sophismata* (1335), a treatise giving methods for 'solving' sophisms. The *sophis-*

---

[4] Because many of them were associated with Merton College, they are often also known as the 'Merton Calculators'; but because not all were members of Merton, this is not an optimal label.

[5] Kilvington is often cited as the first of the Calculators; however, his methods differed from that of later calculators (Kretzmann 1988, p. 226; Ashworth 1992, p. 520), and it is likely that he left Oxford before the others Calculators really became active (Sylla 1999). Nevertheless, his treatises were enormously influential on the later Calculators, especially on William Heytesbury, student of Kilvington, whose *Regule solvendi sophismata* (1335) is indebted to Kilvington's *Sophismata* (Wilson 1956, p. 7).

mata genre is a specifically philosophical one, with a sophism being a logical or philosophical puzzle whose analysis is either difficult, fallacious, or inconsistent. Many of Heytesbury's rules for generating solutions to sophisms were calculatory in nature, especially sophisms arising from the analysis of statements involving *incipit* 'it begins' and *desinit* 'it ceases', as well as those involving maximal and minimal bounds of capacities as measured on a linear continuum. Heytesbury devotes a chapter to each of these topics (*incipit* and *desinit* are treated in Chap. 4; maxima and minima in Chap. 5). It is in Chap. 5 that the mean speed theorem can be found, but the analysis of sophisms involving maxima and minima is closely related to the analysis of beginning and ceasing, since both involve how we are to understand limits.

The analysis of starting and stopping, given a continuous account of time, was a typical issue that occupied many of the Calculators. It was a central topic because of its relationship to change, as "every change... involves a beginning and a ceasing: the ceasing of one state and the beginning of another" (Kretzmann 1977, p. 4). Change itself is central phenomenon in Aristotelian natural philosophy, as it is required to understand generation and corruption, the topic of Aristotle's treatise *De generatione et corruptione*. Many Calculators either wrote commentaries on this treatise or treatises specifically addressing the question of beginning and ceasing, including Richard Kilvington's *Quaestiones super De generatione et corruptione*, written before he obtained his Masters c. 1324–1325; Thomas Bardwardine's *De incipit et desinit* (Bradwardine 1982) and John Dumbleton's *Summa Logica et Philosophiae Naturalis* (c. 1349?) which includes a commentary on *De generatione et corruptione*.

In the parlance of the Calculators and their contemporaries, terms such as *incipit* and *desinit* are called *exponible*, that is, sentences in which they are used can be decomposed into conjunctions of sentences not containing those terms, and it is these conjunctions which must be analysed in order to understand the terms. Often, a syncategorematic term can be expounded in more than one way, and that is why sentences containing these words can provide puzzles. For example, a sentence of the form *A incipit esse B* "*A* begins to be *B*" can be expounded in two ways:

1. *A* is now *B* and now is the first moment where *A* is *B*.
2. *A* is now not *B* and now is the last moment where *A* is not *B*.

In the first way of expounding *incipit*, the limit is intrinsic; in the second, the limit is extrinsic. The analysis of *desinit* is symmetric. Many of the sophisms rely on conflating these two notions, or interpreting the word in one way in one premise and in the other in another. Thus, every time that these words occur in Kilvington's analyses, one must be careful to identify when the analysis is trading on this ambiguity between the two readings of *incipit* and *desinit*.

The calculatory approach was not restricted to applications in physics and metaphysics, but also merged with computistics in eschatology, the calculation of the timetable for the end times (Oberman 1981, p. 526), thus merging the computational threads of this section and the previous.

3.4 Mechanical Reasoning

In the previous sections, we have looked at aspects of computation that are on the mathematical side of the spectrum. In this section, we move away from mathematical reasoning or calculation to linguistic reasoning, specifically to computation as a means of producing valid arguments. The most notorious medieval attempt to mechanize linguistic reasoning is that of Ramon Llull, one of the most eccentric men in the history of computation. But though Llull is the best known, he was not the first to have such a lofty goal. Writing in the middle of the twelfth century, John of Salisbury tells us that his student, William of Soissons,

> invented a device (*machinam*) to revolutionize the old logic by constructing unacceptable conclusions and demolishing the authoritative opinions of the ancients (John of Salisbury and McGarry 1955, Bk. II, ch. 10, p. 98).[6]

Unfortunately, we do not have any of William's own writings, or any other references to his *machina*, making it difficult (perhaps impossible) to determine what kind of mechanism is being referred to. According to the Kneales, "some people" have thought that it was an actual physical construction, akin to Jevon's logical machine (Kneale and Kneale 1984, p. 201).[7] However, it is more likely that "machine" should be understood here in a metaphorical sense, and that William had in mind some particular method or sort of argument-construction which, given a contradiction or an impossible statement, would return any other statement (Martin 1986, p. 565).

Whether William's machine was physically embodied or merely a procedure for a reasoner to follow, it is an interesting example of a computational method where the user is no longer necessarily the reasoner; rather, it is the "machine" itself which is doing the reasoning. But there is no doubt that Ramon Llull's goal was a physically-implemented mechanical computer.

Ramon Llull (1232/33–1315/16)'s early years were devoted to a secular life as a courtier and troubadour-lyric writer. In 1263 he underwent a religious conversion and turned his attentions to theological and philosophical pursuits, including missionary travel.[8] One of Llull's goals was to develop a mechanical system of argumentation or demonstration which could be used to show the Jew and the Muslim the error of their ways, and the correctness of Christian theology, and that could not be disputed. This mechanism is best witnessed in two of Llull's works: The *Ars demonstrativa* (c. 1283–1289, hereafter referred to as AD) and the *Ars brevis* (1308, hereafter referred to as AB), which was his single most influential work.

[6]Adamson (Adamson 1919, p. 27) translates John of Salisbury's *machinam* as "method", and the Kneales translate it as "engine" (Kneale and Kneale 1984, p. 201).

[7]The Kneales do not say who these "some people" are, and I have had no success in determining this.

[8]For biographical information, see Llull and Bonner (1985, vol. 1, pp. 3–52), which includes extensive excerpts from Llull's autobiography.

	Fig. A	Fig. T	Questions & Rules	Subjects	Virtues	Vices
B	goodness	difference	whether?	God	justice	avarice
C	greatness	concordance	what?	angel	prudence	gluttony
D	eternity	contrariety	of what?	heaven	fortitude	pride
E	power	beginning	why?	man	temperance	pride
F	wisdom	middle	how much?	imaginative	faith	accidie
G	will	end	of what kind?	sensitive	hope	envy
H	virtue	majority	when?	vegetative	charity	ire
I	truth	equality	where?	elementative	patience	lying
K	glory	minority	how? and with what?	instrumentative	pity	inconstancy

Fig. 3.1 The alphabet of the *Ars brevis* (Llull and Bonner 1985, p. 581)

The AB both builds upon and simplifies the AD, and together these works are referred to as simply the 'Art'. The Art is a mechanism for abstract reasoning in a restricted domain based a system of constants, each representing different concepts. In AD, the alphabet is two-tiered, with 16 symbols representing basic concepts and seven symbols representing what we might call meta-concepts. This two-tiered alphabet is simplified in AB to just nine symbols, whose meaning depends on their usage. Figure 3.1 gives the interpretation of the alphabet of AB in different contexts.

The Art consisted in combinatorial arrangements of these alphabets of letters, resulting in the mechanistic computation of new combinations, and hence new concepts or conclusions. The allowed combinations of the constant symbols in the alphabet are illustrated by various tables and diagrams.[9] Figure 3.2 of the *Ars brevis* consisted in three concentric circles, the outermost of which was fixed to the manuscript and the two inner ones being mobile (see Fig. 3.2 for a redrawing of Fig. 3.2 as it occurs in one of the manuscripts (Llull and Bonner 1985, Plate XVIII). For further reproductions of Llull's tables and diagrams, see Yates (1954), between pages 117 and 118.). By rotating the moving circles in various ways, one can extract all of the valid Aristotelian syllogisms, where the term on the middle circle is the middle term relating the major and minor terms, located on the outer and inner circles. This illustrates how the Art "became a method for 'finding' all the possible propositions and syllogisms on any given subject and for verifying their truth or falsehood" (Llull and Bonner 1985, p. 575).

The physical nature of the movable circles results in a crude mechanism for computing new concepts (the output) on the basis of a given set of concepts (the input). The mechanistic aspects of this computation cannot be overemphasized; but the process was quite crude and primitive. Because the Art starts from a finite alphabet, there are only finitely many combinations that can be computed, and it has difficulty moving beyond the calculation of intersection (Styazhkin 1969, p. 12). But we should not allow the primitiveness of the mechanism to detract from its novelty:

[9]The diagrams of the first, second, third, and fourth figures of the *Ars brevis* as found in the Escorial MS are reproduced in Llull and Bonner (1985) between pages 582 and 583.

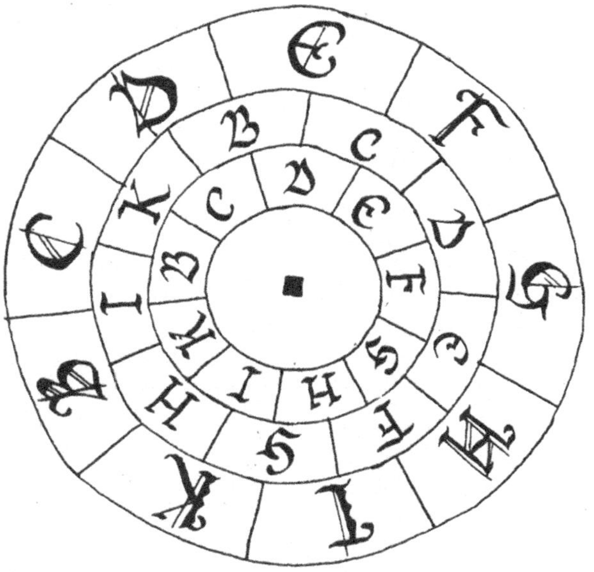

Fig. 3.2 The fourth figure

It is the first known attempt in Western Europe to provide a physical implementation of a mechanistic method of reasoning.[10]

As with many figures whom history eventually identifies as ahead of their time, Llull's combinatorics were little appreciated in his own time, or in the succeeding century. But by the end of the fifteenth century, Llullism was revived, especially among the Franciscans, and by the early sixteenth century it had become quite fashionable, especially in Paris where the Basque Franciscan Bernard de Lavinheta was invited to introduce Llullism to the Sorbonne in 1514 (Mertens 2009, p. 513). De Lavinheta's *Explanatio compendiosaque applicatio artis Raymundi Lulli* was published in Lyon in 1523 (Bonner 1993, p. 65). In 1518, Pietro Mainardi published the *Opusculum Raymundinum de auditu kabbalistico*, picking up on the link between Llull's methods and the kabbalah that was originally asserted by Pico della Mirandola 30 years earlier (Mertens 2009, p. 514), a link grounded in the combinatory nature of both Llull's methods and the Hebrew mysticism. At the end of the sixteenth century, Italian philosopher and mathematician Giordano Bruno wrote a number of treatises both on Llull's views directly and incorporating Llullism into his own views on memory. Llullism was one of the "major forces in the Renaissance" and it remained "enthusiastically cultivated in Paris throughout the seventeenth century", influencing Descartes and others (Yates 1954, p. 166). It was revived again in Germany in the eighteenth-century, where its end product was Leibniz's combinatorial systems (Yates 1954, p. 167). And thus, Llull's trajectory

[10]For further discussion of Llull's system, see Bonner (2007), Llull and Bonner (1985), and Uckelman (2010).

takes us out of the Middle Ages and into the Early Modern era, and it is time to draw our discussion to an end.

3.5 Conclusion

In order to discuss the history of computing and computation in the Middle Ages, we must widen what we mean by 'computation' to cover a broader conception than mere mindless mechanistic practices. When we do so, we can see that medieval Europe, far from being computer-less, was the site of a variety of developments in computation ranging from the arithmetic to the linguistic, of which we have focused on three: Irish and English computistics in the seventh to ninth centuries; the calculatory and arithmetic turn in natural philosophy in the thirteenth and fourteenth centuries; and the use of mechanical methods in linguistic reasoning in the twelfth and thirteenth centuries. These developments are all closely tied to advances in education more generally, both secular and ecclesiastical. We saw how the insular computistic treatises were embedded into the Carolingian educational structure and disseminated across the continent, as well as the importance of the concentration of mathematical philosophers for the development of physics in Oxford at the beginning of the fourteenth century. Llull's own project was less concerned with formal education and more outward facing—taking the benefits of traditional scholastic learning and using them to convert the heathens—but by the end of the Middle Ages his developments were integrated into the university education of Europe.

References

Adamson, J. W. (1919). *A short history of education*. Cambridge: Cambridge University Press.

Ashworth, E. J. (1992). New light on medieval philosophy: The Sophismata of Richard Kilvington. *Dialogue, 31*, 517–521.

Bisagni, J. (2013–2014). A new citation of a work of Columbanus in BnF lat. 6400B. *Peritia, 24–25*, 116–122.

Bonner, A. (Ed.). (1993). *Doctor illuminatus: A Ramón Llull reader*. Princeton: Princeton University Press.

Bonner, A. (2007). *The art and logic of Ramon Llull: A user's guide* (Studien und Texte zur Geistesgeschichte des Mittelalters 95). Leiden/Boston: Brill.

Borst, A. (1993). *The ordering of time: From the ancient computus to the modern computer*. Chicago: University of Chicago Press.

Borst, A. (2006). *Schriften zur Komputistik im Frankenreich von 721 bis 818*. Hannover: Hahnsche Buchhandlung.

Bradwardine, T. (1982). De incipit et desinit. *Cahiers de l'Institut du Moyen Age Grec et Latin, 42*, 1–83 (Nielsen, L. O., Ed.).

Chabas, J., & Goldstein, B. R. (Eds.). (2014). *Essays on medieval computational astronomy*. Leiden: Brill.

Costa, D. (2012). Review of C. Philipp E. Nothaft. In *Dating the passion: The life of Jesus and the emergence of scientific chronology (200–1600)* (Time, astronomy, and calendars: Texts and studies, Vol. 2, x + 320 pp.). Leiden: Brill; *Time's books/KronoScope, 13*(2), 298–302 (2013).

Cróiní, D. (1982). Mo-Sinnu moccu Min and the computus of Bangor. *Peritia, 1*, 281–295.

Folkerts, M., & Lorch, R. (2007). The Arabic sources of Jordanus de Nemore. *Foundation for Science, Technology, and Civilization*, 1–15.

Freely, J. (2013). *Before Galileo: The birth of modern science in medieval Europe*. New York: Overlook Press.

Graff, E. (2010). Review of Immo Warntjes. In *The Munich computus: Text and translation. Irish computistics between Isidore of Seville and the venerable Bede and its reception in Carolingian times* (Sudhoffs Archiv, Zeitschriften für Wissenschaftsgeschichte, heft 59, pp. ccxxi, 402). Stuttgart: Franz Steiner Verlag; *Journal of Medieval Latin, 22*, 326–329 (2012).

Hannam, J. (2010). *God's philosophers*. Duxford: Icon Books.

Hawk, B. W. (2012). A Greco-Latin numerical list in a St. Gall fragment. *Manuscripta, 56*(1), 28–46.

John of Salisbury, & McGarry, D. D. (Ed. and Trans.). (1955). *The metalogicon of John of Salisbury: A twelfth-century defense of the verbal and logical arts of the trivium*. Berkeley: University of California Press.

Kneale, W., & Kneale, M. (1984). *The development of logic* (Revised ed.). Oxford: Clarendon.

Kretzmann, N. (1977). Socrates is whiter than Plato begins to be white. *Noûs, 11*(1), 3–15.

Kretzmann, N. (1988). *Tu Scis Hoc Esse Omne Quod Est Hoc*: Richard Kilvington and the logic of knowledge. In Kretzmann, N. (Ed.), *Meaning and inference in medieval philosophy* (pp. 225–247). Dordrecht: Kluwer Academic Publishers.

Llull, R., & Bonner, A. (Ed. and Trans.). (1985). *Selected works of Ramon Llull (1232–1316)* (Vols. 2). Princeton: Princeton University Press.

Martin, C. J. (1986). William's machine. *Journal of Philosophy, 83*(10), 564–572.

McCarthy, D. (1994). The chronological apparatus of the Annals of Ulster AD 431–1131. *Peritia, 8*, 46–79.

McCluskey, S. C. (1998). *Astronomies and cultures in early medieval Europe*. Cambridge: Cambridge University Press.

MED. (2001–2014). Compotiste, n. In *Middle english dictionary*. University of Michigan. http://quod.lib.umich.edu/cgi/m/mec/med-idx?type=id\&id=MED8760, March 31, 2017.

Mertens, M. (2009). A perspective on Bruno's *De compendiosa architectura et complemento artis Lullii. Bruniana & Campanelliana, 15*(2), 513–525.

Murdoch, J. E. (1959). Review of *The science of mechanics in the middle ages* by Marshall Clagett (pp. xxix, 711; nine plates). Madison: University of Wisconsin Press; *The Philosophical Review, 71*(1), 120–126 (1962).

Oberman, H. A. (1981). The reorientation of the fourteenth century. In A. Maierù & A. P. Bagliani (Eds.), *Studi sul XIV Secolo in Memoria di Anneliese Maier* (pp. 513–530). Edizioni di Storia e Letteratura.

OED. (2017a). Algorithm, *n.* In *Oxford English dictionary* (March 2017 ed.). Oxford University Press. http://www.oed.com/view/Entry/4959, March 29, 2017.

OED. (2017b). Computation, *n.* In *Oxford English dictionary* (March 2017 ed.). Oxford University Press. http://www.oed.com/view/Entry/37968, March 29, 2017.

Palmer, J. (2010). Review of *The Munich computus: Text and translation. Irish computistics between Isidore of Seville and the Venerable Bede and its reception in Carolingian times*. immo warntjes (Sudhoffs Archiv, Beihefte lix, pp. ccxxii + 402). Stuttgart: Franz Steiner Verlag; *Journal for the History of Astronomy, 43*(1), 129–131 (2012).

Philipp, C., & Nothaft, E. (2012). Dating the passion: The life of Jesus and the emergence of scientific chronology (200–1600) (Volume 2 of time, astronomy, and calendars: Texts and studies). Leiden: Brill.

Philipp, C., & Nothaft, E. (2013). Review of Mary Kelly and Charles Doherty (Eds.), *Music and the stars: Mathematics in medieval Ireland* (xx + 249 pp.). Dublin: Four Courts. ISBN:978-1-84682-392-3. *Peritia* (2014).

Philipp, C., & Nothaft, E. (2014). The reception and application of Arabic science in twelfth-century computistics: New evidence from Bavaria. *Journal for the History of Astronomy, 45*(1), 35–60.

Philipp, C., & Nothaft, E. (2015). Roman vs. Arabic computistics in twelfth-century England: A newly discovered source (Collatio Compoti Romani et Arabici). *Early Science and Medicine, 20*, 187–208.

Styazhkin, N. I. (1969). *History of mathematical logic from Leibniz to Peano*. Cambridge/London: MIT Press.

Sylla, E. (1999). Oxford calculators. In R. Audi (Ed.), *The Cambridge dictionary of philosophy* (2nd ed.). Cambridge/New York: Cambridge University Press.

Tabak, J. (2014). *Algebra: Sets, symbols, and the language of thought*. New York: Infobase Publishing.

Uckelman, S. L. (2010). Computing with concepts, computing with numbers: Llull, Leibniz, & Boole. In F. Ferreira, B. Löwe, E. Mayordomo, & L. M. Gomes (Eds.), *Proceedings of the 6th Conference on Computability in Europe Programs, Proofs, Processes, Ponta Delgada*, Azores, June 30–July 4, 2010 (Volume 6158 of lecture notes in computer science, pp 427–437).

Wagner, R., & Briggs, A. (2016). *The Penultimate curiosity: How science swims in the slipstream of ultimate*. Oxford: Oxford University Press.

Warntjes, I. (2005). A newly discovered Irish computus: *Computus Einsidlensis*. *Peritia, 19*, 61–64.

Warntjes, I. (2010). *The Munich computus: Text and translation. Irish computistics between Isidore of Seville and the Venerable Bede and its reception in Carolingian times* (Volume 51 of Sudhoffs Archiv). Stuttgart: Franz Steiner Verlag.

Wilson, C. (1956). *William Heytesbury: Medieval logic and the rise of mathematical physics*. Madison: University of Wisconsin Press.

Yates, F. A. (1954). The Art of Ramon Lull: An approach to it through Lull's theory of the elements. *The Journal of the Courtald and Warburg Institutes, 17*(1/2), 115–173.

Chapter 4
Leibniz and the Calculus Ratiocinator

Wolfgang Lenzen

Abstract This paper deals with the interconnections between mathematics, metaphysics, and logic in the work of Leibniz. On the one hand, it touches upon some practical aspects such as Leibniz's construction of a Four-species calculating machine, a mechanical digital calculating machine, and even a cipher machine. On the other hand, it examines how far Leibniz's metaphysical dreams concerning the "calculus ratiocinator" and its underlying "characteristica universalis" have in fact been realized by the great philosopher. In particular it will be shown that Leibniz not only developed an "intensional" algebra of concepts which is provably equivalent to Boole's "extensional" algebra of sets, but that he also discovered some basic laws of quantifier logic which allowed him to define individual concepts as maximally-consistent concepts. Moreover, Leibniz had the ingenious idea of transforming the basic principles of arithmetical addition and subtraction into a theory of "real" addition and subtraction thus obtaining some important building blocks of elementary set-theory.

4.1 Introduction and Summary

The so-called *calculus ratiocinator* is a bit like the Loch Ness Monster Nessie. Many people talk about it, but nobody seems to know whether it really exists or what it exactly consists of. In a Wikipedia entry it is roughly described as follows:

> The *Calculus ratiocinator* is a [...] universal logical calculation framework, a concept described in the writings of Gottfried Leibniz, usually paired with his [...] *characteristica universalis*, a universal conceptual language. The received point of view in analytic philosophy and formal logic is that the *calculus ratiocinator* anticipates mathematical logic [and that it] is a formal inference engine or computer program which can be designed so as to grant primacy to calculations. [...] From this perspective the *calculus ratiocinator*

W. Lenzen (✉)
Department of Philosophy, Osnabrück University, Osnabrück, Germany
e-mail: lenzen@uos.de

S. O. Hansson (ed.), *Technology and Mathematics*, Philosophy of Engineering and Technology 30, https://doi.org/10.1007/978-3-319-93779-3_4

> is only a part [...] of the universal characteristic and a complete universal characteristic includes a "logical calculus".
>
> A contrasting point of view stems from synthetic philosophy and fields such as cybernetics [...] and general systems theory [...]. The synthetic view understands the calculus ratiocinator as a "calculating machine". The cybernetician Norbert Wiener considered Leibniz's *calculus ratiocinator* a forerunner to the modern day digital computer.
>
> Hartley Rogers saw a link between the two, defining the *calculus ratiocinator* as "an algorithm which, when applied to the symbols of any formula of the *characteristica universalis*, would determine whether or not that formula were true as a statement of science".[1]

The aim of this paper is to unveil some of the mysteries surrounding the *calculus ratiocinator*. First, as regards the "hardware" of such a calculus, it will be shown in Sect. 4.2 that although Leibniz had not the slightest idea of a modern day computer (nor, for that matter, of any other *electronic* device), he successfully invented a *mechanical* computer in the form of a Four-Species Calculating Machine. Furthermore he even made concrete plans for the construction of a (non-electronic) *Dyadic* Calculating Machine.

Second, as regards the "software", Leibniz thought it *possible* to determine the truth-value of *any* proposition by mere calculation. More concretely he believed that such a calculation had once been carried out by God when he set out to decide which possible world – out of an infinite number of alternatives – should become realized. In order to arrive at that decision, God used his infinitely powerful mind to calculate in every detail the consequences which would result if a certain individual X – rather than any other out of an infinite number of alternative possible individuals – were created. This metaphysical vision, which shall be analyzed in more detail in Sect. 4.3, has been summarized by Leibniz in the oft-quoted dictum "Cum Deus calculat et cogitationem exercet fit mundus".

Third, Leibniz was hoping that mankind, although endowed by God only with a *finite* mind, might eventually develop the tools for determining the truth-value of arbitrary propositions by translating them into a precise universal language (the "*characteristica universalis*") which allows calculating the truth in an infallible way. This metaphysical dream, which shall be closer investigated in Sect. 4.4, lies behind Leibniz's even more famous slogan "Calculemus!"

In order to answer the question whether, or how much, of these dreams and visions are realizable (or have in fact been realized by Leibniz himself), a survey of the development of the "calculus ratiocinator" will be given in Sect. 4.5. In Sect. 4.5.1 the background of early seventeenth century syllogistic will be sketched. In Sect. 4.5.2 it will be shown how Leibniz gradually transformed the traditional theory of the syllogism into a much more powerful logic which turned out to be equivalent to so-called *Boolean algebra*. In Sect. 4.5.3 some expansions of Leibniz's algebra of concepts will be considered; in particular it will be shown that the introduction of "indefinite concepts", which function as *quantifiers* ranging

[1] https://en.wikipedia.org/wiki/Calculus_ratiocinator; online access January 3, 2017. The concluding quote comes from Rogers (1963), p. 943.

over concepts, allows the definition of *individual concepts* as maximally-consistent concepts. Section 4.5.4 describes Leibniz's ingenious transformation of some basic laws of elementary arithmetic into a "Calculus of real addition and subtraction" which forms a subsystem of modern *set-theory*.

4.2 Leibniz's Calculating Machines

In 1990, the main curator of the Astronomic-Physical Cabinet of the Hessian State Museum, Ludolf von Mackensen, published an article about the prehistory of calculating machines. After summarizing some early seventeenth century inventions by Wilhelm Schickard and Blaise Pascal, he characterized the role that Leibniz played in this connection as follows.

> The third big universal scientist of the age of baroque, who decisively advanced the invention of a Four-species Calculating Machine, was the philosopher and mathematician Gottfried Wilhelm Leibniz. [. . .] [F]rom the very beginning Leibniz strived for surpassing Schickard and Pascal by creating a machine which was able to make multiplications and divisions. Guided by the idea that a multiplication is a repeated addition and a division a repeated subtraction, Leibniz aimed at a complete mechanization of the first two species so that they could be repeated many times in the shortest possible time. He solved this problem by separating the process of entering the numbers from the process of calculation, i.e. the movement of the counting wheels. Hence the machine was designed by Leibniz to work in two steps, which was achieved by putting special switchgear between the number entry and the calculation device. Such entry/calculation switchgear is a necessary component of each mechanical calculation machine, no matter whether driven by electricity or by means of a hand crank.
>
> In the absence of any example of such switchgear which transmits the entered number into the calculating device, Leibniz invented a completely new element, a gear-wheel, whose effective number of teeth could be varied between 0 and 9 so that if, e.g., the number 5 was set, five teeth would become effective. Leibniz even devised two variants of such a device, a so-called sprocket wheel [. . .] and a so-called stepped drum, i.e. a cylinder which carries nine toothed rings on its circumference. [. . .]
>
> [Therefore] Leibniz may be considered as the first ancestor of a whole line of development of stepped drum machines that ended in 1948 when Curt Herzstack's model "Curta" came to the market.[2]

Leibniz's invention was mainly motivated by the consideration that it is "unworthy to waste the time of excellent people by servile work of calculating when, with the help of a machine, everybody can get the result in a fast and secure way".[3] Figure 4.1 shows the original machine built in 1693.

---

[2]Cf. Mackensen (1990), p. 56–57 (my translation).

[3]In "Machina arithmetica in qua non additio tantum et subtractio sed et multiplicatio nullo, divisio vero paene nullo animi labore peragantur" Leibniz wrote: "Indignum est excellentium virorum horas servili calculandi perire quia Machina adhibita velissimo cuique secure transcribi possit." The translation is taken from "Leibniz on his calculating machine" in Smith (1929), 173–181.

Fig. 4.1 Leibniz's Four-species calculating machine (© Niedersächsische Landesbibliothek Hannover)

Mackensen further pointed out that although Leibniz certainly was not the inventor of the binary number system, in the 1679 paper "De progressione dyadica" he had developed a clear idea of a *binary calculating machine*:

> In the year 1974 [Mackensen] transformed Leibniz's ideas into a drawing and found out that, if one knows his mechanical Four-species-machine and if one adds a few constructive elements from the technique of the time of baroque to the description of Leibniz's dual calculating machine, a functional model can be built. [...] This machine doesn't use wheels or electric impulses but rolling balls. [...] In the calculation process and in the device yielding the result the numbers are not represented by teeth of wheels but by balls: a ball means 1, no ball means 0. Thus for the first time the binary principle is applied for the mechanical representation of data.[4]

Figure 4.2 shows von Mackensen's model which allows to perform additions and multiplications while "subtractions and divisions can only be performed quite cumbersome by way of the complements of the numbers".[5]

Although the practical value of this machine was further restricted by the fact that it presupposed the possibly laborious transformation of decimal numbers into dual numbers, the very idea of a "Machina arithmeticae dyadicae" remained so important for Leibniz that he later devised another version working with gear-wheels rather than with balls. And he also invented a mechanical device to convert

[4]Cf. Mackensen (1990), p. 58 (my translation). A facsimile of Leibniz's manuscript "De progressione dyadica" may be found in Stein and Heinekamp (1990), p. 109.

[5]Cf. Mackensen (1990), p. 59 (my translation).

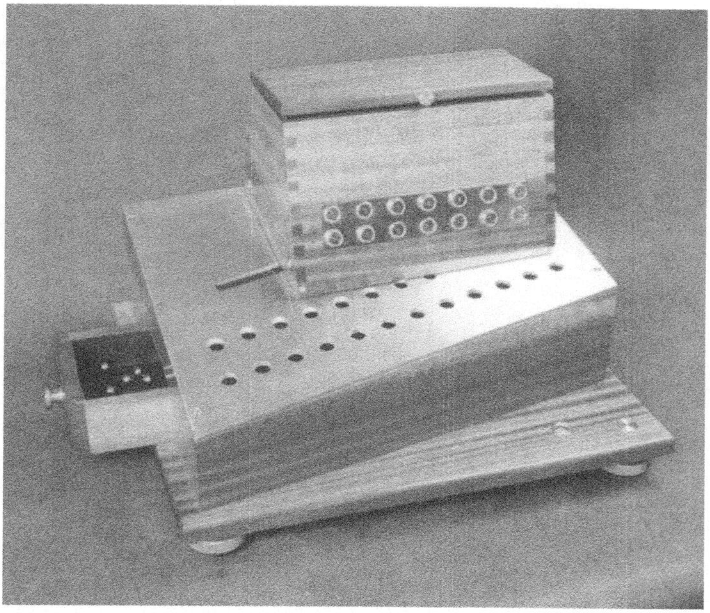

Fig. 4.2 Model of Leibniz's digital calculating machine (© Ludolf von Mackensen)

decimal numbers into dyadic numbers. Figure 4.3 shows a functional model built by Rolf Paland after the construction plans of Ludolf von Mackensen:

To conclude this section let it be pointed out that Nicholas Rescher recently reconstructed Leibniz's ideas of a "Machina Deciphratoria", i.e. a cipher machine, as sort of a byproduct of his calculating machine. In a letter of February 1679 to the Duke of Hanover-Calenberg, Leibniz described his ideas as follows:

> This arithmetical machine led me to conceive another beautiful machine that would serve to encipher and decipher letters, and do this with great swiftness and in a manner indecipherable by others. For I have observed that the most commonly used ciphers are easy to decipher, while those difficult to decipher are generally difficult to use, so that busy people abandon them. But with this machine of mine an entire letter is almost as easy to encipher and decipher for one who uses it as it is to copy it.[6]

Eleven years later, in a memorandum for emperor Leopold I in Vienna, he revealed some further details of this machine:

> It is a smallish mechanism (*machinula*) that is easy to transport. [...] While both encipherment and decipherment is [ordinarily] laborious, there is now a facility enabling one to get at the requisite ciphers or alphabetic-letters as easily as though one were playing

[6]Cf. A I 2, p. 125; the translation is from Rescher (2012), p. 35–36.

Fig. 4.3 Leibniz's machine for converting decimal into binary numbers (© H. Gramann)

> on a clavichord or other [keyboard] instrument. The requisite letters will immediately emerge, and only need to be copied off.[7]

On the basis of these and some other hints, Rescher developed a conjectural reconstruction of Leibniz's cryptographic machine which, with the assistance of several engineers, has meanwhile been physically realized. Figure 4.4 shows the result of this reconstruction.

4.3 Leibniz's Grand Vision of the Creation of the World

The Christian idea that God created the entire world literally out of nothing does not sound very reasonable. Yet Leibniz evidently did believe in this doctrine or, somewhat more exactly, in the slightly weakened claim that God created the world out of nothing *plus one*. In 1981 the "Stadtsparkasse Hannover" edited a commemorative coin (Fig. 4.5):

In the middle of the coin there is a table with the beginning of the binary number system, framed by samples of elementary arithmetical calculations. On top of the coin one can read "Omnibus ex nihilo ducendis sufficit unum", which may be translated as follows: "In order to produce everything out of nothing one [thing] is

[7]Cf. A IV 4, p. 68; translation from Rescher (2012), p. 37.

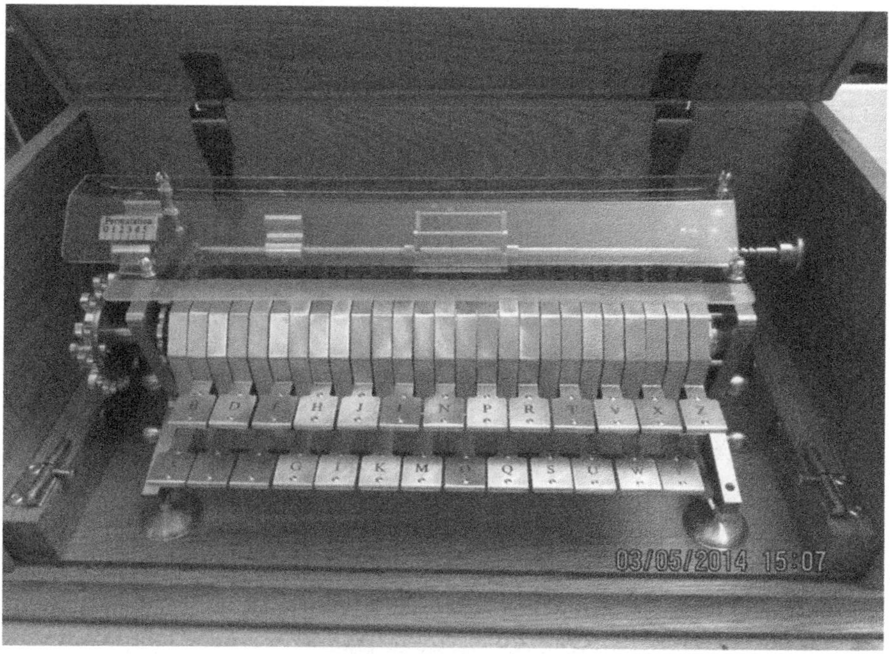

Fig. 4.4 Leibniz's cipher machine (© Nicholas Rescher)

sufficient". The whole picture is said to represent an "imago creationis", i.e. a picture of the creation. Now, a skeptic may want to object that the editors of the coin grossly misunderstood Leibniz's intentions. After all, the diagram mainly illustrates the fact that the set of natural *numbers* can be built up from just two elements, namely from the numerals 0 and 1. Moreover, since Leibniz used to refer to the number zero by the Latin word 'nihil', the quoted dictum can alternatively be translated as saying: "In order to produce every number from 0, the number 1 is sufficient". Thus one might suspect that the Hanover savings bank mistakenly charged Leibniz with holding the Christian view of the creation of the world while in fact he only wanted to put forward the much more modest claim that the world of numbers can be created from zero plus one. In 1697, however, Leibniz himself had painted the picture shown in Fig. 4.6.

Again we are told to see a "Bild der Schöpffung", a picture of the creation, which contains drawings of the sun, the moon, and other celestial objects. On top one can read "Einer hat alles aus nichts gemacht", which means 'One [namely God] has made everything out of nothing'. The ambiguous statement at the bottom "Eins ist noht" can be interpreted as saying either that *one thing* or that *the number one* is necessary. In what follows it will be argued that Leibniz did not only have the trivial arithmetical interpretation in mind, but rather the Christian doctrine of the creation of the world. Somewhat more exactly, Leibniz thought it possible for God

Fig. 4.5 Commemorative coin Stadtsparkasse Hannover

to construct the world – or better: the *idea* of the world – out of the ideas or the *concepts* of Nothing and One in seven steps.

1. Starting with the numerals 0 and 1, one obtains the set of natural numbers.
2. Each of these numbers is interpreted as representing, or being characteristic of, a specific primitive *concept*.
3. By way of logical combination the larger set of *general concepts* is obtained.
4. *Individual-concepts*, i.e. the "ideas" corresponding to individuals, will then be defined as *maximally consistent concepts*.
5. Among the set of all possible individuals the relation of *compossibility* is introduced.
6. Possible worlds are defined as certain *maximal* collections of pairwise compossible individuals.
7. The real world is distinguished from its rivals by being the richest, i.e. most numerous and, perhaps, also in some other respect the *best of all possible worlds*.

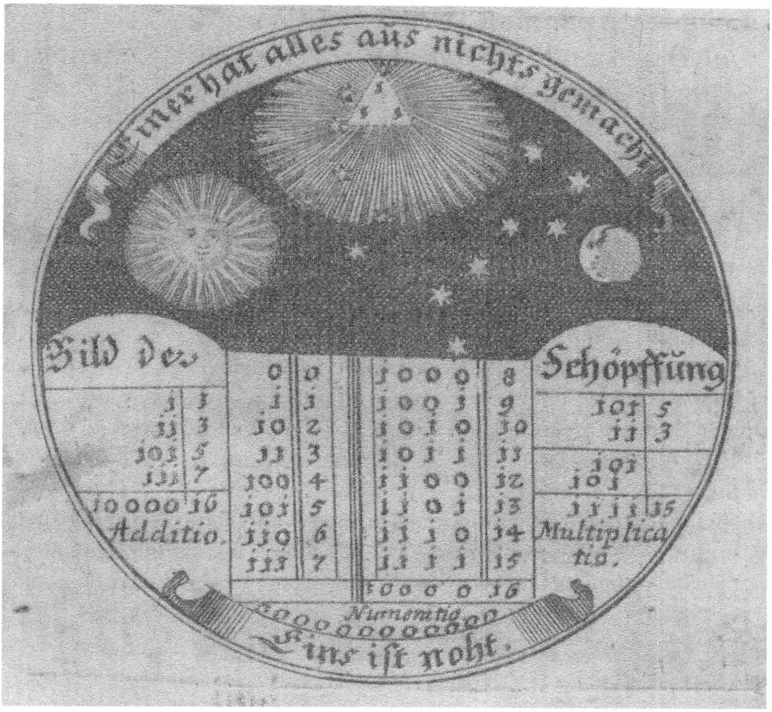

Fig. 4.6 Leibniz's drawing of the creation (© Niedersächsische Landesbibliothek Hannover)

These seven steps try to capture what Leibniz had in mind with his famous remark "Cum Deus calculat et cogitationem exercet fit mundus", which means "While God is calculating and carrying out his deliberations the world comes into existence".[8]

In order to support this interpretation, step 2 of the "logical creation of the world", *viz*. the idea of assigning characteristic numbers to concepts, will be closer examined in Sect. 4.4. Steps 3 and 4, i.e. the construction of the algebra of concepts and the definition of individual concepts, will be outlined in Sects. 4.5.2 and 4.5.3. For reasons of space, the remaining steps which deal with the ontological ideas of compossibility, existence, and possible worlds, must stay out of consideration here. The reader is referred to the reconstruction of "The System of Leibniz's Logic" given elsewhere.[9]

[8]This remark, which Louis Couturat chose as motto for his ground-breaking book (1901), was written by Leibniz on the margin of the "Dialogus" of August 1677; cf. GP 7, p. 191. As far as I know, Leibniz nowhere seriously discussed the problem of the proper *creation* of the world, i.e. the transition from the mere *idea* to its *physical actualization*.

[9]Cf. Lenzen (1990), especially Chap. 6.

4.4 Leibniz's Ambitious Dream of a *Characteristica Universalis* and Its Modest Realization as a Semantics for Syllogistic Inferences

Top-ranking among famous quotes from Leibniz certainly is the slogan "Calculemus":

> [...] whenever controversies arise, there will be no more need of disputation between two philosophers than between two calculators. For it would suffice for them to take their pencils in their hand, to sit down at the abacus, and to say to one another [...]: *Let us calculate!*[10]

This vision of the computability of all (scientific) problems rests on two pillars: (i) the invention of a "characteristica universalis", into which the respective question can be translated in an unambiguous way; and (ii) the construction of a "calculus ratiocinator", which in application to this language yields a precisely determined result.[11] This section is devoted to an explanation of task (i) while (ii) will be dealt with in Sects. 4.5.1, 4.5.2, and 4.5.3.

Already in 1666, in his dissertation "De Arte Combinatoria", Leibniz mentioned the possibility of "a universal writing, i.e. one which is intelligible to anyone who reads it, whatever language he knows."[12] More than 10 years later he explained in some more detail:

> Not long ago, some distinguished persons devised a certain universal language or characteristic in which all notions and things are nicely ordered, a language with whose help different nations can communicate their thoughts, and each, in its own language, read what the other wrote. But no one has put forward a language or characteristic which embodies, at the time, both the art of discovery and the art of judgment, that is, a language whose signs or characters perform the same task as arithmetic signs do for numbers.[13]

Leibniz was convinced

> [...] that one can devise a certain alphabet of human thoughts and that, through the combination of the letters of this alphabet and through analysis of words produced from them, all things can both be discovered and judged. [...] Once the characteristic numbers of most notions are determined, the human race will have a new kind of tool, a tool that will increase the power of the mind much more than optical lenses helped our eyes, a tool that will be as far superior to microscopes or telescopes as reason is to vision.[14]

---

[10]Cf. GP 7, p. 200; the translation has been adopted from https://en.wikiquote.org/wiki/Gottfried_Leibniz

[11]Cf. A VI, 4, p. 443: "Itaque profertur hic calculus quidam novus et mirificus, qui in omnibus nostris ratiocinationibus locum habet, et qui non minus accurate procedit, quam Arithmetica aut Algebra".

[12]Parkinson (1966), p.10.

[13]Cf. the fragment "De Numeris Characteristicis ad Linguam universalem constituendam" in GP 7, p. 184–9. The translation has been adopted with some modifications from Ariew & Garber (1989), p. 6–8.

[14]Cf. GP 7, p. 185 and p. 187.

The application of the "true" characteristic numbers would allow reducing the question whether an arbitrary state of affairs holds or not to a mere arithmetical issue. However, as Leibniz soon came to realize, "due to the wonderful interconnection of things, it is extremely difficult to produce the characteristic numbers". Therefore in a series of essays of April 1679 he contented himself with the much more modest task of developing a *formal semantics* by means of which the *logical validity* of syllogistic inferences can be decided:

> I have contrived a device, quite elegant, if I am not mistaken, by which I can show that it is possible to corroborate reasoning through numbers. And so, I imagine that those so wonderful characteristic numbers are already given, and, having observed a certain general property that characteristic numbers have, I meanwhile assume that these numbers, whatever they might be, have that property. By using these numbers I can immediately demonstrate through numbers, and in an amazing way, all of the logical rules and show how one can know whether certain arguments are formally valid.[15]

This semantics was guided by the idea that a term composed of concepts *A* and *B* gets assigned the *product* of the numbers assigned to the components:

> For example, since 'man' is 'rational animal', if the number of 'animal', *a*, is 2, and the number of 'rational', *r*, is 3, then the number of 'man', *m*, will be the same as $a*r$, in this example 2*3 or 6.[16]

Now a universal affirmative proposition like 'All gold is metal' can be understood as maintaining that the concept 'gold' *contains* the concept 'metal' (because 'gold' can be defined, e.g., as 'the heaviest metal'). Therefore it seems obvious to postulate that in general 'Every *S* is *P*' is true if and only if *s*, the characteristic number assigned to *S*, contains *p*, the number assigned to *P*, as a prime factor; or, in other words, *s* must be divisible by *p*. In a first approach, Leibniz thought that the truth-conditions for the *particular* affirmative proposition 'Some *S* are *P*' might be construed analogously by requiring that *either s* can be divided by *p or* conversely *p* can be divided by *s*. But this was a mistake![17] After some trials and errors, Leibniz eventually found the following more complicated solution[18]:

(i) To every term *T*, a *pair* of natural numbers $<+t_1;-t_2>$ is assigned such that t_1 and t_2 are relatively prime, i.e. they don't have a common divisor.

---

[15]Cf. GP 7, p. 189, and Ariew and Garber (1989), p. 9–10.

[16]Cf. "Elementa Calculi" in Couturat (1903), p. 49–57; the translation has been adopted from Parkinson (1966), p. 17–24.

[17]According to Leibniz's condition, the valid mood DARII would become invalid. The assignment of numbers $B = 3$, $C = 6$, $D = 2$ satisfies the premise 'All *C* are *D*', because 6 can be divided by 2; furthermore 'Some *B* are *C*' becomes true because the number of the predicate, $C = 6$, is divisible by the number of the subject, $B = 3$. But the conclusion 'Some *B* are *D*' would result as false since neither $B = 3$ can be divided by $D = 2$, nor conversely *D* by *B*. Thus also Leibniz soon noticed that for the truth of a particular affirmative proposition "it is not necessary that the subject can be divided by the predicate or the predicate divided by the subject"; cf. C., p. 57.

[18]Cf. "Regulae ex quibus de bonitate consequentiarum [...] judicari potest, per numeros", in C. p. 77–84; an English version may be found in Parkinson (1966), p. 25–32.

(ii) 'Every S is P' is true (relative to the assignment (i)) if and only if $+s_1$ is divisible by $+p_1$ and $-s_2$ is divisible by $-p_2$.
(iii) 'No S is P' is true if and only if $+s_1$ and $-p_2$ have a common divisor or $+p_1$ and $-s_2$ have a common divisor.
(iv) 'Some S is P' is true if and only if condition (iii) is not satisfied.
(v) 'Some S isn't P' is true if and only if condition (ii) is not satisfied.
(vi) An inference from premises P_1, P_2 to the conclusion C is logically valid if and only if for each assignment of numbers satisfying condition (i), C becomes true whenever both P_1 and P_2 are true.[19]

As Leibniz himself proved in Theorems 1–8, the "simple" inferences of the theory of the syllogism, i.e. the laws of *opposition*, *subalternation* and *conversion*, are all satisfied by this semantics. Furthermore, as was first shown by Lukasiewicz (1951), the semantics of characteristic numbers satisfies all (and only) those moods which are commonly regarded as valid. Hence it is a model of a syllogistic which dispenses with negative concepts. Although Leibniz repeatedly tried to generalize his semantics so as to cover also negative concepts, he never found a satisfactory solution. This problem has only been solved by contemporary logicians like Sanchez-Mazas (1979) and Sotirov (1999).

Leibniz's much further reaching hope that mankind might once discover the "true" characteristic numbers which enable to calculate the truth of arbitrary propositions, must, however, be assessed as an illusion:

> When we have the true characteristic numbers of things, we will be able to judge without any mental effort or danger of error whether arguments are materially [!] sound.[20]

One reason why such "true" characteristic numbers are bound to remain a chimera consists in the fact that the result of a mathematical calculation always is *necessary*, while the obtaining or not-obtaining of an arbitrary state of affairs may be *contingent*.

4.5 The Development of Leibniz's Universal Calculus

In the seventeenth century, logic was still dominated by syllogistic, i.e. the theory of the four categorical forms:

Universal affirmative proposition (UA)	Every *S* is *P*	*SaP*
Universal negative proposition (UN)	No *S* is *P*	*SeP*
Particular affirmative proposition (PA)	Some *S* is *P*	*SiP*
Particular negative proposition (PN)	Some *S* isn't *P*	*SoP*

[19]Cf. C., p. 25–28; condition (vi) was put forward only in another fragment. Cf. C., p. 245–7: "Ex hoc calculo omnes modi et figurae derivari possunt per solas regulas Numerorum. Si nosse volumus an aliqua figura procedat vi formae, videmus an contradictorium conclusionis sit compatibile cum praemissis, id est an numeri reperiri possint satisfacientes simul praemissis et contradictoriae conclusionis; quodsi nulli reperiri possunt, concludet argumentum vi formae."
[20]Cf. GP 7, p. 189, or Ariew and Garber (1989), p. 10.

A typical textbook of that time is the well-known "Logique de Port Royal"[21] which, apart from an introductory investigation of ideas, concepts, and propositions, basically consists of a theory of the so-called "simple" laws (of subalternation, opposition, and conversion) and a theory of the syllogistic moods which are classified into four different figures. The following summary of this theory does not rely, however, on Arnaud & Nicole's presentation but rather on Leibniz's reception of the traditional logic. For the sake of preciseness, we use the modern symbols $\neg$, $\wedge$, $\vee$ for the negation, conjunction, and disjunction of propositions and $\rightarrow$, $\leftrightarrow$ for (strict) implication and (strict) equivalence.

4.5.1 Early Seventeenth Century Syllogistic

As Leibniz explains, "a subalternation takes place whenever a particular proposition is inferred from the corresponding universal proposition",[22] i.e.:

SUB 1 $SaP \rightarrow SiP$
SUB 2 $SeP \rightarrow SoP$.

According to the modern analysis of the categorical forms in terms of first order logic, these laws are not strictly valid but hold only under the assumption that the subject term S is not empty. This problem of so-called existential import will be discussed further below.

The theory of opposition first has to determine which propositions are *contradictories* of each other in the sense that they can neither be together true nor be together false. Clearly, the PN is the contradictory, or negation, of the UA, while the PA is the negation of the UN:

OPP 1 $\neg SaP \leftrightarrow SoP$
OPP 2 $\neg SeP \leftrightarrow SiP$.

The next task is to determine which propositions are *contraries* to each other in the sense that they cannot be together true, while they may well be together false. As Leibniz states in C., p. 82: "Theorem 6: The universal affirmative and the universal negative are contrary to each other". Finally, two propositions are said to be *subcontraries* if they cannot be together false while it is possible that are together true. As Leibniz notes in another theorem, the two particular propositions, SiP and SoP, are logically related to each other in this way. The theory of subalternation and opposition is often summarized in the familiar Square of Opposition (Fig. 4.7):

[21]Cf. Arnauld and Nicole (1683).

[22]Cf. C., p. 80. In Arnauld & Nicole (1683) the principle of subalternation is put forward informally as follows: "Les propositions particulières sont enfermés dans les générales de même nature, et non les générales dans les particulières, I dans A, et O dans E, et non A dans I, ni E dans O".

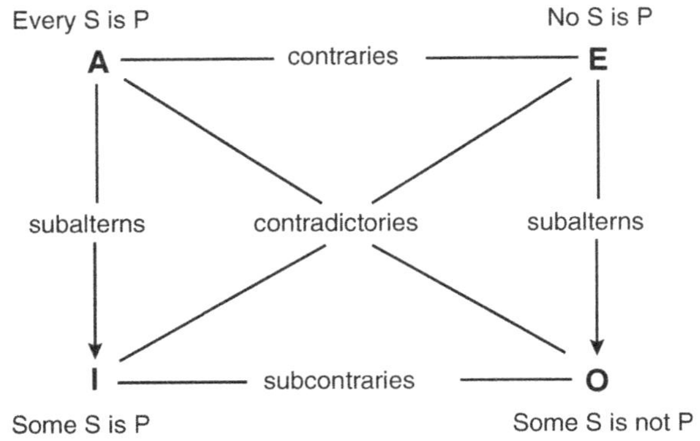

Fig. 4.7 Square of opposition

In the paper "De formis syllogismorum Mathematice definiendis" written around 1682 Leibniz tried to axiomatize the theory of the syllogistic figures and moods by reducing them to a small number of basic principles. The "Fundamentum syllogisticum", i.e. the axiomatic basis of the theory of the syllogism, is the "Dictum de omni et nullo":

> If a total *C* falls within another total *D*, or if the total *C* falls outside *D*, then whatever is in *C*, also falls within *D* (in the former case) or outside *D* (in the latter case).[23]

These laws warrant the validity of the following "perfect" moods of the First Figure:

BARBARA	$CaD, BaC \rightarrow BaD$
CELARENT	$CeD, BaC \rightarrow BeD$
DARII	$CaD, BiC \rightarrow BiD$
FERIO	$CeD, BiC \rightarrow BoD.$

On the one hand, if the second premise of the affirmative moods BARBARA and DARII is satisfied, i.e. if *B* is either totally or partially contained in *D*, then, according to the "Dictum de Omni", also *B* must be either totally or partially contained in *D* since, by the first premise, *C* is entirely contained in *D*. Similarly the negative moods CELARENT and FERIO follow from the "Dictum de Nullo":

> *B* is either totally or partially contained in *C*; but the entire *C* falls outside *D*; hence also *B* either totally or partially falls outside *D*.[24]

Next Leibniz derives the laws of subalternation from DARII and FERIO by substituting '*B*' for '*C*' and '*C*' for '*D*', respectively. This derivation (and hence also the validity of the laws of subalternation) tacitly presupposes the following principle which Leibniz considered as an "identity":

SOME BiB.

[23]Cf. C., p. 410–411.

[24]Cf. C., p. 411.

With the help of the laws of subalternation, BARBARA and CELARENT may be weakened into

BARBARI $CaD, BaC \rightarrow BiD$
CELARO $CeD, BaC \rightarrow BoD.$

Thus the First Figure altogether has six valid moods, from which one obtains six moods of the Second and six of the Third Figure by means of the logical inference-scheme "Regressus":

REGRESS If a conclusion Q logically follows from premises P_1, P_2, but if Q is false, then one of the premises must be false

When Leibniz carefully carries out these derivations, he presupposes the laws of opposition, OPP 1, OPP 2. Finally, six valid moods of the Fourth Figure can be derived from corresponding moods of the First Figure with the help of the laws of conversions. According to traditional doctrines, the PA and the UN may be converted "simpliciter", while the UA can only be converted "per accidens":

CONV 1 $BiD \rightarrow DiB$
CONV 2 $BeD \rightarrow DeB$
CONV 3 $BaD \rightarrow DiB.$

As Leibniz shows, these laws can in turn be derived from some previously proven syllogisms with the help of the "identical" proposition:

ALL $BaB.$

Furthermore one easily obtains another law of conversion according to which the UN can also be converted "accidentally":

CONV 4 $BeD \rightarrow DoB.$

The announced derivation of the moods of the Fourth Figure was not carried out in the fragment "De formis syllogismorum Mathematice definiendis" which breaks off with a reference to "Figura Quarta". It may, however, be found in the manuscript LH IV, 6, 14, 3 which, unfortunately, was only partially edited by Couturat. At any rate, Leibniz managed to prove that all valid moods can be reduced to the "Fundamentum syllogisticum" in conjunction with the laws of opposition, the inference scheme "Regressus", and the "identical" propositions SOME and ALL.

Now while ALL really is an identity or theorem of first order logic, $\forall x(Bx \rightarrow Bx)$, SOME is nowadays interpreted as $\exists x(Bx \wedge Bx)$. This formula is equivalent to $\exists x(Bx)$, i.e. to the assumption that there exists at least one x such that x is B. Hence the laws of subalternation presuppose that each concept B which can occupy the position of the subject of a categorical form is »non-empty«. Leibniz discussed this problem of existential import in the paper "Difficultates quaedam logicae" (GP 7, pp. 211–217) where he distinguished two kinds of existence: *Actual* existence of the individuals inhabiting our real world vs. merely *possible* subsistence of individuals "in the region of ideas". According to Leibniz, logical inferences should always be evaluated with reference to "the region of ideas", i.e. the larger set of all possible individuals. Therefore all that is required for the validity of subalternation is that the term *B* has a non-empty extension within the domain of possible individuals.[25]

4.5.2 From the 1678 "Calculus Ratiocinator" to an Algebra of Concepts

In this section it will be shown how Leibniz's algebra of concepts gradually evolves from the traditional theory of the syllogism in four steps: First, by distilling an abstract operator of *conceptual containment* out of the informal proposition 'Every *S* is *P*'. Second, by inventing or discovering the operator of *conceptual conjunction* inherent in the operation of juxtaposition of concepts like 'rational animal'. Third, by a thoroughgoing elaboration of the laws of *conceptual negation*, which goes hand in hand, fourth, with the invention of the operator of *possibility*, or *self-consistency*, of *concepts*.

4.5.2.1 Conceptual Containment and Conceptual Coincidence

By the end of the 1670s, Leibniz has come to realize that, in the traditional formulation of the UA, the quantifier expression 'every' is basically superfluous. Instead of 'Every *A* is *B*' one may simply say '*A is B*'. Thus in an early draft of a "Calculus ratiocinator" he abbreviates 'Omnis homo est animal' by the formula '*A* est *B*' because "the subject is always supposed to be preceded by a universal sign".[26] Similarly, in the "Specimen Calculi Universalis" of 1679, he explains:

> (1) A universal affirmative proposition will be expressed here as follows: *A* is *B*, or (every) man is an animal. We shall, therefore, always understand the sign of universality to be prefixed.[27]

By the year 1686 at the latest, when elaborating his "General Inquiries about the Analysis of Concepts and of Truths", Leibniz uses to express the UA not only with the help of the "copula" 'is' as '*A is B*', but alternatively also as '*A contains B*' or '*B* is contained in *A*':

> (28) I usually take as universal a term which is posited simply: e.g. '*A* is *B*', i.e. 'Every *A* is *B*', or 'In the concept *A* the concept *B* is contained'.[28]

---

[25]As will turn out below, this weak condition of existential import is tantamount to the assumption that concept *B* is *self-consistent*!

[26]Cf. A VI, 4, p. 274: "Subjectum *a* in exemplo praecedenti, *Omnis homo*. Semper enim signum universale subjecto praefixum intelligatur".

[27]Cf. GP 7, p. 218 or the translation in Parkinson (1966), p. 33. For the sake of uniformity, Leibniz's small letters '*a*', '*b*' have been replaced by capitals '*A*', '*B*'.

[28]Cf. C., p. 367 or the translation in Parkinson (1966), p. 57.

Leibniz's terminology is based upon the traditional distinction between the *extension* and the "*intension*" (or comprehension) of a concept. Thus in the "Elementa Calculi" of April 1679 he wrote:

> (11) [...] For example, the concept of gold and the concept of metal differ as part and whole; for in the concept of gold there is contained the concept of metal and something else – e.g. the concept of the heaviest among metals. Consequently, the concept of gold is greater than the concept of metal.
>
> (12) The Scholastics speak differently; for they consider, not concepts, but instances which are brought under universal concepts. So they say that metal is wider than gold, since it contains more species than gold, and if we wish to enumerate the individuals made of gold on the one hand and those made of gold on the other, the latter will be more than the former, which will therefore be contained in the latter as a part in the whole. [...] However, I have preferred to consider universal concepts, i.e. ideas, and their combinations, as they do not depend on the existence of individuals. So I say that gold is greater than metal, since more is required for the concept of gold than for that of metal and it is a greater task to produce gold than to produce simply a metal of some kind or other. Our language and that of the Scholastics, then, is not contradictory here, but it must be distinguished carefully.[29]

Similarly, in the "New Essays on Human understanding" the two opposing points of view are explained as follows:

> The common manner of statement concerns individuals, whereas Aristotle's refers rather to ideas or universals. For when I say *Every man is an animal* I mean that all the men are included among all the animals; but at the same time I mean that the idea of animal is included in the idea of man. 'Animal' comprises more individuals than 'man' does, but 'man' comprises more ideas or more attributes: one has more instances, the other more degrees of reality; one has the greater extension, the other the greater intension.[30]

From these considerations it follows quite generally that the *extension* of concept *A* is *contained in* the extension of *B* if and only if the *intension* of *A contains* the intension of *B*:

REZI 1 $\text{Ext}(A) \subseteq \text{Ext}(B) \leftrightarrow \text{Int}(A) \supseteq \text{Int}(B)$.

Leibniz defended this principle of reciprocity in a paper of August 1690 as follows:

> If I say 'Every man is an animal' I want the notion of animal to be contained in the idea of man. And the method of notions is contrary to the method of individuals: just as [...] all men are contained in all animals, so conversely the notion of animal is contained in the notion of man. And just like there are more animals besides the men, so something must be added to the idea of animal in order to get the idea of man. For by augmenting the conditions, the number decreases.[31]

Now the law REZI 1 immediately entails that two concepts with the same extension must also have the same intension:

REZI 2 $\text{Ext}(A) = \text{Ext}(B) \rightarrow \text{Int}(A) = \text{Int}(B)$.

[29] Cf. C., p. 53 or the translation in Parkinson (1966), p. 20–21. A similar distinction may also be found in Arnauld & Nicole (1683), p. 51–2.

[30] Cf. GP 5, p. 469.

[31] Cf. C., p. 235.

According to our modern understanding of "intensionality", this principle is clearly invalid because one can find concepts or predicates *A* and *B* which have the same extension but not the same "intension". To quote a famous example of Quine, it seems biologically plausible to assume that all animals with a heart have a kidney, and vice versa.[32] Therefore the predicates 'animal with heart' and 'animal with kidney' have the same extension, while their "intensions" or "meanings" differ widely. However, "intension" in the sense of traditional logic must not be mixed up with "intension" in the modern sense. While for contemporary modal logic the intension of an expression is always considered as something dependent of the respective *possible world*, according to the traditional view the intension of a concept *A* is not to be taken relative to different possible worlds. Instead it only *mirrors* the extension of *A* in the actual world. Furthermore, in Leibniz's view, the extension of concept *A* is not just the set of *actually existing* individuals, but rather the set of all *possible* individuals that fall under that concept. Anyway, in what follows the containment-relation between concepts *A*, *B* shall be symbolized as:

$A \in B$ "*A* is *B*"; "*A* contains *B*".

The logical properties of this relation are easily determined. Already in "De Elementis cogitandi" of 1676, Leibniz had put forward the "absolute identical proposition *A* is *A*" together with the "hypothetical identical proposition: If *A* is *B*, and *B* is *C*, then *A* is *C*".[33] Hence the containment-relation is both *reflexive* and *transitive*.

CONT 1 $A \in A$
CONT 2 $A \in B \wedge B \in C \rightarrow A \in C$.

Furthermore Leibniz soon recognized that the *identity* or *coincidence* of concepts may be defined as *mutual containment*. Thus in the "Elementa ad calculum condendum" of around 1678 he notes that "If *A* is *B* and *B* is *A*, then one can be substituted for the other salva veritate", where a few lines later he defines that "*A* and *B* are the same, if one can everywhere be substituted for the other".[34] With the help of the symbol '=', the former definition may be rendered as follows:

IDEN 1 $A = B \leftrightarrow A \in B \wedge B \in A$.

The famous "Leibniz law of identity", i.e. the principle of the substitutivity of identicals, can be formalized by the following inference scheme (where α is an arbitrary proposition):

IDEN 2 If $A = B$, then $\alpha[A] \Longleftrightarrow \alpha[B]$.

---

[32] Cf. Quine (1953), p. 21.
[33] Cf. A VI, 3, p. 506.
[34] Cf. A VI, 4, p. 154.

By means of these two basic principles, the subsequent corollaries, according to which the identity-relation is reflexive, transitive, and symmetric, can easily be derived:

IDEN 3 $A = A$

IDEN 4 $A = B \wedge B = C \rightarrow A = C$

IDEN 5 $A = B \rightarrow B = A$.[35]

4.5.2.2 Conceptual Conjunction

The operator of conceptual *conjunction* combines, e.g., 'animal' and 'rational' by mere *juxtaposition* to 'rational animal'. More generally, two concepts A and B may be conjunctively combined into AB. As Leibniz points out in an early draft of a logical calculus, it follows from the very meaning of conjunctive juxtaposition that AB contains A (and similarly AB contains B) because "AB wants to express just this, namely that which is A and which also is B"[36]:

CONJ 1 $AB \in A$

CONJ 2 $AB \in B$.

In the "Addenda to the specimen of the Universal Calculus" of 1679, Leibniz noted that the operation of conceptual conjunction is *symmetric* and *idempotent*:

> It must also be noted that it makes no difference whether you say *AB* or *BA*, for it makes no difference whether you say 'rational animal' or 'animal rational'.
>
> The repetition of some letter in the same term is superfluous, and it is enough for it to be retained once; e.g. *AA* or 'man man'.[37]

In our symbolism these laws take the form:

CONJ 3 $AB = BA$

CONJ 4 $AA = A$.

Furthermore Leibniz recognized that in addition to principles CONJ 1, 2, which show that a "compound predicate can be divided into several", also conversely:

> Different predicates can be joined into one; thus if it is agreed that *A* is *B*, and (for some other reason) that *A* is *C*, then it can be said that *A* is *BC*. For example, if man is an animal, and if man is rational, then man will be a rational animal.[38]

Hence one gets the further law of conjunction:

CONJ 5 $A \in B \wedge A \in C \rightarrow A \in BC$.

---

[35] Leibniz stated these laws especially in the "Generales Inquisitiones". Cf. A VI, 4, p. 751 "Propositio per se vera est *A* coincidit ipsi *A*"; p. 750: "(6) Si *A* coincidit ipsi *B*, *B* coincidit ipsi *A* [...] (8) Si *A* coincidit ipsi *B*, et *B* coincidit ipsi *C*, etiam *A* coincidit ipsi *B*".

[36] Cf. A VI, 4, p. 148: "*AB* est *A* pendet a significatione huiusmodi compositionis literarum. Hoc ipsum enim vult *AB*, nempe id quod est *A*, itemque *B*".

[37] Cf. GP 7, p. 221–2, or the translation in Parkinson (1966), p. 40.

[38] Cf. Parkinson (1966), p. 40.

In Leibniz's riper calculi this law will usually be strengthened into an *equivalence*: "That A contains B and A contains C is the same as that A contains BC"[39]:

CONJ 6 $A \in BC \leftrightarrow A \in B \wedge A \in C$.

To conclude this section, let it be pointed out that just as the identity operator can be defined by '∈' (according to IDEN 1), so conversely the ∈-operator might be defined (with the additional help of the operator of conceptual conjunction) by '=', namely according to the following law which was put forward by Leibniz, e.g., in § 83 of the "General Inquiries"[40]:

CONT 3 $A \in B \leftrightarrow (A = AB)$.

4.5.2.3 Conceptual Negation

Leibniz always used one and the same word, 'not' (in Latin 'non'), to designate the negation either of a *proposition* or of a *concept*. Here we will use, however, two different symbols, namely '¬' for the negation of a proposition, and '∼' for the negation of a concept. The logic of the propositional connective is quite straightforward. If one defines the negation of a proposition in the traditional way such that these "two propositions neither can be together true, nor can be together false", one obtains the following truth-conditions:

> If the affirmation is true, then the negation is false; if the negation is true, then the affirmation is false [...] If it is true that it is false, or if it is false that it is true, then it is false; if it is true that it is true, and if it is false that it is false, then it is true. All these are usually subsumed under the name of the *Principle of contradiction*.[41]

While Leibniz had an absolute clear understanding of the logic of *propositional* negation, during his research into the laws for *conceptual* negation he faced serious problems. From the tradition, he was acquainted with the law of double negation, "Not-not-$A = A$"[42]:

NEG 1 $\sim(\sim A) = A$.

Also Leibniz easily transformed the Scholastic principle of *contraposition* into a corresponding law of his "Universal calculus", *viz.*: "In general, 'A is B' is the same as 'Not-B is Not-A'"[43]:

NEG 2 $A \in B \leftrightarrow \sim B \in \sim A$.

[39]Cf. Parkinson (1966), p. 58, fn. 4.

[40]Cf. C., p. 378, or the translation in Parkinson (1966), p. 67.

[41]The first quotation is from April 1679, the second from around 1686; cf. A VI, 4, p. 248 and p. 804.

[42]Cf. § 96 of the "General Inquiries", e.g., A VI, 4, p. 767.

[43]Cf. § 77 of the "General Inquiries", e.g. A VI, 4, p 764, or the translation in Parkinson (1966), p. 67.

Furthermore Leibniz discovered the following variants of the law of *consistency* where the symbols '$\neq$' and '$\notin$', of course, are meant to abbreviate the *negation* of '$=$' and of '$\in$', respectively:

NEG 3 $A \neq \sim A$
NEG 4 $A = B \rightarrow A \neq \sim B$.
NEG 5* $A \notin \sim A$
NEG 6* $A \in B \rightarrow A \notin \sim B$.[44]

Principles NEG 5, 6 have been marked with a '*' in order to indicate that the laws are not *absolutely* valid. As will be explained below, they have to be restricted to *self-consistent* terms.

The cardinal mistake of Leibniz's theory of negation, however, consists in the frequent assumption that the one-way implication NEG 6 might be *strengthened* into the *equivalence*:

NEG 7* $A \notin B \leftrightarrow A \in \sim B$.[45]

This error is a bit surprising because in general Leibniz was well aware of the fact that the formula '$A \in B$' expresses the *universal affirmative* proposition while, on the background of the traditional principle of obversion, '$A \in \sim B$' formally represents the *universal negative* proposition. In view of the laws of *opposition*, the negated formulae '$A \notin B$' and '$A \notin \sim B$' therefore represent the *particular* negative and the *particular* affirmative proposition, respectively. Hence all four categorical forms can uniformly be expressed in Leibniz's algebra of concepts as follows:

(UA) $A \in B$
(UN) $A \in \sim B$
(PA) $A \notin \sim B$
(PN) $A \notin B$.

From this it follows that the basically (but not entirely) correct principle NEG 6* is just the formal counterpart of the law of *subalternation*, SUB 1, and this inference clearly must never be *converted*! Thus, e.g., in § 92 of the "General Inquiries", Leibniz emphasized that the inference from $A \notin \sim B$ to $A \in B$ (or, similarly, from $A \notin B$ to $A \in \sim B$) is *invalid*.[46] On the other hand, a little bit earlier in the same work, namely

[44] In the "General Inquiries", the above principles had been formulated as follows: "A proposition false in itself is '*A* coincides with not-*A*'" (§ 11); "If $A = B$, then $A \neq$ not-*B*" (§ 171, Seventh); "It is false that *B* contains not-*B*, that is, *B* doesn't contain not-*B*" (§ 43); and "*A* is *B*, therefore *A* isn't not-*B*" (§ 91). Cf. A VI, 4, p. 751, p. 783, p. 755, and p. 766, or the translation in Parkinson (1966), p. 56, p. 83, p. 59, and p. 68.

[45] Cf. GRUA, p. 536.

[46] Cf. A VI, 4, p. 766: "Non valet consequentia: Si *A* non est non-*B*, tunc *A* est *B*, seu Omne animal esse non hominem falsum est, quidem; sed tamen hinc non sequitur Omne animal esse hominem."

in § 82, he had maintained that "'*A* isn't *B*' is the same [!] as '*A* is not-*B*'", and this error was repeated again and again in many other fragments.[47]

The root of Leibniz's notorious problem of mixing up '$A \notin B$' and '$A \in \sim B$' is closely connected with the distinction between *singular* and *general* terms! If *A* is the name of some *individual*, or, as one could also say, if *A* is an *individual-concept*, then the two propositions $\neg$(*A* is *B*) and (*A* is $\sim B$), though being syntactically different, are semantically equivalent because it seems reasonable to assume that for each individual *x*, *x* has the negative property $\sim B$ iff *x* fails to have the positive property *B*. Thus in the "Calculi universalis investigationes" Leibniz explained:

> Two *terms* are *contradictory* if one is positive and the other the negation of this positive, as 'man' and 'not man'. About these the following rule must be observed: If there are two propositions with *exactly the same singular subject* of which the first has the one and the second the other of the contradictory terms as predicate, then necessarily one proposition is true and the other false. But I say: *exactly the same [singular] subject*, for instance if I say 'Apostle Peter was a Roman bishop' and Apostle Peter was not a Roman bishop'.[48]

The crucial law NEG 7* is indeed *valid* for the special case where the subject *A* is an *individual concept*. Unfortunately, Leibniz failed to realize with sufficient clarity that this principle may *not be generalized* to the case where *A* is an *arbitrary* concept. Thus, after the just quoted passage, he temporarily considered that of the pair of propositions 'Every man is learned', 'Every man is not learned', always exactly one would be true and the other false, but soon afterwards he noticed this error and remarked that the *generalized* rule is wrong.[49] However, a few lines later he considered the rule once again in a more abstract way (omitting the informal quantifier expression 'Every') and *then* he repeated the mistake of postulating not only the (basically) correct principle NEG 6: "I If the proposition '*A* is *B*' is true, then proposition '*A* is not-*B*' is false", but also the incorrect conversion: "III If the proposition '*A* is *B*' is false, then the proposition '*A* is not-*B*' is true".[50]

4.5.2.4 Conceptual Possibility

Fortunately, the partial confusions and errors of Leibniz's theory of negation (as described in the preceding section) are highly compensated by an ingenious discovery, namely the invention of the operator of *possibility* or *self-consistency* of concepts. This operator shall here be symbolized by

---

[47]Cf., e.g., § 21 of "Specimina calculi rationalis" in A VI, 4, p. 813: "*A* non est *B* idem est quod *A* est non *B*."

[48]Cf. A VI, 4, p. 218; the quoted example of Apostle Peter only appears in the critical apparatus of variants; Leibniz later replaced it by the less fortunate example 'this piece of gold is a metal' vs. 'this piece of gold is a non-metal'.

[49]Cf. A VI, 4, p. 218; critical apparatus, variant (d): "Imo hic patet me errasse, neque enim procedit regula."

[50]Cf. A VI, 4, p. 218; in order to avoid confusions, I have interchanged Leibniz's symbolic letters '*B*' and '*A*'.

P(*A*) ("*A* is possible").

Leibniz himself used many different locutions to express the self-consistency of a concept. Instead of '*A* is possible' he often says '*A* is a thing' ("*A* est res"), or '*A* is being' ("*A* est ens"), or simply '*A* is' ("*A* est"). In the opposite case of an impossible concept, he sometimes also calls *A* a "false term". Now in Leibniz's "Universal calculus", one can consider, in particular, the *inconsistent* concept $A{\sim}A$ ("*A* Not-*A*"), and therefore one may define that a concept *B* is *possible* if and only if *B doesn't contain* a contradiction like $A{\sim}A$:

Poss 1 $\mathrm{P}(B) =_{\mathrm{df}} (B \notin A{\sim}A)$.[51]

In order to get a clearer understanding of the truth-condition of the proposition P(*B*), let it be noted that the extension of the negative concept $\sim A$ must always be conceived as the set-theoretical *complement* of the extension of *A*, because an object *x* has the negative property $\sim A$ just in case that *x* fails to have property *A*. Furthermore, the extension of a *conjunctive* concept *BC* generally is the *intersection* of the extension of *B* and the extension of *C*, because *x* has the property *BC* if and only if *x* has both properties. From these conditions it follows that the extension of $A{\sim}A$ is the intersection of Ext(*A*) and its complement, i.e. the *empty set*! Hence a concept *B* is possible if and only if its extension is *not contained* in the empty set, which in turn means that Ext(*B*) itself *is not empty*!

At first sight, this requirement appears inadequate, since there are certain concepts – such as that of a unicorn – which happen to be empty but which may nevertheless be regarded as possible, i.e. not involving a contradiction. However, as Leibniz explained, e.g., in a paper on "Some logical difficulties",[52] the universe of discourse underlying the extensional interpretation of his logic should not be taken to consist of *actually existing* objects only, but instead to comprise all *possible* individuals. Therefore the non-emptiness of Ext(*B*) is both necessary and sufficient for guaranteeing the self-consistency of *B*. Clearly, if *B* is possible, then there must be at least one *possible individual x* that falls under concept *B*.

The following two laws describe some characteristic relations between the possibility-operator P and other operators of the algebra of concepts:

Poss 2 $A \in B \wedge \mathrm{P}(A) \rightarrow \mathrm{P}(B)$
Poss 3 $A \in B \leftrightarrow \neg\, \mathrm{P}(A{\sim}B)$.

Leibniz's own formulation of principle Poss 2: "If *A* contains *B* and *A* is true, *B* also is true" *prima facie* sounds a bit strange, but he goes on to explain:

[51]Cf. A VI, 4, p. 749, fn 8: "*A* non-*A* contradictorium est Possibile est quod non continet contradictorium seu *A* non-*A*. Possibile est quod non est *Y*, non-*Y*".

[52]Cf. GP 7, p. 211–217, or the translation in Parkinson (1966), p. 115–121.

> By a false letter I understand either a false term (i.e. one which is impossible, or not-being) or a false proposition. And in the same way by [a true letter] I understand either a possible term or a true proposition.[53]

Hence, if the term (or concept) A contains B and if A is "true", i.e. possible, then also B must be possible. This law, incidentally, might be proved as follows. Assume that $A \in B$ and that P(A); then also P(B) must hold, because otherwise B would contain a contradiction like $C{\sim}C$. But from $A \in B$ and $B \in (C{\sim}C)$ it would follow by CONT 2 that $A \in (C{\sim}C)$ which contradicts the assumption P(A).

The important law POSS 3, in contrast, cannot be derived from the remaining laws for containment, negation, and conjunction, but has to be adopted as a fundamental *axiom* of the algebra of concepts.[54] The systematic importance of POSS 3 is evidenced by the fact that in the "General Inquiries" Leibniz stated no less than six different versions of this law. Leibniz hit upon this crucial axiom by his investigation of propositions "*secundi adjecti*" vs. "*tertii adjecti*" which culminated in the discovery:

> (151) We have, therefore, propositions *tertii adjecti* reduced as follows to propositions *secundi adjecti*:
>
> 'Some *A* are *B*' gives '*AB* is a thing'
> 'Some *A* are not *B*' gives '*A* not-*B* is a thing'
> 'Every *A* is *B*' gives '*A* not-*B* is not a thing'
> 'No *A* is *B*' gives '*AB* is not a thing'.[55]

With the help of the possibility-operator, the two (almost valid) laws of consistency NEG 5* and NEG 6* can be improved as follows:

NEG 5 $\quad \mathrm{P}(A) \rightarrow A \notin {\sim}A$
NEG 6 $\quad \mathrm{P}(A) \rightarrow (A \in B \rightarrow A \notin {\sim}B)$.

As NEG 6 shows, the validity of the principle of *subalternation*, i.e. the inference from the UA '$A \in B$' to the PA '$A \notin {\sim}B$', only presupposes that the subject term A is self-consistent (and hence has a non-empty extension within the set of all possible individuals).

Note also that axioms POSS 2, POSS 3 admit a proof of the following counterpart of what in propositional logic is called "ex contradictorio quodlibet":

---

[53] Cf. § 55 of the "General Inquiries", e.g. A VI, 4, p. 757, or the translation in Parkinson (1966), p. 60.

[54] More exactly, this holds only for the implication $\neg\mathrm{P}(A{\sim}B) \rightarrow A \in B$, while the converse $A \in B \rightarrow \neg\mathrm{P}(A{\sim}B)$ is easily proven: If $A \in B$, then (by CONT 3) $A = AB$, hence (by IDEN 6) $A{\sim}B = AB{\sim}B$, and thus $(A{\sim}B) \in (B{\sim}B)$, i.e. $\neg\mathrm{P}(A{\sim}B)$. Cf. A VI, 4, p. 863: "*Vera propositio categorica affirmativa universalis* est: *A* est *B*, si *A* et *AB* coincidat et *A* sit possibile, et *B* sit possibile. Hinc sequitur, si *A* est *B*, vera propositio est, *A* non-*B* implicare contradictionem, nam pro *A* substituendo aequivalens *AB* fit *AB* non-*B* quod manifeste est contradictorium".

[55] Cf. Parkinson (1966), p. 81.

NEG 8 $(A(\sim A))\in B$.[56]

Just as the contradictory proposition $\alpha\wedge\neg\alpha$ entails any other proposition β, so the contradictory concept $A(\sim A)$ contains any other concept B![57]

It has often been criticized that Leibniz's logic lacks the operator of conceptual *disjunction*. Although this is by and large correct, it doesn't imply any defect or any incompleteness of his algebra of logic because the "missing" operator may simply be introduced by *definition*:

DISJ 1 $A\vee B =_{df} \sim(\sim A\sim B)$.[58]

On the background of the above axioms of *negation*, the standard laws for *disjunction*, e.g.

DISJ 2 $A\in(A\vee B)$

DISJ 3 $B\in(A\vee B)$

DISJ 4 $A\in C \wedge B\in C \rightarrow (A\vee B)\in C$,

may easily be *derived* from corresponding laws of *conjunction*. More generally, as was shown in Lenzen (1984), Leibniz's "intensional" logic of concepts turns out to be provably equivalent, or isomorphic, to Boole's extensional algebra of sets, and in this sense Leibniz really managed to transform the theory of the syllogism into a complete and sound algebra of concepts.

4.5.3 Indefinite Concepts, Quantifiers, and Individual Concepts

Leibniz's quantifier logic emerges from the algebra of concepts by the introduction of so-called "indefinite concepts". These concepts are symbolized by letters from the end of the alphabet *X*, *Y*, *Z* ..., and they function as quantifiers ranging over *concepts*. Thus in the "General Inquiries" Leibniz explained:

> (16) An affirmative proposition is '*A* is *B*' or '*A* contains *B*' [...]. That is, if we substitute the value for *A*, one obtains '*A* coincides with *BY*'. For example, 'Man is an animal', i.e. 'Man' is the same as 'a ... animal' (namely, 'Man' is 'rational animal'). For by the sign '*Y*' I mean

[56]Consider the concept $A(\sim A(\sim B))$ which contains $A(\sim A)$. Since $A\sim A$ is contradictory, it follows by POSS 2 that $A(\sim A(\sim B))$ is also impossible; but from $\neg P(A(\sim A(\sim B)))$ it immediately follows by POSS 3 that $A(\sim A)\in B$!.

[57]The inference from a contradictory pair of premises, α, $\neg\alpha$ to an arbitrary conclusion β was well known in Medieval logic, but Leibniz wasn't convinced of its validity. In his excerpts from Caramuel's *Leptotatos* (A VI 4, p. 1334–1343) he considered the "argumentatio curiose" by means of which, e.g., the conclusion 'Circulus habet 4 angulos' is derived from the premises 'Petrus currit' and 'Petrus non currit'. Although the deduction is based on two impeccable formal principles, Leibniz annotated: "Videtur esse sophisma".

[58]Leibniz knew quite well that the corresponding *propositional* connective $(\alpha\vee\beta)$ can similarly be defined as $\neg(\neg\alpha \wedge\neg\beta)$. For a closer discussion cf. Lenzen (1983), p. 132–133.

> something undetermined, so that '*BY*' is the same as 'Some *B*', or 'a ... animal' [...], or 'a certain animal'. So '*A* is *B*' is the same as '*A* coincides with some *B*', i.e. '$A = BY$'.[59]

With the help of the modern symbol for the existential quantifier, ∃, the latter law can be expressed more precisely as follows:

CONT 4 $A{\in}B \leftrightarrow \exists Y(A = BY)$.

As Leibniz himself noted, the formalization of the UA according to CONT 4 is provably equivalent to the simpler representation according to CONT 3.[60] On the one hand, according to the rule of existential generalization,

EXIST 1 If $\alpha[A]$, then $\exists Y\alpha[Y]$,

$A = AB$ immediately entails $\exists Y(A = YB)$. On the other hand, if there exists some Y such that $A = YB$, then according to IDEN 6, $AB = YBB$, i.e. $AB = YB$ and hence (by the premise $A = YB$) $AB = A$.[61]

Next observe that Leibniz often used to formalize the PA 'Some *A* is *B*' by means of an indefinite concept as '$YA{\in}B$'. In view of CONT 3, this representation might be transformed into the (elliptic) equation $YA = ZB$. However, both formalizations are somewhat *inadequate* because they are *theorems* of Leibniz's quantifier logic! According to CONJ 4, $BA{\in}B$, hence by EXIST 1:

CONJ 6 $\exists Y(YA{\in}B)$.

Similarly, since, according to CONJ 3, $AB = BA$, a twofold application of EXIST 1 yields:

CONJ 7 $\exists Y \exists Z(YA = ZB)$.

These *tautologies*, of course, cannot adequately represent the PA which for an appropriate choice of concepts *A* and *B* may become *false*! In order to resolve these difficulties, consider a draft of a calculus probably written between 1686 and 1690, where Leibniz proved principle:

NEG 9* $A{\notin}B \leftrightarrow \exists Y(YA{\in}{\sim}B)$.[62]

On the one hand, it is interesting to see that after first formulating the right hand side of the equivalence, "as usual", in the *elliptic* way '*YA* is Not-*B*', Leibniz later paraphrased it by means of the *explicit* quantifier expression "there exists a *Y* such that *YA* is Not-*B*". On the other hand, Leibniz discovered that NEG 8* has to be improved by requiring more exactly that there exists a *Y* such that *YA* contains ~*B* *and YA is possible*, i.e. *Y* must be *compatible* with *A*:

---

[59] Cf. A VI, 4, p. 751 or the translation in Parkinson (1966), p. 56

[60] Cf. A VI, 4, p. 751, fn. 13, or Parkinson (1966), p. 56, fn. 1: "It is noteworthy that for '$A = BY$' one can also say '$A = AB$' so that there is no need to introduce a new letter".

[61] This proof was given by Leibniz himself in the important paper "Primaria Calculi Logic Fundamenta" of August 1690; cf. C., 235.

[62] Cf. C., 259–261, or the text-critical edition in A VI, 4, p. 807–814.

NEG 9 $A \notin B \leftrightarrow \exists Y(\mathrm{P}(YA) \wedge YA \in {\sim}B)$.

Leibniz's proof of this important law is quite remarkable:

> (18) [. . .] to say '*A* isn't *B*' is the same as to say 'there exists a *Y* such that *YA* is Not-*B*'. If '*A* is *B*' is false, then '*A* Not-*B*' is possible by [POSS 2]. 'Not-*B*' shall be called '*Y*'. Hence *YA* is possible. Hence *YA* is Not-*B*. Therefore we have shown that, if it is false that *A* is *B*, then *YA* is Not-*B*. Conversely, let us show that if *YA* is Not-*B*, '*A* is *B*' is false. For if '*A* is *B*' would be true, '*B*' could be substituted for '*A*' and we would obtain '*YB* is Not-*B*' which is absurd.[63]

To conclude the sketch of Leibniz's quantifier logic, let us consider some of the few passages where an indefinite concept functions as a *universal* quantifier. In the above quoted draft, Leibniz put forward principle "(15) '*A* is *B*' is the same as 'If *L* is *A*, it follows that *L* is *B*'", i.e.:

CONT 5 $A{\in}B \leftrightarrow \forall Y(Y{\in}A \rightarrow Y{\in}B)$.[64]

Furthermore, in § 32 GI, Leibniz at least vaguely recognized that just as $A{\in}B$ (according to CONJ 6) is equivalent to $\exists Y(A = YB)$, so the negation $A{\notin}B$ means that, for *any* indefinite concept Y, $A \neq BY$:

CONT 6 $A \notin B \leftrightarrow \forall Y(A \neq YB)$.

According to the text-critical edition, Leibniz had written:

> (32) The negative proposition '*A* does not contain *B*', or it is false that *A* is (or contains) *B*, or *A* does not coincide with *BY*.[65]

Unfortunately, the passage 'or *A* does not coincide with *BY*' ("seu *A* non coincidit *BY*") had been overlooked by Couturat and it is therefore also missing in Parkinson's translation! Anyway, with the help of '∀', one can formalize Leibniz's conception of *individual concepts* as *maximally-consistent* concepts as follows:

IND 1 $\mathrm{Ind}(A) \leftrightarrow_{\mathrm{df}} \mathrm{P}(A) \wedge \forall Y(\mathrm{P}(AY) \rightarrow A \in Y)$.

Thus *A* is an individual concept iff *A* is self-consistent and *A* contains every concept *Y* which is compatible with *A*. The underlying idea of the *completeness* of individual concepts had been formulated in § 72 of the "General Inquiries" as follows:

> So if *BY* is ["being"], and the indefinite term *Y* is superfluous, i.e., in the way that 'a certain Alexander the Great' and 'Alexander the Great' are the same, then *B* is an *individual*. If the term *BA* is ["being"] and if *B* is an individual, then *A* will be superfluous; or if $BA = C$, then $B = C$.[66]

---

[63]Cf. C., p. 261.

[64]Cf. C., p. 260.

[65]Cf. A VI, 4, p. 753: "(32) *Propositio Negativa*. *A* non continet *B*, seu *A* esse (continere) *B* falsum est., seu *A* non coincidit *BY*."

[66]Cf. A VI, 4, p. 762 or the translation in Parkinson (1966), p. 65, § 72 including fn. 1; for a closer interpretation of Leibniz's logical criteria for individual concepts cf. Lenzen (2004).

Note, incidentally, that IND 1 might be simplified by requiring that, for each concept Y, A either contains Y or contains $\sim Y$:

IND 2 $\quad \mathrm{Ind}(A) \leftrightarrow \forall Y(A \in \sim Y \leftrightarrow A \notin Y)$.

As a corollary it follows that the *invalid* principle NEG 7*, which Leibniz again and again had considered as valid, in fact holds only for *individual concepts*:

NEG 7 $\quad \mathrm{Ind}(A) \rightarrow (A \notin B \rightarrow A \in \sim B)$.

Already in the "Calculi Universalis Investigationes" of 1679, Leibniz had pointed out:

> [...] if two propositions are given with exactly the same singular [!] subject, where the predicate of the one is contradictory to the predicate of the other, then necessarily one proposition is true and the other is false. But I say: *exactly the same [singular] subject*, for example, 'This gold is a metal', 'This gold is a not-metal'.[67]

The crucial issue here is that NEG 7 holds only for an individual concept like, e.g., 'Apostle Peter', but not for *general concepts* as, e.g., 'man'. The text-critical apparatus of the Academy Edition reveals that Leibniz was somewhat diffident about this decisive point. He began to illustrate the above rule by the correct example "if I say 'Apostle Peter was a Roman bishop', and 'Apostle Peter was not a Roman bishop'" and then went on, erroneously, to generalize this law for arbitrary terms: "or if I say 'Every man is learned' 'Every man is not learned'." Finally he noticed this error "Here it becomes evident that I am mistaken, for this rule is not valid."[68]

4.5.4 "Real Addition and Subtraction": Some Building Blocks of Elementary Set-Theory

The so-called Plus-Minus-Calculus was mainly developed in the "Specimen Calculi Coincidentium et Inexistentium" and in the "Non inelegans specimen demonstrandi in abstractis" of around 1687.[69] Strictly speaking, it is not a *logical* calculus but rather a much more general calculus which admits of different applications and interpretations. In its abstract form, it should be regarded as a theory of set-theoretical *containment*, set-theoretical "*addition*", and set-theoretical "*subtraction*". Unlike modern systems of set-theory, however, Leibniz's calculus has no counterpart of the relation 'x is an element of A'; and it also lacks the operator of "negation", i.e. set-theoretical *complement*! The complement of set A *might*, though, be *defined*

---

[67]Cf. A VI, 4, p. 217–218.

[68]Cf. A VI, 4, p. 218, lines 3–6, variant (*d*). The long story of Leibniz's cardinal mistake of mixing up '*A* isn't *B*' and '*A* is not-*B*' is analyzed in detail in Lenzen (1986).

[69]Cf. the text-critical edition in A VI, 4, p. 830–845 and 845–855; English translations may be found in Parkinson (1966), p. 122–130, 131–144.

with the help of the subtraction operator as (U-A) where the constant 'U' designates the *universe of discourse*. But, in Leibniz's calculus, this additional logical element is lacking.

Leibniz's drafts exhibit certain inconsistencies which result from the experimental character of developing the laws for "*real*" addition and subtraction in close analogy to the laws of *arithmetical* addition and subtraction. The genesis of this idea is described in detail in Lenzen (1989, 2000). The inconsistencies might be removed basically in two ways. First, one might restrict A-B to the case where B is contained in A; such a conservative reconstruction of the Plus-Minus-Calculus has been developed in Dürr (1930). The second, more rewarding alternative consists in admitting the operation of »real subtraction« A-B also if B is not contained in A. In any case, however, one has to give up Leibniz's idea that subtraction might yield "privative" entities which are "less than nothing".

In the following reconstruction, Leibniz's symbols '+' for the addition (i.e. union) and '-' for the subtraction of sets are retained, while his informal expressions 'Nothing' ("nihil") and 'is in' ("est in") are replaced by the modern symbols '∅' and '⊆'. Set-theoretical identity may be treated either as a *primitive* or as a *defined* operator. In the former case, inclusion can be defined either by $A{\subseteq}B =_{\mathrm{df}} \exists Y(A + Y = B)$ or simpler as $A{\subseteq}B =_{\mathrm{df}} (A + B = B)$. If, conversely, inclusion is taken as primitive, identity can be defined as mutual inclusion: $A = B =_{\mathrm{df}} (A{\subseteq}B) \wedge (B{\subseteq}A)$.[70] Set-theoretical addition, of course, is symmetric, or, as Leibniz puts it, "transposition makes no difference here"[71]:

PLUS 1 $A + B = B + A$.

The main difference between arithmetical addition and "real addition" is that the addition of one and the same "real" thing (or set of things) doesn't yield anything new:

PLUS 2 $A + A = A$.

As Leibniz puts it:

> A+A = A [. . .] i.e. repetition changes nothing here. (For although four coins and another four coins are eight coins, four coins and the same four already counted are not.)[72]

The "real nothing", i.e. the empty set ∅, is characterized as follows: "It does not matter whether Nothing is put or not, i.e. A+Nih. = A"[73]:

NIHIL 1 $A + \varnothing = A$.

In view of the relation $(A{\subseteq}B) \leftrightarrow (A{+}B = B)$, this law can be transformed into:

---

[70]Cf., e.g., Parkinson (1966), p. 131–144 (Definition 3, Propositions 13, 14 and 17).

[71]Cf. Parkinson (1966), p. 132 (Axiom 1).

[72]Cf. Parkinson (1966), p. 132.

[73]Cf. C., p. 267.

NIHIL 2 $\varnothing \subseteq A$.

"Real" subtraction may be regarded as the converse operation of addition: "If the same is put and taken away [...] it coincides with Nothing. I.e. *A* [...] – *A* [...] = *N*"[74]:

MINUS 1 $A - A = \varnothing$.

Leibniz also considered the following principles which in a stronger form express that subtraction is the converse of addition:

MINUS 2* $(A + B) - B = A$
MINUS 3* $(A + B) = C \to C - B = A$.

But he soon recognized that these laws do not hold in general but only in the special case where *A* and *B* are "uncommunicating".[75] The new operator of "communicating" sets has to be understood as follows:

> If something, *M*, is in *A*, and the same is in *B*, this is said to be 'common' to them, and they will be said to be 'communicating'.[76]

Hence two sets *A* and *B* have something in common if and only if there exists some set *Y* such that $Y \subseteq A$ and $Y \subseteq B$. Now since, trivially, the empty set is included in every set (cf. NIHIL 2), one has to add the qualification that *Y* is not empty:

COMMON 1 $\text{Com}(A, B) \leftrightarrow_{\text{df}} \exists Y(Y \neq \varnothing \wedge Y \subseteq A \wedge Y \subseteq B)$.

The necessary restriction of MINUS 2* and MINUS 3* can then be formalized as follows:

MINUS 2 $\neg\text{Com}(A, B) \to ((A + B) - B = A)$
MINUS 3 $\neg\text{Com}(A, B) \wedge (A + B = C) \to (C - B = A)$.

Similarly, Leibniz recognized that from an equation $A + B = A + C$, *A* may be subtracted on both sides provided that *C* is "uncommunicating" both with *A* and with *B*, i.e.:

MINUS 4 $\neg\text{Com}(A, B) \wedge \neg\text{Com}(A, C) \to (A + B = A + C \to B = C)$.[77]

Furthermore Leibniz discovered that the implication in MINUS 2 may be converted (and hence strengthened into a biconditional). Thus one obtains the following criterion: Two sets *A*, *B* are "uncommunicating" if and only if the result of first *adding* and then *subtracting* *B* coincides with *A*. Inserting negations on both sides of this equivalence one obtains:

[74]Cf. Parkinson (1966), p. 124 (Axiom 2).

[75]Cf. C., p. 267, # 29: "Itaque si $A + B = C$, erit $A = C\text{-}B$ [...] Sed opus est *A* et *B* nihil habere commune".

[76]Cf. Parkinson (1966), p. 123, who misleadingly inserts the word '*term*' before the entities *M*, *A*, *B*, while Leibniz himself spoke more neutrally of "*aliquid M*"!

[77]Cf. Parkinson (1966), p. 132 (Axiom 1).

COMMON 2 $\text{Com}(A,B) \leftrightarrow ((A+B)-B) \neq \text{A}$.

Whenever two sets A, B are communicating or "have something in common", the *intersection* of A and B, in modern symbols $A \cap B$, is not empty, i.e.:

COMMON 3 $\text{Com}(A,B) \leftrightarrow A \cap B \neq \varnothing$.[78]

Furthermore, "What has been subtracted and the remainder are uncommunicating",[79] i.e.:

COMMON 4 $\neg\text{Com}(A-B,B)$.

Leibniz further discovered the following formula which permits to "calculate" the *intersection* or "commune" of A and B by a series of additions and subtractions:

INTER 1 $A \cap B = B - ((A+B)-A)$.

In a small fragment (C., p. 250) he explained:

> Suppose you have A and B and you want to know if there exists some M which is in both of them. Solution: combine those two into one, $A+B$, which shall be called L [. . .] and from L one of the constituents, A, shall be subtracted [. . .] let the rest be N; then, if N coincides with the other constituent, B, they have nothing in common. But if they do not coincide, they have something in common which can be found by subtracting the rest N [...] from B [. . .] and there remains M, the commune of A and B, which was looked for.

In this way Leibniz gradually transformed the theory of arithmetical addition and subtraction into a fragment of the theory of sets. It is interesting to see how the incompatibility between the characteristic axiom of set-theoretical union, PLUS 1, and certain laws which hold only for numbers lead him to the discovery of new operators like 'Com' and '∩' which have no counterpart in elementary arithmetic.

References

Leibniz's Works and Manuscripts

(A) Deutsche Akademie der Wissenschaften (Ed.). *Sämtliche Schriften und Briefe*, quoted according to series (I–VIII) and volume. Cf. in particular series VI, *Philosophische Schriften*. Darmstadt 1930, Berlin 1962 ff.

(C.) Couturat, L. (Ed.). *Opuscules et fragments inédits de Leibniz*. Paris 1903, reprint Hildesheim 1961.

(GP) Gerhardt, C. I. (Ed.). *Die philosophischen Schriften von Gottfried Wilhelm Leibniz*, 7 volumes, Berlin 1875–1890, reprint Hildesheim 1965; quoted according to volume and page.

---

[78] Cf. Parkinson (1966), p. 127, Case 2 of Theorem IX: "Let us assume meanwhile that E is everything which A and G have in common – if they have something in common, so that if they have nothing in common, E = Nothing".

[79] Cf. Parkinson (1966), p. 128, Theorem X.

(GRUA) Grua, G. (Ed.). *G. W. Leibniz – Textes inédits*, Paris 1948 (2 volumes).
(LH) Bodemann, E. *Die Leibniz-Handschriften der Königlichen öffentlichen Bibliothek zu Hannover*, Hannover & Leipzig 1895, reprint Hildesheim 1966.

Secondary Literature

Ariew, R., & Garber, D. (Eds.). (1989). *G. W. Leibniz – philosophical essays*. Indianapolis: Hackett.
Arnauld, A., & Nicole, P (1683) *La Logique ou L'Art de Penser* (5th ed.), reprint Paris 1965.
Couturat, L. (1901). *La Logique de Leibniz d'aprés des documents inédits*. Paris, reprint Hildesheim 1985.
Dürr, K. (1930). *Neue Beleuchtung einer Theorie von Leibniz – Grundzüge des Logikkalküls*. Darmstadt: O. Reichl.
Lenzen, W. (1983). Zur extensionalen und ≫intensionalen≪ Interpretation der Leibnizschen Logik. *Studia Leibnitiana, 15*, 129–148.
Lenzen, W. (1984). Leibniz und die Boolesche Algebra. *Studia Leibnitiana, 16*, 187–203.
Lenzen, W. (1986). 'Non est' non est 'est non' – Zu Leibnizens Theorie der Negation. *Studia Leibnitiana, 18*, 1–37.
Lenzen, W. (1989). Arithmetical vs. 'Real' addition – a case study of the relation between logic, mathematics, and metaphysics in Leibniz. In N. Rescher (Ed.), *Leibnizian inquiries – a group of essays* (pp. 149–157). Lanham: University Press of America.
Lenzen, W. (1990). *Das System der Leibnizschen Logik*. Berlin: de Gruyter.
Lenzen, W. (2000). Guilielmi Pacidii Non plus ultra – oder – Eine Rekonstruktion des Leibnizschen Plus-Minus-Kalküls. *Philosophiegeschichte und logische Analyse, 3*, 71–118.
Lenzen, W. (2004). Logical criteria for individual(concept)s. In M. Carrara, A. M. Nunziante, & G. Tomasi (Eds.), *Individuals, minds, and bodies: Themes from Leibniz* (pp. 87–107). Stuttgart: Franz Steiner Verlag.
Lukasiewicz, J. (1951). *Aristotle's syllogistic – From the standpoint of modern formal logic*. Oxford: Clarendon Press.
Mackensen, L. v. (1990). Die ersten dekadischen und dualen Rechenmaschinen. In Stein/Heinekamp, pp. 52–61.
Parkinson, G. H. R. (Ed.). (1966). *Leibniz logical papers – A selection*. Oxford: Clarendon Press.
Quine, W. V. O. (1953). *From a logical point of view*. New York: Harper and Row.
Rescher, N. (2012). *Leibniz and cryptography – An account on the occasion of the initial exhibition of the reconstruction of Leibniz's Cipher Machine*. Pittsburgh: University Library System, University of Pittsburgh.
Rogers, H., Jr. (1963). An example in mathematical logic. *The American Mathematical Monthly, 70*, 929–945.
Sanchez-Mazas, M. (1979). Simplification de l'arithmétisation leibnitienne de la syllogistique par l'expression arithmétique de la notion intensionelle du ≫non ens≪. *Studia Leibnitiana Sonderheft, 8*, 46–58.
Smith, D. E. (Ed.). (1929). *A source book in mathematics* (Vol. 1). New York: Dover Publications.
Sotirov, V. (1999). Arithmetizations of Syllogistic à la Leibniz. *Journal of Applied Non-Classical Logics, 9*, 387–405.
Stein, E., & Heinekamp, A. (1990). *Gottfried Wilhelm Leibniz – Das Wirken des großen Philosophen und Universalgelehrten als Mathematiker, Physiker, Techniker*. Hannover: Gottfried-Wilhelm-Leibniz-Ges.

Chapter 5
Mathematics and Mechanical Computation

Doron Swade

> *what cannot be investigated and understood mechanically, cannot be investigated and understood at all – Thomas Carlyle (1829)*

Abstract The interplay of mathematics and machine is explored through early physical aids from pebbles to the 'analytical machines' of the nineteenth century. The earliest speculations on the nature and potential of computing machines are traced through the work of Charles Babbage for whom calculating engines represented a new technology for mathematics. Babbage's Analytical Engine, a mechanical embodiment of mathematical analysis, and his Mechanical Notation, a universal language of signs and symbols, are described. Ideas prompted by the intersection of mathematics and machine are discussed: the physicalisation of memory and the implications for coding, algorithmic programming, machine solution of equations, heuristics, computation as systematic method, halting, and numerical analysis. A brief Epilogue links this material to the modern era.

5.1 Introduction

Computing looks to its origins in early counting systems, and from earliest times practitioners have sought, through the use of physical aids, to offset human deficiencies of memory, computational ability, and trust.

As a medium of record physical aids have a long history. The use of knotted cords dates back to Biblical times in Old Testament references to knots as religious reminders as well as a record of dimensions of the temple to be. Roman tax collectors used knotted strings to record tax liabilities and payments. A system of knotted cords, *quipu,* were used by American Indians in the fifteenth and sixteenth

D. Swade (✉)
Department of Computer Science, Royal Holloway University, London, UK
e-mail: doron.swade@rhul.ac.uk

S. O. Hansson (ed.), *Technology and Mathematics*, Philosophy of Engineering and Technology 30, https://doi.org/10.1007/978-3-319-93779-3_5

centuries to record numerical data and also as a prompt for the recollection of historical events (Williams 1985: 35–6).

Material artefacts were enlisted to ensure accountability when trust and probity were at issue. Notched Tally Sticks to record sums of money, usually debts, date back some 8000 years. Such sticks were used in Medieval Europe to record tax liabilities or the amount of a debt. Split Tally Sticks were adopted in the thirteenth to nineteenth centuries by the English exchequer to record tax liabilities. Notches scored across the stick represented the sum owed or deposited, and the stick was split lengthwise, along the grain, into two matching pieces similarly scored. The slimmer piece was called the 'foil' and was held by the bank or Exchequer while the larger piece, the 'stock' was held by the depositor or debtor. The device served not only as a record of a loan and its partial repayments, but also as protection against one or another of the parties swindling the other as fraudulently modifying the marks by one party would create a mismatch when the two halves were later compared. The words 'counterfoil' and 'stock holder' are legacies of this practice.

The transition from counting to calculation can be found in *calculi*, small pebbles, used as markers or tokens freely placed on Roman counting boards. In the abacus with beads threaded on wires in a frame we find incipient mechanism – motion under mechanical constraint. Like the *quipu* and the counting board, the abacus uses a positional system of value in which the placement of the bead represents numerical value.

The European Enlightenment saw a surge in calculating aids. Analog devices with graduated scales for calculation and measurement were the mainstay of calculation from the seventeenth century onwards. The quadrant, sector and proportional compass are some. John Napier's eponymous Napier's Bones, described in 1617, consisting of a set of inscribed slats or rods, was a device to aid mainly paper-and-pen multiplication and division. There is nothing macabre in the name which derives from the fact that upmarket versions of the device were made from bone, horn or ivory, rather than wood.

Slide rules, with logarithmically graduated scales, were publicised in the 1630s following the introduction of logarithms by Napier in 1614. The most favoured of these for general use were 'universal' slide rules for multiplication, division, logarithmic and trigonometric functions. These had many variants some exotically specific: estimating excise duties (conversions scales for cubic inches to bushels, finding the mean diameter of a cask), calculating the volume of timber, the weight of cattle, estimating varieties of interest rates, and scales for a host of specialised engineering applications (Baxandall 1926). Slide rules offered the convenience of portability and robustness, and were widely used for the next 350 years for rapid calculation where accuracy of two to four digit places would suffice (Horsburgh 1914).

But the algorithmic rule still resided in the human operator upon whom the execution of the calculation depended. The seventeenth century saw several early stirrings to incorporate computational rule in mechanism. Savants in Continental Europe sought to produce mechanical devices for simple arithmetic. Wilhelm Schickard built his 'Calculating Clock' (1623), Blaise Pascal, his 'Pascaline'

(1642), and Gottfried Wilhelm Leibniz his ‘Reckoner’ (1674) (Martin [1925] 1992). Leibniz’s Reckoner introduced an innovative stepped drum that was the basis of calculator designs for the next two centuries. Decimal numbers were entered, sometimes using a stylus, on circular dials or sliders, and results, the outcome mainly of simple addition with carriage of tens, were displayed on engraved or annotated discs. The Reckoner is celebrated more for its ambition than for any practical accomplishment: the ‘carriage of tens’ failed to work as intended and only one, a largely unsuccessful prototype, appears to have been made (Morar 2015). The Pascaline stimulated philosophical debate about the mechanisation of mental process. Models were paraded before royalty, and demonstrated in the drawing rooms of merchants, government officials, aristocrats, and university professors. Most were ornate and expensive, philosophical novelties, insufficiently robust for daily use, and not many were made.

For all the ingenuity of their makers and their seriousness of purpose, mechanical calculators prior to the nineteenth century were largely *objets de salon*, many exquisite and delicate, sumptuous testaments to the instrument maker’s art, but unsuited to daily use in trade, finance, commerce, science or engineering.

The mechanical calculator that made a serious bid for widespread take-up was the arithmometer, patented and made public by Thomas de Colmar in 1820, and became the vanguard of mechanical calculator development in the nineteenth century (Johnston 1997). This was a desk-top device with sliders for entering numbers, numbered dials to display results, a moveable carriage for shifting decades, and a rotary crank handle. While often described as the first successful commercial calculator, the arithmometer was far from an instant success. It took over fifty years of modification and improvement before commanding even a small market. A contemporary government report evaluating utility of arithmometers records that even in the 1870s, they were troublesome, noisy, subject to derangement, imprecisely made, and in frequent need of repair (Mowatt 1893; Henniker 1893; Swade 2003a: 35–9). Arithmometers went on to sell in the tens of thousands but it had taken the better part of a century for them to mature as a product (Johnston 1997; Mounier-Kuhn 1999).

The function of these devices depends on the ability of the mechanism to manipulate physical representations of numerical value, and their mathematical capabilities were bounded by the state of contemporary mechanics. Unreliability was one issue, digit precision another. Arithmometers, for example, typically featured no more than six or eight digits. Here the limiting mechanism was the carriage of tens. In the worst case of a 1, say, being added to a row of 9s, the carriage of tens needed to propagate across each digit position as it altered 9s to 0s. The action followed a digit-to-digit causal sequence and to effect this domino or ripple-through carry the force required to advance all the digit wheels is derived from a single motion – the addition of 1 to the least significant digit. With calculators made of wood, ivory and soft workable metals digit precision was limited by the strength of the material transferring force from the manual dial, knob, or handle to the all of the digit positions in the same action.

There were other generic deficiencies in manual calculators that further inhibited their use. Multiplying two numbers using an arithmometer is accomplished by the accumulation of partial products. The operator enters the digits of the multiplier on sliding dials, rotates a handle the correct number of times for the current digit of the multiplier, lifts and correctly repositions the moveable carriage, and repeats this process for each next digit of the multiplier. Use of the device requires the continuous informed intervention of the operator and the correctness of the final result relies, not only on the repeated correct mechanical functioning of the device, but on the faultless execution by an operator of a sequence of physical manipulations. Only a limited part of the computational process is embodied in the mechanism (addition and the carriage of tens) with the overall computational algorithm provided not by the device but by the operator.

A further limitation was the absence of a permanent record of the outcome. Each new calculation replaces the last set of numbers in the mechanism, and the only way of retaining a record of prior or intermediate results is for the operator to note the contents of the registers by writing them down. Such transcription was again dependent on human agency with each manual operation in the sequence susceptible to error.

In the face of such constraints the mathematical ambitions of the calculator makers were modest confined as they were to four-function arithmetic at best, and while the struggle to produce viable devices that were more than aspirational novelties continued, practitioners, needing to perform other than elementary calculations relied for the most part on printed mathematical tables, or the slide rule. By a curious twist it was the reliance on printed tables that led to the game-changing episode. And it was the promise of mechanised mathematics that played a decisive role in subsequent change.

5.2 Mechanical Computation

The event that lifted the prospects for computational machines from the hands of struggling instrument makers is captured in the increasingly well-known vignette in which the English mathematician, Charles Babbage (1791–1871) and the astronomer, John Herschel, met in 1821 to check the accuracy of newly prepared manuscripts of astronomical tables.

An established practice of the day was double computation in which two sets of results were prepared independently by human computers without collaboration (Swade 2003a: 74–77). The manuscripts of the separately computed sets of results would then be compared for discrepancies (Lardner 1834: 278). The reduced likelihood of two independent computers making the same mistake increased confidence in the integrity of the results. The effectiveness of the technique relies on the independence of the computers from each other. Two computers were instantly dismissed when, hired to assist in the preparation of the British Nautical Almanac for 1771–2, they were found copying from each other (Forbes 1965: 394).

The technique of double computation was not foolproof: it was not unknown for computers who, despite insulation from each other, produced the same incorrect result, and these would elude detection using coincidence checks (Swade 2003b: 151–3).

Herschel and Babbage met to compare the two newly computed tables. During the process, Babbage, increasingly dismayed by the many discrepancies, exclaimed 'I wish to God these calculations had been executed by steam' (Hyman 1988: 46). 'Steam' can be read as a metaphor for the infallibility of machinery, as well as for the model of industrial production to solve the problem of supply. With machine as factory and number as product, tables, like manufactured goods, could be produced at will. In Babbage's invocation of steam we have an essential extension of the model of industrial production from goods to information, from physical to mental, from matter to mind (Schaffer 1994).

There are three accounts by Babbage of the meeting with Herschel, dating from 1822, 1834 and 1839 (Collier 1990: 14–8). The first account leaves it open as to whether it was Babbage or Herschel that made the suggestion of solution by machine. In the second and third accounts Babbage claims ownership of the suggestion for himself. The third account is the most dramatic and is the only one to include direct speech. All three accounts refer to steam. Babbage may well have dramatised the episode or aggrandised his role appropriating more credit with each retelling. But that the episode occurred, and was the jumping-off point for the half-century that followed in which Babbage's devoted the major part of his efforts to design and build automatic calculating engines, is well evidenced by published accounts that Herschel, who was party to the original event, neither questioned nor contradicted.

Even with an automatic calculating engine the production of tables would not eliminate human agency in its entirety. The practices in the production of printed tables had remained unchanged for centuries and involved five essential stages: calculation by hand of each tabular result, the transcription of these results into a tabular format suitable for typesetting, typesetting in loose type by a compositor, printing copies in a conventional printing press and, finally, verification and proof-checking results. In Babbage's aspirational world the 'unerring certainty of mechanical agency' (Lardner 1834: 311) would ensure error-free calculation; having the machine typeset results automatically would eliminate transcription and typesetting errors; the automatic production of stereotype plates during the calculating cycle would serve as moulds for the production of printing plates and would eliminate printing errors – the displacement of loose type by adhesion to sticky ink, for example – and automatically printing a checking copy would assist proof reading. 'It is only by the *mechanical fabrication of tables* that such [human] errors can be rendered impossible' asserted Dionysius Lardner in his grandiloquent advocacy of Babbage's Engine (Lardner 1834: 282). So at least in prospect Babbage's intended machine would, at a stroke, eliminate the risk of human error to which each of the manual processes was prone, and error-free tables would be available on demand. Astronomers were one group of potential beneficiaries. No longer would they need to petition a reluctant Astronomer Royal to compile tables for the trajectory of newly observed celestial bodies and incur inevitable delay. They would be able to produce tables on demand.

Fired up by the meeting with Herschel, Babbage was seized by the idea of automatic machine calculation, and he immediately began drafting exploratory designs for his difference engine, so called because of the mathematical principle on which it was based, the method of finite differences – an established method of manual calculation used by table-makers. Rather than evaluating the required function *ab initio* for each successive value of the table by repeated substitution of the argument uniformly incremented for each new entry, the method consisted of first finding the value of the function for relatively widely space intervals of the argument to yield a set of 'pivotal values', and then finding intermediate values by interpolation. The favoured technique was interpolation by subtabulation using the method of finite differences. The first use of the technique is not known. It may have originated with Henry Briggs who described it in 1624 though the term 'method of differences' appears not to have been adopted till the nineteenth century (Lindgren 1990: 311).

Examining more exactly the processes and division of labour involved in pre-mechanised tabulation helps to clarify the role a difference engine was intended to play. Tabulation by differences started with mathematicians who chose the formulae for the function to be tabulated, chose the particular form (typically a series expansion), fixed the range of the table (the start and end values of the independent variable), decided the number of decimals to be worked to, and calculated the pivotal values. The mathematicians also calculated the initial differences required to start the process and these, together with the pivotal values, the starting line of initial values, and a set of procedural instructions, were given to the human computers. Starting with the first pivotal value the computers calculated each next tabular value by the repeated addition of differences. The nth difference was added to the $(n\text{-}1)^{th}$ difference, the $(n\text{-}1)^{th}$ to the $(n\text{-}2)^{th}$ and so on, until the first difference was added to the last tabular result to yield the next tabular value. Each repetition of the train of additions generated the next tabular value and the process of repeated additions continued until the new pivotal value was reached. Subtabulation runs of as many as one hundred to two hundred values between pivotal values were not uncommon.

The hierarchy of skills involved is exemplified by the great French cadastral tables project led by Gaspard Francois de Prony in the late eighteenth century (Grattan-Guinness 2003; Swade 2003a: 56–62). The tables project, directed by de Prony, which aimed, amongst other things, to monumentalise the French metric system, was the most ambitious single tabulation project undertaken to that time. de Prony, France's leading civil engineer, was charged with preparing a set of trigonometric and logarithmic tables of unprecedented scope, scale and precision. He distributed the work to three groups reflecting the hierarchy of mathematical skills involved. The preparatory mathematical work was split between two groups of mathematicians, five or six high ranking mathematicians notable amongst whom were Legendre and Carnot, and seven or eight lesser mathematicians who calculated the pivotal values and starting values for the computers. The third group was the largest and consisted of sixty to eighty computers. These had no more than an elementary knowledge of arithmetic and carried out the most laborious and repetitive part of the process. The guillotining of the aristocracy had hit the hairdressing trade and the market for elaborate coiffures was in recession. The hairstyles of the aristocracy became a loathed symbol of the defunct pre-revolutionary regime and many of the computers were unemployed hairdressers who turned their hands to rudimentary arithmetic (Grattan-Guinness 1990).

Babbage was familiar with de Prony's project and greatly admiring of it. His engine would be used exclusively for interpolation using the method of differences and he calculated that interpolation by machine would reduce de Prony's workforce from ninety five to twelve, the greatest savings being made by replacing the entire group of computers. The machine would replace only the largely 'mechanical' work of the human computers. The role of the mathematicians was largely unaffected.

While tabular errors feature prominently in the historical narrative of Babbage's efforts it would be a mistake to take the elimination of error as the enduring motive for Babbage's interest in machine computation let alone its sole purpose. There is clear evidence that the primacy of errors in Babbage's motivational landscape has been over-emphasised, and a close reading of his earliest writings on his expectations for his machines demonstrates a parallel and possibly superordinate interest in the mathematical potential of mechanised computation that has been largely overlooked (Swade 2003a: 164–173, 2011: 246–8).

5.3 Mathematics and Machines

By the spring of 1822 Babbage had made a small working model of a Difference Engine powered by a falling weight. The model has come to be known as 'Difference Engine 0' (DE0) as it predates the later Difference Engine No. 1. The machine has never been found but from Babbage's descriptions we know that numerical value was represented by the rotation of geared wheels called 'figure wheels' inscribed with decimal numbers 0 through 9, and that multi-digit numbers were represented by figure wheels stacked in vertical columns. DE0 was capable of automatically tabulating quadratics using a repeated cycle that added the second difference to the first, and the first to the current tabular value to generate the next tabular value in the sequence, a mechanised version of the manual method practised by the table makers.

The capacity of DE0 was modest featuring a three-digit tabular value, a two-digit first difference and a single-digit second difference which was, in the case of native polynomials, constant (Hyman 1988: 47). However, the little machine has historical significance that transcends its modest capabilities. This was the first automatic calculating device that incorporated mathematical rule in mechanism and the computational algorithm was, for the first time, embodied in an autonomous machine. Through the physical agency of a falling weight, results could be achieved that up to that point in time were achievable only by mental effort.

Up to this point Babbage's main interest and experience had been in mathematics. He went up to Cambridge in 1810 at the age of seventeen, already a moderately accomplished mathematician. His published output at the time of his mechanical epiphany consisted entirely of mathematical papers, some thirteen in all between 1813 and 1821, of which the most interesting to modern mathematicians are those on the calculus of functions (Dubbey 1978). So in 1822, when experimenting with his new model, we have a mathematician aged twenty nine running the first practical automatic computing machine and reflecting on the implications for machine computation. With his first trials fresh in his mind he articulated these early reflections in five papers written in between June and December that year. The ideas and speculations contained in these writings are remarkable and evidence a two-way relationship between mathematics and computational analysis.

While Difference Engine 0 has never been found, a larger demonstration piece, completed in 1832, representing one-seventh of the calculating section of the complete Difference Engine No. 1 (Illustration 5.1), has all the essential features of its lost predecessor and is used here to illustrate Babbage's earliest recorded reflections on machine computation. The 'beautiful fragment', as the piece was referred to by Babbage's son, is the oldest surviving automatic computational machine (Babbage 1889: Preface).

Here, as in the first small model, number values are represented by the rotation of geared wheels inscribed with the decimal numbers 0–9 arranged in columns with the least significant digit in the lowermost position. The right-most column represents the tabular value, the middle column the first difference, and the left-most column the second difference. Initial values from a table, precalculated for the specific function being tabulated, are entered by rotating individual figure wheels by hand to the required digit value in a fixed setting-up sequence. For a given function the initial values fix the start value of the argument and the fixed increment of the argument for each next result. The Engine is then operated by cranking to and fro the handle above the top plate. Each cycle of the Engine produced, by repeated addition, the next value of the mathematical expression in the table with the tabular appearing on the on the right.

5.4 Computation as Systematic Method

New mathematical implications of machine computation are articulated in Babbage's open letter of 3 July 1822, to Sir Humphrey Davy, President of the Royal Society. Here Babbage advertised the use of his Engine to solve equations with no known analytical solution: 'Another and very remarkable point in the structure of this machine is ... that it will solve equations for which analytical methods of solution have not yet been contrived' (Babbage 1822: 4). This claim lifts machine computation from the relatively mundane context of tabulation, to computation as a systematic method of solution (Swade 2003a: 137).

(a)

(b)

Illustration 5.1 (**a**) Difference Engine No. 1 demonstration piece (1832). (**b**) Top view showing crank handle

The roots or solutions of an equation are the values of the independent variable at which the function passes through zero. The standard analytical technique for solving equations was to equate the expression to zero and to solve for the unknown. There was no systematic process for doing this and the success of the process depends on ingenuity, creativity, and often an ability to manipulate the problem into a recognisable form that has a known class of solution. Not only was there no guarantee of solution using such techniques, but there was no way of determining whether or not the equation in question was soluble in principle. If analytical methods failed, then trial and error substitution could be tried. This involves substituting trial values of the independent variable and repeating this process to see if values of the argument can be found for which the function converges to zero. But the technique was hit and miss, was regarded as 'inelegant' by mathematicians, and did not guarantee success.

Starting with an initial value of the independent variable, each cycle of the engine generates each next tabular value, and the machine has found a 'solution' when the figure wheels giving the tabular result are all at zero. So finding a solution reduces to detecting the all-zero state, and the number of machine cycles taken to achieve this represents the value of the independent variable, which is the solution sought. If two adjacent tabular results straddle zero (i.e. if the argument does not exactly coincide with the root but the value of the function passes through zero) the solution will be signalled by a change of sign. To remove the reliance on visual detection of the all zero state Babbage first incorporated a bell that would ring to alert the operator to the occurrence of specific conditions in the column of tabular values (Lardner 1834: 311). The operator would then halt the machine and read off the number of cycles the machine had run to give the first root of the equation (the machine had facilities for automatic cycle counting). If there were multiple roots the operator keeps cranking until the bell rings again. In the event that there are no roots, the machine continues *ad infinitum*. To further remove reliance on a human operator, provision was later made for the machines to halt automatically (Swade 2011: 249–50). Babbage wrote explicitly of the machine halting on finding a root.

Pre-echoes of Turing's 1936 paper 'On computable numbers' are unmistakable. What later became known as the 'halting problem', though not explicitly referred to as such by Turing, is inseparably associated with him. For those interpreting Turing's 'circular machine' halting became a logical determinant of whether or not machines could decide whether a certain class of problems was soluble. While Babbage himself did not claim any special *theoretical* significance for the halting criterion the resonances with decidability and solubility are unmistakeable.

The internal organisation and spatial layout of the engine suggested to Babbage new series for which there was a generative rule but no general expression for the nth term and Babbage speculates on the heuristic value of machine computation to mathematics (Babbage 1822: 312–3).

> On considering the arrangements of its parts, I observed that a different mode of connecting them would produce tables of a new species altogether different from any with which I was acquainted. I therefore computed with my pen a small table such as would have

been informed by the engine had it existed in this new shape and I was much surprised at discovering that no analytical method was yet known for determining its *n*th term.

Using external gearing any given figure wheel (representing units, tens, hundreds etc. of a given number) could add its value to a wheel in another column during the execution of a standard machine cycle. The technique allows feedback, feed-forward or cross-feeding of individual digits in a way that influences the step-wise generation of successive results. Cycling the machine would produce a new series for which there was a clear generative rule for each next value, but for which there was no known analytical formula: 'we are not in possession of methods of determining its *n*th term, without passing through all the previous ones' (Babbage 1823: 123). By cycling the machine to calculate each next value in turn, any requisite term can be reached. Machine computation offered solutions where formal analysis failed.

The portion of Difference Engine No. 1, assembled in 1832, shows additional axes and gearing that allow such cross-coupling and which were added when Babbage later returned to these ideas. He was intrigued by the general question of finding general laws for empirically generated series and he provides an example of how, using an inductive process, he derived a general expression for a new series suggested by the engine. Later he considered recurrence relationships in which each next term in the series is defined in terms of the current term and a few prior terms. In these cases there is a general expression for the *n*th term, but one that does not allow it to be calculated by direct substitution. A machine capable of iterating the requisite sequence of operations to calculate each new result in turn could again provide computational solutions that resisted analytical process.

This line of thinking fell outside the comfort zone of the traditions of the time. The appeal of analytical formulation derived from its generality, that is, the ability to represent, in a single symbolic statement, any and all specific instances of the relations expressed. A silent premise of contemporary mathematics and philosophy was that example was inferior to generalisation, induction inferior to deduction, empirical truths to analytical truths, and the synthetic to the analytic. Generality and universality were elevated above example and instantiation. Calculation, which involves specific numerical example was, in the prevailing culture, implicitly inferior to formal analysis. The existence of a series that could be produced by computational rule for which no formal law was known was at odds with prevailing attitudes. This was new territory and Babbage shows awareness that his enquiries were off *piste* when he wrote that he would desist from a general investigation of methods to determine general laws for series defined only by generative rules because the techniques involved (essentially induction) were 'not so much in unison with the taste which at present prevails in that science' (Babbage 1826b: 218). So computation and computational theory were cast as the methodological poor relation of mathematics and mathematical logic. The stigma of the contingent still rankles. Computer scientists still baulk at the imputation that their discipline might not be as well anchored intellectually as its elite campus neighbours, and they tend to shift in their seats at the suggestion that their subject has its historical roots in engineering rather than something more rarefied.

Babbage foresaw that computation by machine would give rise to new forms of mathematical analysis. One such was what we would now call numerical analysis that would *precede* computation by machine. In a manuscript unpublished in his lifetime he predicted that the need for techniques to optimise efficiency by tailoring the problem for computation by machine. He used as an example an expression that required 35 multiplications and 6 additions to evaluate and showed that manipulating it into an alternative but mathematically identical form reduced the computational load to 5 multiplications and 1 addition. (Babbage 1837a). In the case of tabulation, new techniques included 'best fit' approximation methods, and the preparatory analysis necessary to ensure that the approximation remains valid to the requisite accuracy in the interval to be tabulated. While neither of these elements was new to table-makers, Babbage articulates the need for such analysis in terms of the stimulus to mathematics to formalise and systematise computational method, and the value of this analysis in rendering practically useful otherwise abstract forms.

Early on Babbage anticipated that without machine computation, or an alternative, science would stultify:

> I will yet venture to predict that a time will arrive, when the accumulating labour which arises from the arithmetical applications of mathematical formulae, acting as a constantly retarding force, shall ultimately impede the useful progress of the science, unless this or some equivalent method is devised for relieving it from the overwhelming incumbrance of numerical detail. (Babbage 1823: 128)

The need to evaluate mathematical formulae for practical purposes stimulated the engine project and with it the development of mechanical logic. The detailed designs of the Difference Engine contain the earliest embodiment of fundamental principles of machine computation recognisable in the modern era. The use of terms familiar to us in the list below is entirely anachronistic and is for expository purposes only. Principles and features of mechanical logic explicitly detailed in the Difference Engine designs include: autonomy (transfer of rule from human to machine eliminating the need for human intervention in algorithmic process); digital operation through the discretisation of motion; parallel operation (the simultaneous operation on each digit of multi-digit numbers); non-destructive addition; carriage of tens including secondary carries; 'microprogramming'; 'pipelining'; 'pulse-shaping' (cleaning up degraded transitions to ensure digital integrity); error prevention, error correction and error detection; 'latching' (one-bit memory); 'polling' (sequenced interrogation of a series of logical states); and manual input of initial values, printed and/or stereotyped output (Swade 2005).

These earliest forays into machine computation evidence a two-way relationship between mathematics and machine. In one direction, the need for reliable mathematical tables was the original stimulus for inventing automatic calculating machines, pursuing their early development, and pioneering basic principles and potential of computational engines. In the reverse direction, we have the earliest articulation of some of the implications of machine computation for mathematics: computation as systematic method, heuristic potential, and the need for new forms of analysis tailored and contrived to serve new computational needs.

What is perhaps remarkable is that the springboard for the nineteenth-century speculations mentioned so far was not a computer as we would now understand the term but an automatic calculator. Difference engines are strictly calculators in that they crunch numbers the only way they know how – by repeated addition according the method of finite differences. They execute a single fixed algorithm on whatever initial values are supplied. While they are capable of conditional action (whether or not to execute a carriage of tens, for example, or whether to halt or not), they are not capable of branching i.e. they cannot deviate from a default operational sequence to pursue an alternative algorithmic trajectory. They have no generality even as a four-function calculator.

The essential feature on which Babbage's speculations are founded was that the machine was *automatic*. Mathematical rule was embodied in mechanism, the algorithm was contained in wheelwork and, by physical effort, results could be achieved which up to that point in time could only be achieved by mental effort. The idea that the machine was 'thinking' was not lost on Babbage or his contemporaries. In 1833 Lady Byron (Ada Lovelace's mother) referred to the 1832 demonstration piece as 'the thinking machine' (Toole 1992: 51). A junior colleague of Babbage's wrote that Babbage had 'taught wheelwork *to think*, or at least to do the office of thought' (Hyman 1988: 48–9). The machine being autonomous was the first step towards machine intelligence and it was this feature more than any other that served as the basis for early speculation about the implications and potential of machine computation.

5.5 From Computation to Computing

A practical advantage of the method of differences is that it requires only addition and this eliminates the need for multiplication and division that would ordinarily be needed to evaluate terms in a series, and addition is significantly simpler to realise in mechanism than direct multiplication and division. The method of finite differences allowed the calculation and tabulation of polynomial functions, and what generality this has to mathematics, science and engineering derives from the capacity to express functions in the form of series expansions. For all the ingenuity of its conception and design, difference engines are still calculators confined as they are to a fixed cycle of unvarying mechanical operations.

By a series of undocumented steps Babbage was led from mechanised calculation by differences to fully-fledged general purpose computing. His Analytical Engine, incipiently conceived in 1834 and on which he worked for the best part of the rest of his life, was a machine of unprecedented generality. Buxton, doubtless ventriloquising Babbage, wrote that 'the powers of the Analytical Engine are coextensive with analysis itself' (Hyman 1988: 150). Babbage himself maintained that two contemporary descriptions of his Analytical Engine (Lovelace 1843;

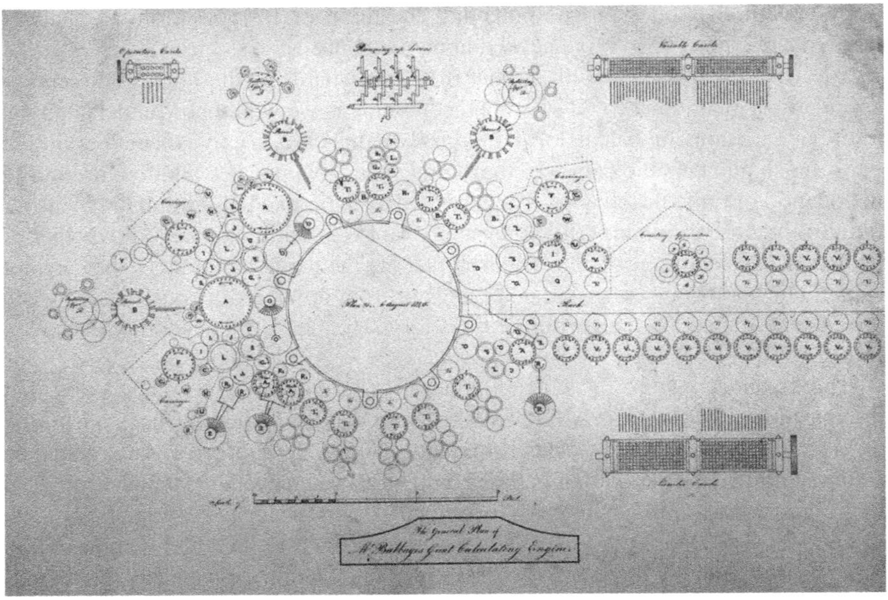

Illustration 5.2 Analytical engine, Plan 25 (1840)

Lardner 1834) demonstrated 'that the whole of the developments and operations of analysis are now capable of being executed by machinery' (Babbage 1864: 136).

The design as it stood in 1840 is shown in Illustration 5.2. In this view the circles represent registers (columns of figure wheels), stud-programmable 'barrels' ('microprograms'), and other mechanisms as seen from above. The cluster of circles around the central circle is the Mill (central processor) and the large central circle is a parallel bus consisting of large toothed wheels that allow transfer of data within the Mill. The Store (memory) is shown as two rows of registers extending indefinitely to the right. Each register or Variable (annotated $\mathbf{V}_n$), so called because the numerical value of its contents is not fixed, consists of a vertical column of up to 50 decimal figure wheels. Provision was made for double precision operation allowing for 100-digit results. The strip between the two rows of Variables (annotated *Rack*) is a stack of independently moveable toothed slats. The Racks act as a parallel data bus that transfers information between the Store and the Mill via buffer registers (Ingress axis (**I**), and Egress axis (**"A**) (Bromley 1982).

The machine described is physically massive. The central wheels of the Mill are some 5 ft 6 in. in diameter and some 15 ft high and the engine as shown would be about 10 ft long. However the Store is truncated in the drawing for reasons of drafting convenience and only 17 Variables are shown. Babbage's minimum engine which would have had some 100 Variables which would stretch it to over 20 ft long and he talks of talks of machines with 1000 Store Variables. A hundred-Variable machine would call for an estimated 50,000 parts.

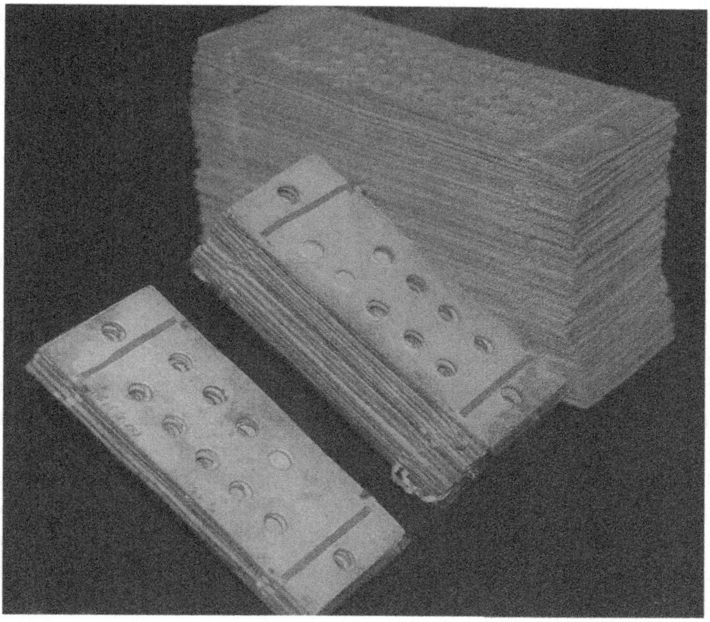

Illustration 5.3 Punched cards for analytical engine (1840). Smaller cards are operation cards. Larger cards are variable cards

The Analytical Engine (an abstraction of design, as it too was never built), offered the prospect of a new technology for mathematics, allowing as it did the evaluation of any definable function of arbitrary generality. At an operational level the Engine was capable of conditional branching, iterative looping, and microprogramming, though neither Babbage nor his contemporaries used these terms. At systems level it had a separate Store and Mill, a serial fetch-execute cycle, punched card input for data and instructions, output through print, punched card, or graph plotter, an internal repertoire of automatically executed operations including direct multiplication, division, addition and subtraction, parallel processing using multiple processors, look-ahead carry, buffering, and pipelining. At user-level it was programmable using punched cards of which there were several kinds, chief amongst which were Operation Cards which contained instructions, Number Cards contained input data, and Variable Cards specified where in the Store the operand was to be found, and the destination location for the result. Cards were made from paste-board and loosely stitched together with ribbon (Illustration 5.3). Repeating an instruction sequence, invaluable for iterative processes in recurrence relationships, was achieved by automatically winding back a pre-specified number of times (determined at one stage of development by a Combinatorial Card) a fixed sequence of Operation Cards (Bromley 2000).

The internal architecture of the machine pre-echoes the signature features of von Neumann's model described in 1945 and which dominated computer design since: the separation of memory and central processor, serial fetch-execute cycle, and input/output. There is no internal stored program in the Analytical Engine. The sequence of operations is stored on and executed from the Operation Cards stitched together in a sequential train.

Between 1836 and 1840 Babbage wrote 25 'programs' for the solution of a variety of mathematical problems, and the nature and scope of these are revealing (Babbage 1836, 1837b). The format of the programs is essentially tabular and similar but not identical for all. Typically the first column features a line number that indicates the order in which the sequence of operations is to be performed. Other columns specify the operation to be performed at each step, the location in store of the operands to be retrieved, which Store Variables are acted upon, where in the Store the results of each operation are to be placed and, for clarification, the changing contents of each Store Variable at each step as the computation progresses. These descriptions are not strictly programs as we now understand the term. For one, there is no control information. Babbage called them Notations of Calculation. 'Traces' or 'walkthroughs' have been suggested as more appropriate descriptions (Bromley 1991: L-1). But I hazard that they are sufficiently algorithmic in intention to exempt us from censure should we continue to refer to them as programs though it is admittedly anachronistic to do so.

Ten of Babbage's twenty four programs are concerned with the solution of simultaneous equations. There are separate programs for the successive reduction of n simultaneous equations in n variables to find the solution for one variable, with separate programs for $n = 3$, 4, and 5. There are several programs for the reduction of a number of n simultaneous equations in n variables to n-1 equations in n-1 variables, as well as the direct solution of simultaneous equations for each of the variables. Four of the programs are for tabulation by differences of quadratics, cubics, and quartrics. There is a program for the computation of the coefficients of a polynomial from those of another polynomial divided by a linear or quadratic term, and for the coefficients in the product of two polynomials. There are three examples of recurrence relationships requiring iterative looping. The most complex program in the suite is for the solution of a problem in astronomy – computing the radius vector of a body from the eccentricity and mean anomaly of its orbit.

The most celebrated program for the Analytical Engine is for the calculation of Bernoulli numbers written by Ada, Countess of Lovelace and published in 1843 (Lovelace 1843). With the possible exception of Babbage's vector radius calculation the Bernoulli calculation (Illustration 5.4) is the most complex program written for the machine requiring as it does nested looping and a form of memory addressing that allows all prior Bernoulli numbers to be retained and available for the calculation of each next term in the series (Glaschick 2016), a requirement not explicit in Babbage's examples of recurrence relations programs. These sample walkthroughs were intended to demonstrate both the computational power and generality of the machine.

Number of Operation	Nature of Operation	Variables acted upon	Variables receiving results	Indication of change in the value on any Variable	Statement of Results	Data			Working Variables										Result Variables			
						$^1V_1$ ○ 0 0 0 1 [1]	$^1V_2$ ○ 0 0 0 2 [2]	$^1V_3$ ○ 0 0 0 4 [n]	$^0V_4$ ○ 0 0 0 0	$^0V_5$ ○ 0 0 0 0	$^0V_6$ ○ 0 0 0 0	$^0V_7$ ○ 0 0 0 0	$^0V_8$ ○ 0 0 0 0	$^0V_9$ ○ 0 0 0 0	$^0V_{10}$ ○ 0 0 0 0	$^0V_{11}$ ○ 0 0 0 0	$^0V_{12}$ ○ 0 0 0 0	$^0V_{13}\cdots$ ○ 0 0 0 0	$^1V_{21}$ ○ B in a dec. fract. [B_1]	$^1V_{22}$ ○ B in a dec. fract. [B_3]	$^1V_{23}$ ○ B in a dec. fract. [B_5]	$^0V_{24}\cdots$ ○ 0 0 0 [B_7]
1	×	$^1V_2 \times {}^1V_3$	$^1V_4, {}^1V_5, {}^1V_6$	$\{^1V_2 = {}^1V_2;\ {}^1V_3 = {}^1V_3\}$	$= 2n$		2	n	2n	2n	2n											
2	−	$^1V_4 - {}^1V_1$	$^2V_4$	$\{^1V_4 = {}^2V_4;\ {}^1V_1 = {}^1V_1\}$	$= 2n - 1$	1			2n − 1													
3	+	$^1V_5 + {}^1V_1$	$^2V_5$	$\{^1V_5 = {}^2V_5;\ {}^1V_1 = {}^1V_1\}$	$= 2n + 1$	1				2n + 1												
4	÷	$^2V_5 \div {}^2V_4$	$^1V_{11}$	$\{^2V_5 = {}^0V_5;\ {}^2V_4 = {}^0V_4\}$	$= \frac{2n-1}{2n+1}$				0	0						$\frac{2n-1}{2n+1}$						
5	÷	$^1V_{11} \div {}^1V_2$	$^2V_{11}$	$\{^1V_{11} = {}^2V_{11};\ {}^1V_2 = {}^1V_2\}$	$= \frac{1}{2} \cdot \frac{2n-1}{2n+1}$		2									$\frac{1}{2} \cdot \frac{2n-1}{2n+1}$						
6	−	$^0V_{13} - {}^2V_{11}$	$^1V_{13}$	$\{^2V_{11} = {}^0V_{11};\ {}^0V_{13} = {}^1V_{13}\}$	$= -\frac{1}{2} \cdot \frac{2n-1}{2n+1} = A_0$											0		$= -\frac{1}{2} \cdot \frac{2n-1}{2n+1} = A_0$				
7	−	$^1V_3 - {}^1V_1$	$^1V_{10}$	$\{^1V_3 = {}^1V_3;\ {}^1V_1 = {}^1V_1\}$	$= n - 1 (= 3)$	1		n							n − 1							
8	+	$^1V_2 + {}^0V_7$	$^1V_7$	$\{^1V_2 = {}^1V_2;\ {}^0V_7 = {}^1V_7\}$	$= 2 + 0 = 2$		2					2										
9	÷	$^1V_6 \div {}^1V_7$	$^3V_{11}$	$\{^1V_6 = {}^1V_6;\ {}^0V_{11} = {}^3V_{11}\}$	$= \frac{2n}{2} = A_1$						2n	2				$\frac{2n}{2} = A_1$						
10	×	$^1V_{21} \times {}^3V_{11}$	$^1V_{12}$	$\{^1V_{21} = {}^1V_{21};\ {}^3V_{11} = {}^3V_{11}\}$	$= B_1 \cdot \frac{2n}{2} = B_1 A_1$											$\frac{2n}{2} = A_1$	$B_1 \cdot \frac{2n}{2} = B_1 A_1$		B_1			
11	+	$^1V_{12} + {}^1V_{13}$	$^2V_{13}$	$\{^1V_{12} = {}^0V_{12};\ {}^1V_{13} = {}^2V_{13}\}$	$= -\frac{1}{2} \cdot \frac{2n-1}{2n+1} + B_1 \cdot \frac{2n}{2}$												0	$\left\{-\frac{1}{2} \cdot \frac{2n-1}{2n+1} + B_1 \cdot \frac{2n}{2}\right\}$				
12	−	$^1V_{10} - {}^1V_1$	$^2V_{10}$	$\{^1V_{10} = {}^2V_{10};\ {}^1V_1 = {}^1V_1\}$	$= n - 2 (= 2)$	1									n − 2							
13	−	$^1V_6 - {}^1V_1$	$^2V_6$	$\{^1V_6 = {}^2V_6;\ {}^1V_1 = {}^1V_1\}$	$= 2n - 1$	1					2n − 1											
14	+	$^1V_1 + {}^1V_7$	$^2V_7$	$\{^1V_1 = {}^1V_1;\ {}^1V_7 = {}^2V_7\}$	$= 2 + 1 = 3$	1						3										
15	÷	$^2V_6 \div {}^2V_7$	$^1V_8$	$\{^2V_6 = {}^2V_6;\ {}^2V_7 = {}^2V_7\}$	$= \frac{2n-1}{3}$						2n − 1	3	$\frac{2n-1}{3}$									
16	×	$^1V_8 \times {}^3V_{11}$	$^4V_{11}$	$\{^1V_8 = {}^0V_8;\ {}^3V_{11} = {}^4V_{11}\}$	$= \frac{2n}{2} \cdot \frac{2n-1}{3}$								0			$\frac{2n}{2} \cdot \frac{2n-1}{3}$						
17	−	$^2V_6 - {}^1V_1$	$^3V_6$	$\{^2V_6 = {}^3V_6;\ {}^1V_1 = {}^1V_1\}$	$= 2n - 2$	1					2n − 2											
18	+	$^1V_1 + {}^2V_7$	$^3V_7$	$\{^2V_7 = {}^3V_7;\ {}^1V_1 = {}^1V_1\}$	$= 3 + 1 = 4$	1						4										
19	÷	$^3V_6 \div {}^3V_7$	$^1V_9$	$\{^3V_6 = {}^3V_6;\ {}^3V_7 = {}^3V_7\}$	$= \frac{2n-2}{4}$						2n − 2	4		$\frac{2n-2}{4}$								
20	×	$^1V_9 \times {}^4V_{11}$	$^5V_{11}$	$\{^1V_9 = {}^0V_9;\ {}^4V_{11} = {}^5V_{11}\}$	$= \frac{2n}{2} \cdot \frac{2n-1}{3} \cdot \frac{2n-2}{4} = A_3$									0		$\left\{\frac{2n}{2} \cdot \frac{2n-1}{3} \cdot \frac{2n-2}{4}\right\} = A_3$						
21	×	$^1V_{22} \times {}^5V_{11}$	$^0V_{12}$	$\{^1V_{22} = {}^1V_{22};\ {}^0V_{12} = {}^2V_{12}\}$	$= B_3 \cdot \frac{2n}{2} \cdot \frac{2n-1}{3} \cdot \frac{2n-2}{4} = B_3 A_3$											0	$B_3 A_3$			B_3		
22	+	$^2V_{12} + {}^2V_{13}$	$^3V_{13}$	$\{^2V_{12} = {}^0V_{12};\ {}^2V_{13} = {}^3V_{13}\}$	$= A_0 + B_1 A_1 + B_3 A_3$												0	$(A_0 + B_1 A_1 + B_3 A_3)$				
23	−	$^2V_{10} - {}^1V_1$	$^3V_{10}$	$\{^2V_{10} = {}^3V_{10};\ {}^1V_1 = {}^1V_1\}$	$= n - 3 (= 1)$	1									n − 3							
					Here follows a repetition of Operations thirteen to twenty-three																	
24	+	$^4V_{13} + {}^0V_{24}$	$^1V_{24}$	$\{^4V_{13} = {}^0V_{13};\ {}^0V_{24} = {}^1V_{24}\}$	$= B_7$																	B_7
25	+	$^1V_1 + {}^1V_3$	$^1V_3$	$\{^1V_1 = {}^1V_1;\ {}^1V_3 = {}^1V_3;\ {}^5V_6 = {}^0V_6;\ {}^5V_7 = {}^0V_7\}$	$= n + 1 = 4 + 1 = 5$; by a Variable-card. by a Variable-card.	1		n + 1			0	0										

Illustration 5.4 Ada Lovelace's 'walkthrough program' for calculation of Bernoulli numbers (1843)

Babbage's writing on the engines is largely technocentric and he offers little in the way of speculation on the broader significance of his work. It was Giovanni Plana, an Italian mathematician, who most clearly situated Analytical Engine and machine computation in the context of mathematics. Plana, writing to Babbage in 1840, made the distinction between the 'legislative' and the 'executive' aspects of analysis positing that 'hitherto the *legislative* department of our analysis has been all powerful – the *executive* all feeble'. He goes on to say that the Analytical Engine redressed this imbalance by giving us 'the same control over the executive which we have hitherto only possessed over the legislative ...' (Babbage 1864: 129). Babbage was much taken with this distinction as one that for him exactly conveyed the role of machine computation in relation to mathematics. The distinction endured till the end. In 1869 two years before he died, he set out finally to write a general description of the Analytical Engine. He made three separate attempts none of which was completed. Each opened with a statement of the purpose of the Analytical Engine. The first of these (Babbage 1869: 134) dated 4 May 1869 reads:

The object of this Engine is to execute by *machinery*

1. All the operations of arithmetic
2. All the operations of Analysis
3. To print any or all of the calculated results.

The greatest generality here is contained in the reference to 'Analysis' which in terms of the engine was a reference to symbolic algebra with number an instantiation of quantity (Priestley 2011: 32). (Lovelace hints at the prospect of an engine capable of implementing arbitrary rules for arbitrary symbols but the ideas are tantalisingly undeveloped). The expressed purpose of the Analytical Engine as that of finding the numerical value of algebraic formulae is, at face value, unexpectedly modest but less so when viewed in the context of the times. In 1841 Lovelace wrote, 'Mathematical Science shows what *is*. It is the language of unseen relations between things' (Huskey and Huskey 1980: 308). And Buxton, reflecting the rationalist credo of the time, wrote:

> There is in fact no question that can be conceived, which does not come within the category of number, or which is not finally reducible to a question to be solved by the investigation of quantities, by one another, according to certain relations.... . all our ideas of quality are reducible to the ideas of quantity. (Hyman 1988: 153)

Mathematics was seen to be uniquely privileged in its descriptive and explanatory powers, and for the rationalists, the world was reducible to number. A generalised machine able to map abstract mathematical description into number was the essential instrument without which the 'language of unseen relations' would remain mute. What emerges from Lovelace's description of the Analytical Engine is the idea that the potential utility of computing machines lies in its ability to manipulate according to rules representations of the world contained in symbols. The machine was the bridge between symbolic abstraction and contingency in the world. We have in these two statements (Lovelace's and Buxton's) a reflection of Plana's two departments of mathematics, legislative and executive. One was not the other's rival, but an essential complement, and Babbage was evidently more than content with this.

5.6 The Mechanical Notation

The unprecedented intricacy the mechanisms and long causal chains of action posed difficulties both for the design process and for modes of representation. Holding in one's mind the multitude of parts and long trains of action in the mechanisms 'would have baffled the most tenacious memory' wrote Babbage in 1864 (Babbage 1864: 113). The solution was his Mechanical Notation, a language of signs and symbols devised to describe the complex mechanisms of his calculating engines. He described the genesis of the language in 1826:

> The difficulty of retaining in the mind all the contemporaneous and successive movements of a complicated machine, and the still greater difficulty of properly timing movements which had already been provided for, induced me to seek for some method by which I might at a glance of the eye select any particular part, and find at any given time its state of motion or rest, its relation to the motions of any other part of the machine, and if necessary trace back the sources of its movement through all its successive stages to the original moving power. I soon felt that the forms of ordinary language were far too diffuse to admit of any expectation of removing the difficulty, and being convinced from experience of the vast power which analysis derives from the great condensation of meaning in the language

> it employs, I was not long in deciding that the most favourable path to pursue was to have recourse to the language of signs. (Babbage 1826a: 250)

The importance of notation to mathematical reasoning is a running motif in Babbage's mathematical work. He wrote that an advantage of symbolic language over common language was lack of ambiguity between sign and signified: unlike words 'an arbitrary symbol can neither convey, nor excite any idea foreign to its original definition' (Babbage and Herschel 1813: i; Babbage 1827: 327–8). A further virtue was that of conciseness which he framed as a mental aid to keeping track of long operational sequences (Babbage and Herschel 1813: i–ii; Babbage 1830: 395).

As a mathematics undergraduate at Cambridge (he graduated in 1814) Babbage had been a vigorous advocate for the superiority of Leibniz's notation for differential calculus over Newton's system of dots, this in defiance of the prevailing orthodoxy. A suite of mathematical papers with the general title 'The Philosophy of Analysis' includes a paper on notation that predates his involvement in the calculating engines i.e. his ideas on notation are rooted in mathematics and were well developed before his mechanical epiphany in 1821. Edward Bromhead, mathematician and friend of Babbage, in a letter to Babbage commenting on the paper in March of that year, endorsed the importance of notation: 'I have always considered Notation as the *Grammar* of symbolic language' (Dubbey 1978: 93). The Mechanical Notation can be seen as an extension to machines of his ideas on the role and importance of notation in mathematics.

The Mechanical Notation provided an abstract form for the nature, timing and causal action of parts, and Babbage used it to specify and describe the structural complexity of mechanisms and their time-dependent behaviour. He also used it as a design aid to optimise timing and eliminate redundancy (Babbage 1851b, 1855, 1856).

In its mature form there are three federated elements that combine to form the Mechanical Notation, each indispensible to the whole. *Forms* refer to mechanical drawings depicting the shape and size of parts and their organisation into mechanisms. The drawings use familiar drafting conventions of plan views, front and end elevations, and sectional views in mainly third angle projections. There is nothing radical or revolutionary in these. They conform to contemporary representational conventions and capture what are essentially spatial relations.

Trains are diagrams that describe the complete causal chain from the first mover to the end result. They show the path of the transmission of motion by parts acting on other parts (Illustration 5.5). Each part in the *Forms* was assigned a capital letter of the alphabet in one of a number of typefaces – italicised letters for moving parts and upright letters for fixed framing pieces, and a variety of typeface families were used including Etruscan, Roman, and Script. Each letter identifying a part had up to six indices – superscripts and subscripts (Swade 2017: 420). Four of the indices were numerical (index of identity, index of circular position, of linear position, and an index to extend the use of a typeface family in the event of running out of letters). The four numerical indices indicated the spatial relationship to other parts and which parts formed functional groups.

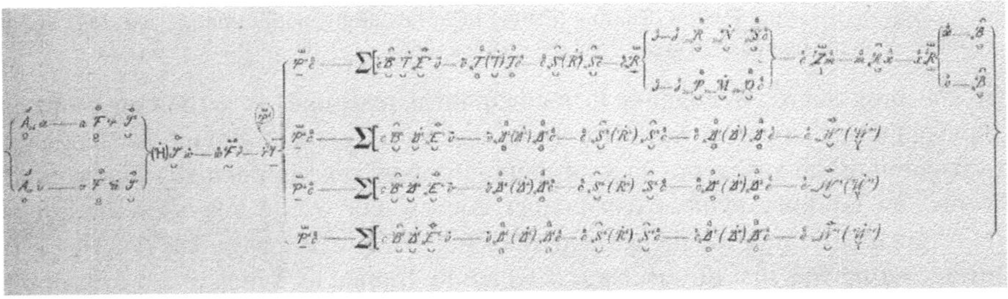

Illustration 5.5 Mechanical notation. *Train* for circular motion of figure wheels, Difference Engine No. 2 (detail) (1848)

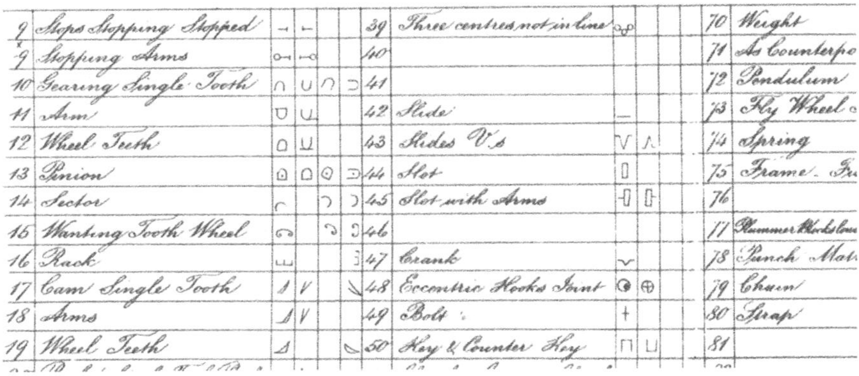

9	Stops Stopping Stopped	39	Three centres not in line	70	Weight
9	Stopping Arms	40		71	As Counterpo[illegible]
10	Gearing Single Tooth	41		72	Pendulum
11	Arm	42	Slide	73	Fly Wheel [illegible]
12	Wheel Teeth	43	Slides V's	74	Spring
13	Pinion	44	Slot	75	Frame. Fr[illegible]
14	Sector	45	Slot with Arms	76	
15	Wanting Tooth Wheel	46		77	Plummer Blocks [illegible]
16	Rack	47	Crank	78	Punch Mat[illegible]
17	Cam Single Tooth	48	Eccentric Hooks Joint	79	Chain
18	Arms	49	Bolt	80	Strap
19	Wheel Teeth	50	Key & Counter Key	81	

Illustration 5.6 Mechanical notation, signs of form (detail) (1851)

The two non-numerical indexes are the Sign of Form, and Sign of Motion. The Sign of Form gave functional specificity to an otherwise arbitrary symbol. It indicated the species of part – rack, gear wheel, cam, pinion, arm, crank, handle etc. – using symbols that are partial pictograms indicating generic function (Illustrations 5.5 and 5.6). The specific purpose of this portrayal was to enable the chain of action to be followed using mental images of parts without the distraction of reference to the mechanical drawings (*Forms*). The Sign of Motion described the nature of motion – circular, linear, curvilinear, or reciprocating – as depicted in a particular view in the *Forms*, plan, elevation, or end view. Signs of Motion could be used in combination in the annotation of a particular part. In 1851 Babbage proposed ten symbols in the Alphabet of Motion and some eighty in the Alphabet of Form and speculated that as many as 200 might be needed (Babbage 1851a: 138).

A *Train* is formed by combining these indexed letters into statements using syntactical rules. Illustration 5.5 shows a portion of the *Train* for the circular motion of the even difference figure wheels for Difference Engine No. 2. The lower case letters indicate 'working points' – points or surfaces on a part that act on or are acted upon by other parts.

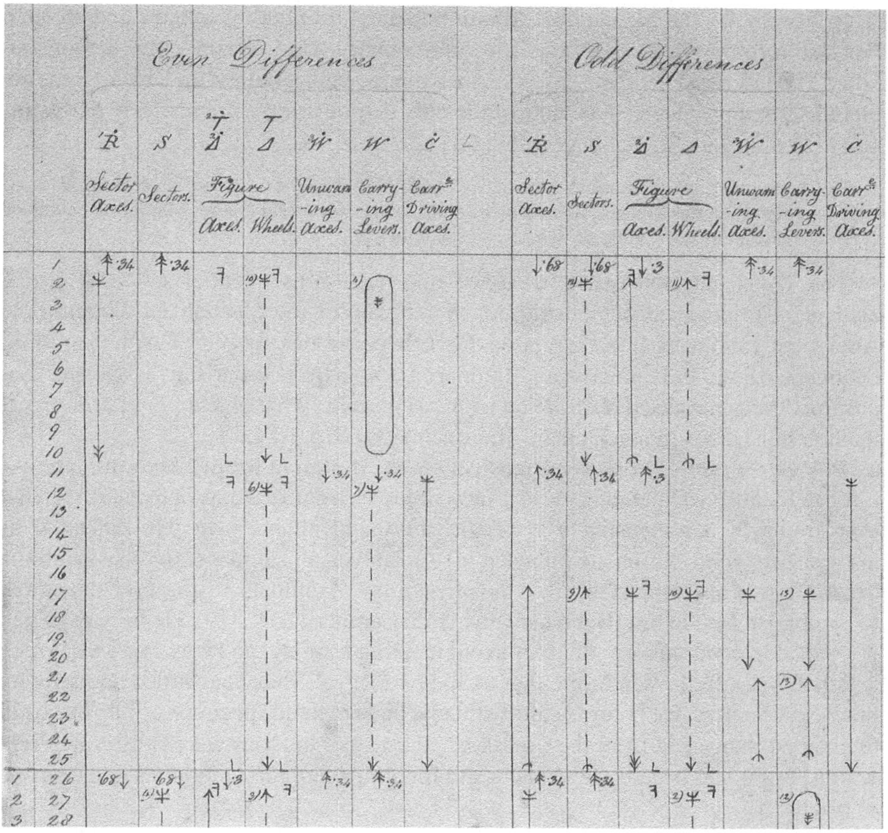

Illustration 5.7 Mechanical notation. *Cycle* for addition and carriage, Difference Engine No. 2 (detail) (1848)

While the direction of flow in the *Trains* is generally from left to right (except where there is feedback) neither the *Forms* nor the *Trains* describe time-dependent behaviour in any detail. The third element in the triptych are *Cycles*, essentially timing diagrams that depict at any point in the cycle what action each designated part is performing, and its timed relation to all other contemporaneous motions (Illustration 5.7).

Cycles show the orchestration of motions of individual parts into a functioning whole. For this a new set of notational conventions was introduced. Annotations at the head or tail of an arrow indicate linear or circular motion, whether rotation is positive or negative, and whether or not the motion depicted returns the part to its rest position. Other conventions indicate whether the motion is conditional or unconditional, continuous or intermittent and the time window in which the motion may or may not occur.

Advocacy for the Mechanical Notation likened it to both geometry and algebra. For Babbage the distinctive properties of geometry were certainty and demonstrability. He described geometry as 'a science of absolute certainty' in which 'signs are pictures', and privileges it as 'almost the only demonstrative science'. He places his Notation alongside it (Babbage 1860: 381–2).

> By these aids the science of constructive machinery becomes simple. It is reduced to mathematical certainty, and I believe now stands by the side of geometry as far as the nature of its reasoning, and that these two sciences stand alone.

The *Trains* allowed one to visualise and trace consequential action, a virtual equivalent to physical demonstration. 'By the aid of the Mechanical Notation the Analytical Engine became a reality: for it became susceptible of demonstration' (Babbage 1864: 113). Dionysius Lardner, a colourful populariser of science, emphasises the generalised abstraction of the Notation: 'what algebra is to arithmetic, the notation ... is to mechanism' (Lardner 1834: 315, 319).

Babbage used the Notation extensively in the design of his machines to optimise timing, identify redundancy, derive new motions from existing ones, and marshal long trains of action using a symbolic shorthand all his own. He ranked it as his greatest contribution to knowledge – a universal language for the symbolic description of anything at all from the physiology of animals, respiration, digestion, to combat on land or sea (Babbage 1864: 145; Lardner 1834: 319). He fully expected it to be adopted as an essential tool in engineering training, and was aggrieved when it failed, in 1826, to win him either of the two Royal Medals awarded annually by the Royal Society. He records sulkily that he had received specimens of its use from the United States and from the Continent, and two of his sons were fully conversant with it. But it was used by few others and for all his faith in it merits its fate was one of obscurity.

The Mechanical Notation can be seen as a response to the unprecedented levels of complexity of the engines' mechanisms and is not unlike the 'hardware description languages' (HDLs) the like of which emerged again in the early 1970s in computer circuit and system design, and especially in the design of integrated circuits. HDLs provide a higher-order representation to manage otherwise unmanageable detail at component level – the same solution to the same need 150 years apart:

> I succeeded in mastering trains of investigation so vast in extent that no length of years ever allotted to one individual could otherwise have enabled me to control. (Babbage 1864: 113)

5.7 A Coding Problem

Machines compute by manipulating, according to rules, physical representations of numbers. Logical relations in a mathematical statement can be seen as timeless or even atemporal, but once 'physicalised' in a machine they are subject to physics and mechanics in ways that logic is not: actions need to be phased in time, measures taken to ensure the integrity of representation and control, and the algorithmic sequence needs to be a correct encoding of the problem. The time-dependence of computational process had unexpected implications for programming that were evidenced in the earliest computer programs. As mentioned earlier, the Variables in the Store are numbered sequentially, $\mathbf{V}_1$, $\mathbf{V}_2$, $\mathbf{V}_3 \ldots$ and represent the locations in memory for operands and results. Conventional mathematical notation was immediately problematic. The statement $\mathbf{V}_4 - \mathbf{V}_2 = \mathbf{V}_4$ meant 'subtract the contents of $\mathbf{V}_2$ from the contents of $\mathbf{V}_4$ and place the result in $\mathbf{V}_4$'. The original value in $\mathbf{V}_4$ is overwritten by the result. The statement $\mathbf{V}_4 - \mathbf{V}_2 = \mathbf{V}_4$ was problematic for mathematicians. At face value it appears to violate the notion of mathematical identity being trivially true for $\mathbf{V}_2 = 0$, and manifestly false for non-zero $\mathbf{V}_2$. Babbage's solution was to add a leading index so the statement then read ${}^1\mathbf{V}_4 - {}^1\mathbf{V}_2 = {}^2\mathbf{V}_4$. The leading superscript, the 'index of alteration', indicated that the contents of the Variable had changed during the operations. The trailing index, the 'index of location' remained as before denoting the location in the Store of the Variable in question. Each reuse of the Variable incremented the index of alteration and the history of the Variable's changing contents could be traced back through the chain of programming steps.

The issue arose because memory, for the first time in a computing machine, in virtue of being 'physicalised', had spatial location, and instructions expressed in standard mathematical notation did not reflect time-dependence. One of the earliest programs Babbage wrote, dated 4 August 1837, has a sequence of instructions for the solution of two simultaneous equations and features the double index, though he used Roman numerals for the index of alteration (Babbage 1837a: L-1), later changed to more familiar Arabic numbers (Babbage 1864: 127). The need for a new index to reflect time-dependence in an instruction sequence signalled a more general finding – that coding would require new notational conventions.

5.8 Epilogue

The practical fate of Babbage's engines was a wretched one. Famed as he is for their invention he is no less famed for failing to build any of them in their entirety. The largest of the few experimental mechanisms he assembled was the demonstration piece for Difference Engine No. 1, 'the finished portion of the unfinished engine' (Babbage 1864: 150) which represents one-seventh of the calculating section of the whole machine (Illustration 5.1). The first complete Babbage engine was built in the modern era. Difference Engine No. 2, designed between 1847 and 1849 was built to the original plans and completed 2002 (Illustration 5.8). It weighs 5 tonnes, consists of 8000 parts, measures 11 ft long and 7 ft high, and calculates and tabulates any seventh-order polynomial to 30 decimal places. It works exactly as Babbage intended (Swade 2005).

The reasons for Babbage's failures are a cocktail of factors: fierce pride, poor management, social organisation of labour, absence of production techniques with inherent repeatability to make hundreds of near-identical parts, abrasive diplomacy that alienated those whose support he needed, and loss of credibility through

Illustration 5.8 Difference Engine No. 2 (2002)

delay, to name but some. Achievable precision in manufacture, and the availability of funding, were indirectly relevant but not critical to the mix (Swade 2000). Others built tabulating difference engines in the nineteenth century stimulated by Babbage's work – Alfred Deacon in London, Martin Wiberg in Sweden, the Scheutz father-and-son team also in Sweden, and Barnard Grant in the United States (Lindgren 1990). Not all the machines were technically flawless. All were commercial failures.

With Babbage's death in 1871 the movement to mechanise calculation lost its most visible protagonist and its major impetus. Leslie Comrie, spoke of the 'dark age of computing history that lasted 100 years' referring to the period between the early 1830s and Comrie's revival in the early 1930s of automatic tabulation by differences using commercial adding machines (Cohen 1988: 180). There were sporadic flickers in the early twentieth century to design and build 'analytical machines' (program controlled calculators) for general calculation, notably by Percy Ludgate and by Torres y Quevedo (Randell 1971, 1982). These were isolated episodes and developmental *culs-de-sac*.

The influence of Babbage's work on the modern era is tenuous at best. The engine designs were not studied in technical detail until the 1970s (Bromley 1982, 1987, 2000) and while his exploits were known to almost all the pioneers of modern computing, they effectively reinvented the principles of computation largely in ignorance of the detail of Babbage's work.

Mechanical computation was not yet entirely defunct. Mechanical devices were used in the transition to fully electronic systems. Konrad Zuse's early machines from the late 1930s relied on mechanical memory in the form of sliding plates, and IBM's Harvard Mark I, completed in 1943, was a hybrid electromechanical system with elements of mechanical logic. One of its early tasks was calculating and printing mathematical tables.

Mechanical analog computation, routinely underrepresented in the canon, had several significant successes in providing computational solutions to mathematically modelled physical phenomena. The prediction of tidal behaviour using techniques of harmonic analysis first introduced by William Thomson (later Lord Kelvin) in the 1860s were the basis of several mechanical analog tide predictors. One of these in service in the United States was not replaced until the 1960s by an electronic computer. Michael Williams reports that using a 37-term formula the mechanical predictor could calculate the tidal heights to a tenth of a foot for each minute of the year (Williams 1985: 209). In the late 1920s Vannevar Bush, frustrated by the tedium and difficulty of analytical methods, developed 'differential analysers' for the solution of differential equations by integration. The analysers were mechanical analog machines using wheel-and-disc integrators as their essential computational element. Differential analysers were used extensively during WWII for the calculation of artillery firing tables.

Tide predictors and differential analysers are problem-specific calculators and in this they are unlike the general purpose programmable 'analytical' machines discussed earlier. But like the earlier machines they exemplify the idea that mathematics and technology intersect where symbols and the rules of their manipulation are physicalised in material form.

Acknowledgement This work is supported in part by the Leverhulme Trust.
Picture Credits:
Illustrations:
2, 5, 7 – Science Museum/Science and Society Picture Library
1a, 1b, 3, 6, 8 – Doron Swade
4 – Rainer Glaschick

References

Babbage, C. (1822). *A letter to Sir Humphrey Davy, Bart., President of the Royal Society, on the application of machinery to the purpose of calculating and printing mathematical tables*. London: Cradock and Joy. Reprinted in, Campbell-Kelley (1989), Vol. 2, pp. 5–14.

Babbage, C. (1823). On the theoretical principles of the machinery for calculating tables. *Edinburgh Philosophical Journal, 8*, 122–128. Reprinted in, Campbell-Kelley (1989), Vol. 2, pp. 39–43.

Babbage, C. (1826a). On a method of expressing by signs the action of machinery. *Philosophical Transactions of the Royal Society, 2*, 250–265. Reprinted in, Campbell-Kelley (1989), Vol. 3, pp. 209–23.

Babbage, C. (1826b). On the determination of the general term of a new class of infinite series. *Transactions of the Cambridge Philosophical Society, 2*, 217–225. Reprinted in, Campbell-Kelley (1989), Vol. 2, pp. 61–68.

Babbage, C. (1827). On the influence of signs in mathematical reasoning. *Transactions of the Cambridge Philosophical Society, 2*, 325–77. Reprinted in, Campbell-Kelley (1989), Vol. 1, pp. 371–408.

Babbage, C. (1830). Notation. In *The Edinburgh Encyclopaedia*.

Babbage, C. (1836). Notations of the analytical engine [manuscript], Science Museum, The Babbage Papers 1821–1905, BAB/F/57x. London.

Babbage, C. (1837a). On the mathematical powers of the calculating engine. Museum of the History of Science, Oxford, *Buxton MSS collection*, in Campbell-Kelley (1989), *3*, 15–61.

Babbage, C. (1837b). Notations of calculations for the analytical engine [manuscript], Science Museum, The Babbage Papers 1821–1905, BAB/L London.

Babbage, C. (1851a). *The exposition of 1851 or views on the industry, the science, and the government of England*. London: John Murray. Reprinted in, Campbell-Kelley (1989), Vol. 10.

Babbage, C. (1851b). Laws of mechanical notation. *Privately printed*. Reprinted in, Campbell-Kelley (1989), *3*, 224–8.

Babbage, H.P. (1855). On mechanical notation, as exemplified in the Swedish calculating machine of Messrs. Scheutz. British Association. Glasgow. Reprinted in, Babbage, H. P. 1889. In *Babbage's calculating engines: A collection of papers by Henry Prevost Babbage*, (pp. 246–247). London: Spon.

Babbage, H. P. (1856). Scheut's difference engine, and Babbage's mechanical notation. *Minutes of proceedings of Civil Engineers*. Reprinted in, Babbage, H. P. 1889. In *Babbage's calculating engines: A collection of papers by Henry Prevost Babbage,* (pp. 248–257). London: Spon.

Babbage, C. (1860). Contribution to the discussions. *Proceedings of the Fourth International Statistical Congress*. London: Wiley. Reprinted in, Campbell-Kelley (1989), Vol. 5, pp. 169–175.

Babbage, C. (1864). *Passages from the life of a philosopher*. London: Longman, Green, Longman, Roberts & Green. Reprinted in, Campbell-Kelley 1989, Vol. 11.

Babbage, C. (1869). Science museum library. London, *The papers of Charles Babbage*, S15, f. 134.

Babbage, H. P. (1889). *Babbage's calculating engines: A collection of papers by Henry Prevost Babbage*. London: Spon. Reprinted in, Bromley, A. G. (ed.) 1982. Babbage's Calculating Engines: a collection of papers by Henry Prevost Babbage, Los Angeles: Tomash.

Babbage, C., & Herschel, J. F. W. (1813). *Memoirs of the analytical society*. Cambridge: Deighton & Sons et al. Preface, Reprinted in, Campbell-Kelley (1989), Vol. 1, pp. 37–60.

Baxandall, D. (1926). *Catalogue of the collections in the science museum: Mathematics: Calculating machines and instruments*. London: Her Majesty's Stationery Office (HMSO)

Bromley, A. G. (1982). Charles Babbage's analytical engine, 1838. *Annals of the History of Computing, 4*, 196–217.

Bromley, A. G. (1987). The evolution of Babbage's calculating engines. *Annals of the History of Computing, 9*, 113–136.

Bromley, A. G. (1991). *The Babbage papers in the science museum: A cross-referenced list*. London: Science Museum.

Bromley, A. G. (2000). Babbage's analytical engine plans 28 and 28a – The programmer's interface. *IEEE Annals of the History of Computing, 22*, 4–19.

Campbell-Kelly, M. (Ed.). (1989). *The works of Charles Babbage*. London: William Pickering.

Carlyle, T. (1829). Signs of the times. *Edinburgh Review, 49*, 439–459.

Cohen, I. B. (1988). Babbage and Aiken. *Annals of the History of Computing, 10*, 171–191.

Collier, B. (1990). *The little engines that could've: The calculating engines of Charles Babbage*. New York: Garland.

Dubbey, J. M. (1978). *The mathematical work of Charles Babbage*. Cambridge: Cambridge University Press.

Forbes, E. G. (1965). The foundation and early development of the Nautical Almanac. *Journal of the Institute of Navigation, 18*, 391–401.

Glaschick, R. (2016). *Ada Lovelace's calculation of Bernoulli's numbers* [Online]. Available: http://rclab.de/rclab/_media/analyticalengine/aal_noteg_glaschick_v1.1.pdf. Accessed 5 Feb 2017.

Grattan-Guinness, I. (1990). Work for Hairdressers: The production of de Prony's logarithmic and trigonometric tables. *Annals of the History of Computing, 12*, 177–185.

Grattan-Guinness, I. (2003). The computation factory: de Prony's project for making tables in the 1790s. In Campbell-Kelly, M. Campbell-Kelly, M. Croarken, R. Flood, & E. Robson (Eds.), *The history of mathematical tables: From sumer to spreadsheets* (pp. 104–121). Oxford: Oxford University Press.

Henniker, B. (1893). Letter to treasury, 2 Feb. National Archive, Kew, RG29-3, Vol, 3, f. 300.

Horsburgh, E. M. (Ed.). (1914). *Napier tercentenary celebration: Handbook of the exhibition of Napier Relics and of books, instruments, and devices for facilitating calculation*. Edinburgh: Royal Society of Edinburgh.

Huskey, V. R., & Huskey, H. D. (1980). Lady Lovelace and Charles Babbage. *Annals of the History of Computing, 2*, 299–329.

Hyman, A. (Ed.). (1988). *Memoir of the life and labours of the late Charles Babbage Esq. F.R.S. by H. W. Buxton*. Cambridge, MA: Tomash.

Johnston, S. (1997). Making the arithmometer count. *Bulletin of the Scientific Instrument Society, 52*, 12–21.

Lardner, D. (1834). Babbage's calculating engine. *Edinburgh Review, 59*, 263–327. Reprinted in, Campbell-Kelly (1989), Vol. 2, pp. 118–186.

Lindgren, M. (1990). *Glory and failure: The difference engines of Johann Muller, Charles Babbage, and Georg and Edvard Scheutz*. Cambridge, MA: MIT Press.

Lovelace, A. A. (1843). Sketch of the analytical engine invented by Charles Babbage Esq. By L. F. Menabrea, of Turin, officer of the Military Engineers, with notes upon the memoir by the translator. *Scientific Memoirs, 3*, 666–731. Reprinted in, Campbell-Kelley (1989), Vol. 3, pp. 89–170.

Martin, E. ([1925] 1992). *The calculating machines (Die Rechensmashinen): Their history and development*. Translated and edited by Peggy Aldrich Kidwell and Michael R Williams. *Charles Babbage Institute Reprint Series for the History of Computing*. Cambridge, MA: MIT Press.

Morar, F.-S. (2015). Reinventing machines: The transmission history of the Leibniz calculator. *British Journal for the History of Science (BJHS), 48*(1), 125–146.

Mounier-Kuhn, P. (1999). *L'informatique en France, de la Seconde Guerre mondiale au Plan Calcul: Science, Industrie, Politiques publiques*. PhD, Conservatoire national des arts et métiers.

Mowatt, F. (1893). Letter to Henniker B, 14 January. National Archive, Kew, RG29-7, f. 80.

Priestley, M. (2011). *A science of operations: Machines, logic and the invention of programming*. London: Springer.

Randell, B. (1971). Ludgate's analytical machine of 1909. *Computer Journal, 14*, 317–326.

Randell, B. (1982). From analytical engine to electronic digital computer: The contributions of Ludgate, Torres, and Bush. *Annals of the History of Computing, 4*, 327–341.

Schaffer, S. (1994). Babbage's intelligence: Calculating engines and the factory system. *Critical Enquiry, 21*, 203–227.

Swade, D. (2000). *The Cogwheel brain: Charles Babbage and the quest to build the first computer*. London: Little, Brown.

Swade, D. (2003a). *Calculation and tabulation in the 19th century: George Biddell Airy versus Charles Babbage*. PhD, University College London.

Swade, D. (2003b). The 'unerring certainty of mechanical agency': Machines and table making in the nineteenth century. In M. Campbell-Kelly, M. Coarken, R. Flood, & E. Robson (Eds.), *The history of mathematical tables: From sumer to spreadsheets*. Oxford: Oxford University Press.

Swade, D. (2005). The construction of Charles Babbage's difference engine no. 2. *IEEE Annals of the History of Computing, 27*, 70–88.

Swade, D. (2011). Calculating engines: Machines, mathematics, and misconceptions. In R. Flood, A. Rice, & R. Wilson (Eds.), *Mathematics in Victorian Britain*. Oxford: Oxford University Press.

Swade, D. (2017). 'Photographing the footsteps of time': Space and time in Charles Babbage's calculating engines. In G. G. Shyam Wuppuluri (Ed.), *Space, time, and the limits of human understanding*. London: Springer.

Toole, B. A. (Ed.). (1992). *Ada, the enchantress of numbers: A selection from the letters of Lord Byron's daughter and her description of the first computer*. Mill Valley: Strawberry Press.

Williams, M. R. (1985). *A history of computing technology*. Englewood Cliffs: Prentice-Hall.

Chapter 6
The Mathematical Origins of Modern Computing

Mark Priestley

Abstract Modern computing has been shaped by the problems and practices of mathematics to a greater extent than is often acknowledged. The first computers were built to accelerate and automate mathematical labour, not as universal logical machines. Very specific mathematical objectives shaped the design of ENIAC, the first general-purpose electronic computer, and its successor, EDVAC, the template for virtually all subsequent computers. As well as machine architecture, software development is firmly rooted in mathematical practice. Techniques for planning large-scale manual computation were directly translated to the task of programming the new machines, and specific mathematical practices, such as the use of tables in calculation, profoundly affected the design of programs.

6.1 Introduction

If there is a truth universally acknowledged in the history of computing, it is this: the "modern computer" was invented in the early 1940s and its design was first described in the *First Draft of a Report on the EDVAC* (von Neumann 1945b). In the preceding 3 years, a group at the University of Pennsylvania's Moore School of Electrical Engineering had designed and built ENIAC, a giant machine that among other things demonstrated the feasibility of large-scale electronic calculation. As ENIAC's design neared completion in 1944, the team began to plan a follow-up project, the EDVAC. They recruited the mathematician John von Neumann as a consultant, and in early 1945 he wrote a report describing, in rather abstract terms, the design of the new machine. This was the first systematic presentation of the new ideas, and proved highly influential. By the end of the decade the first machines based on the EDVAC design were operational, marking the first step on a ladder of technological progress leading to the ubiquity of computational machinery today.

M. Priestley (✉)
Independent Researcher, London, UK

S. O. Hansson (ed.), *Technology and Mathematics*, Philosophy of Engineering and Technology 30, https://doi.org/10.1007/978-3-319-93779-3_6

Historian Michael Mahoney (2005) challenged such machine-centric views of computer history. Mahoney urged historians to turn their attention to the history of computing, not just the technical history of computers. He further argued that the history of computing cannot be understood as a single unified narrative. The computer can be many things to different people, generating a multitude of diverse stories. Mahoney supported his argument by appealing to a particular view of the nature of the computer: while acknowledging that the first computers were built to perform scientific calculations, he believed that the machines based on the EDVAC design were something different, not just calculators but "protean machines" that could be bent to any task.

> But making it [the computer] universal, or general purpose, also made it indeterminate. Capable of calculating any logical function, it could become anything but was in itself nothing (well, as designed, it could always do arithmetic). (Mahoney 2005, 123)

A machine which is in itself nothing cannot have much of a history. Instead, Mahoney urged, historians of computing should tell the stories of how the machine was introduced to and transformed, and was itself transformed by, a wide range of existing "communities": the people involved in areas of application such as data processing, management, or military command and control systems (Fig. 6.1).

It is striking that, despite Mahoney's revisionist intentions, this schema retains a prominent place for the traditional origin story involving ENIAC and EDVAC. On Mahoney's account, EDVAC has a dual nature. On the one hand, it is a room-sized mathematical calculator, built for very specific purposes by a particular group

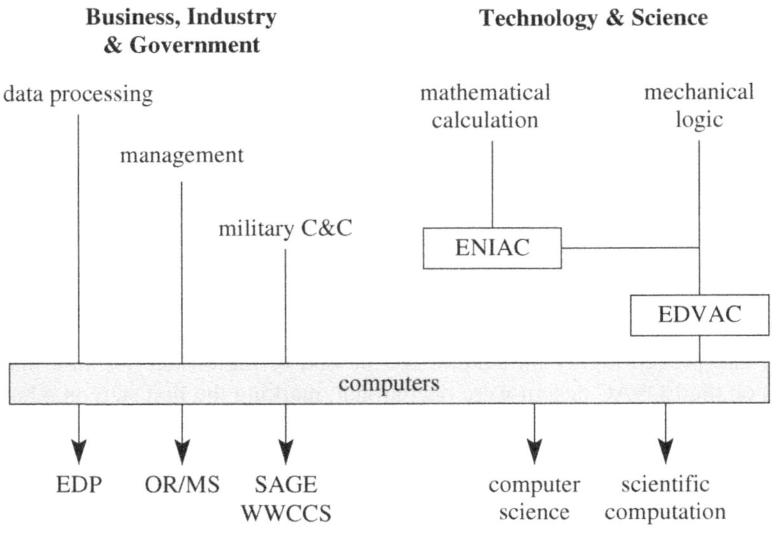

Fig. 6.1 The communities of computing (Redrawn extract from Mahoney 2005, fig. 5, copyright © Institute of Materials, Minerals and Mining, reprinted by permission of Taylor & Francis Ltd, http://www.tandfonline.com, on behalf of Institute of Materials, Minerals and Mining)

of people. But at the same time it is an abstract machine, "in concept a universal Turing machine". According to Mahoney, it is this second, spectral machine which moves between communities. Being universal and general-purpose, its potential for use in different fields can be taken for granted.

From this point of view, the computer's mathematical origins are little more than an historical curiosity. Mahoney followed logician-turned-historian Martin Davis (2000) in seeing the crux of the computer's evolution as being an injection of logic between ENIAC and EDVAC that turned a brute calculator into an ethereal logic machine with, incidentally, the capability to do "arithmetic".[1]

However, the idea that a new technology can transform many application areas is not the novelty that Mahoney seems to suggest, and does not depend on the universal nature of the technology being transferred, as two examples from the prehistory of computing illustrate. In the 1920s, Leslie Comrie began an extended investigation into the use of punched card machinery to support scientific calculation, work that was continued by Wallace Eckert in the USA. Similarly, Tommy Flowers took with him to Bletchley Park the experience that he had gained with electronic switching before World War 2 in the British GPO, and deployed it very effectively in the development of the Robinson and Colossus machines. In this perspective, the idea that the invention of the computer might give rise to different histories of adoption in different areas is simply another example of a regular historical pattern.

The computer remains a special case in its breadth of application, of course, and this is a fact that calls out for explanation. In response, Mahoney appealed to the modern computer's "protean" nature. But how does the computer come to have such a nature? The conventional answer to this is technological: the "stored-program concept", itself said to be derived from Turing's description of a universal machine, is the particular feature that allows a single machine to perform an unlimited variety of tasks.[2] But there is an unsatisfying circularity in the suggestion that it is the "universal" logico-technical properties of the computer that make it inevitable that it will find universal application. It is more illuminating to start with a functional characterization: if the computer is a technology of automation, what was it intended to automate? In his proposal for the ACE, a machine to be built at the UK's National Physical Laboratory, Turing suggested an answer to this question:

> How can one expect a machine to do all this multitudinous variety of things? The answer is that we should consider the machine as doing something quite simple, namely carrying out orders given to it in a standard form which it is able to understand. (Turing 1946, 3)

The modern computer, in other words, is a machine that obeys orders. As a matter of historical contingency, the first such machines were developed to automate the

[1]Similar views were widely canvassed in connection with Turing's centenary celebrations in 2012. An alternative perspective, challenging the view that logic played a central role in the development of EDVAC, has been articulated recently by historians including Tom Haigh (2014) and Edgar Daylight (2015). See also Sect. 6.5, below.

[2]The stored-program concept has been discussed, and its usefulness as an analytical category critiqued, by Haigh et al. (2014).

specific processes involved in large-scale mathematical calculation. This is far from being the incidental detail that Mahoney suggests, however, and deeply affected the ways in which computers could be deployed in areas outside mathematics, as computer scientist Donald Knuth's comments on the problems of carrying out data retrieval with electronic computers illustrate:

> Computers have increased the speed of scientific calculations by a factor of 10^7 or 10^8, but they have provided nowhere near this gain in efficiency with respect to problems of information handling. [. . .] We shouldn't expect too much of a computer just because it performs other tasks so well. (Knuth 1973, 551)

The first half of this chapter describes the effects of the mathematical context of innovation on the ENIAC and EDVAC projects and the machines they developed. The computer's mathematical origins are reflected in more than just its physical characteristics, however. The modern computer automated a certain kind of human labour, that of following a plan of computation in a more or less mechanical way. Many of the established practices of manual calculation were transferred to the new machines and profoundly shaped the ways in which they were used. The second half of this chapter examines how two such practices, the social organization of large-scale computation and the use of mathematical tables, were translated into the context of the automatic computer and the consequences of this for the way the new task of programming was conceived.

6.2 The Organization of Large-Scale Calculation

In the 1790s, the French engineer Gaspard Riche de Prony embarked on a mammoth project to calculate a new set of tables of logarithmic and trigonometric functions (Grattan-Guinness 1990). The undertaking was industrial in scale, and to manage it de Prony employed the principle of the division of labour recently described by Adam Smith in *The Wealth of Nations*, first published in 1776.

De Prony divided his workforce into three sections. The first section consisted of a small number of leading mathematicians who derived the formulas that would be used to calculate the various functions. These formulas were passed on to a second section of skilled but less eminent mathematicians whose job was to work out how to calculate the values of the formulas using the method of differences.

An advantage of the method of differences was that it enabled the functions to be calculated using only the basic operations of addition and subtraction. The third section consisted of relatively unskilled labour, many of them hairdressers who had been made redundant by changing fashions after the French revolution. The workers of the third section carried out sequences of additions and subtraction as specified by calculating sheets prepared by the second section. Rough working was carried out on loose sheets of paper, and the results were transcribed onto the calculating sheets, which were then passed back to the second section for checking.

The third section had little scope for the exercise of judgement or initiative. As Smith had observed, the division of labour often broke a complex task down into activities that were simple enough to be mechanized. Fully aware of de Prony's approach, Charles Babbage took advantage of this when beginning the development of his first Difference Engine in the 1820s:

> If the persons composing the second section, instead of delivering the numbers they calculate to the computers of the third section, were to deliver them to the engine, the whole of the remaining operations would be executed by machinery. (Babbage 1822, 10)

More than a hundred years after Babbage, large-scale computation was still being organized along the lines pioneered by de Prony. David Grier (1998) has described the organization of the Math Tables Project (MTP), a Depression-era project aimed at providing jobs for unemployed office workers in New York. The work of the MTP was directed by a Planning Committee which "developed the mathematical methodology, and prepared the computing instructions" that were passed onto the Computing Floor Division. This consisted of two groups of trained mathematicians who could be trusted to work independently: the "Special Computing Unit", who among other responsibilities helped the project leaders to prepare the worksheets for the "Manual Unit", and the "Testing Section". The Manual Unit were unskilled workers who were trained to perform to perform specific operations, such as multiplication by a single digit. Their work was directed by the worksheets.

Desk calculating machines were widely used in the 1930s to perform arithmetical operations, including multiplication and division. As the MTP grew, it acquired numbers of second-hand calculators and the size of the Manual Unit shrank as its suitably qualified members were promoted to the Machine Unit.

From the French revolution right through to the mid-twentieth century, then, the organization of large-scale calculation took the form of a pyramid resting on the base of a large group of mathematically unsophisticated (human) computers. The computers were expected to perform individual arithmetico-logical operations, with or without mechanical assistance, and to closely follow a plan telling them what operations to perform, in what order, and how and where to record the results. The ability to work independently and the exercise of initiative or judgement were not required.

It was precisely these characteristics that machine developers of the early 1940s were hoping to automate and that George Stibitz, designer of an influential series of machines at Bell Telephone Laboratories, made the defining property of computers understood as machines rather than human beings.[3]

> By "calculator" or "calculating machine", we shall mean a device (mechanical, electrical or what not) capable of accepting two numbers, A and B, and of forming some or any of

[3]The word "computer" before 1945 did not always refer to a human being. From the 1890s onward, "computers" were also computational aids, sometimes booklets containing useful collections of tables and methods (Hering 1891), but more often special-purpose circular slide-rules embodying particular formulas or algorithms (Halsey 1896). David Mindell (2002) has traced the further usage of the word in the 1930s in the field of fire-control systems in the US military.

the combinations $A + B$, $A - B$, $A \times B$, A/B. By "computer", we shall mean a machine capable of carrying out automatically a succession of operations of this kind and of storing the necessary intermediate results". (Stibitz 1945, 1–2)

In the 1830s Babbage had made the first attempt to design such a computer with his work on the Analytical Engine. Around a hundred years later, Konrad Zuse in Germany and Howard Aiken in the USA independently began projects leading to the construction of the first machines capable of automatically carrying out a sequence of operations.

6.3 Automating Calculation

In 1935, Zuse set up a workshop in his parents' Berlin apartment and began work. The following year, he submitted a patent application describing a machine which would automatically execute "frequently recurring computations, of arbitrary length and construction, consisting of an assembly of elementary arithmetic operations" (Zuse 1936, 163). The operations to be performed were described by what Zuse called a "computation plan" which would be presented to the machine in some suitable form, such as a punched tape. As an example, Zuse gave a plan for calculating the determinant of a 3×3 matrix. This involved a total of 17 operations, each with two operands: 12 multiplications, two additions and three subtractions.

Zuse developed a series of machines designed along these lines. The third of these machines, the Z3, was completed in 1941 and is now considered to be the first programmable computer. The Z3 and its predecessors were destroyed in air raids, but Zuse's next machine, the Z4, survived and was moved to Zurich, where it played an important role in the post-war development of European computing.

In 1937, Harvard graduate student and physics instructor Howard Aiken wrote a proposal for "an automatic calculating machine specifically designed for the purposes of the mathematical sciences" (Aiken 1937). He observed that existing punched-card calculating machinery did "the reverse of that required in many mathematical operations", in that it allowed the evaluation of a limited range of formulas on sequences of data read from punched cards. By contrast, Aiken believed that the characteristic of scientific calculation was that it required long and varied sequences of operations to be carried out on relatively small amounts of data. In principle, this could be done on existing machinery by manually switching from one operation to another: it was precisely this manual switching that Aiken planned to automate.

Aiken managed to enlist the help of IBM in building his machine, officially called the IBM Automatic Sequence Controlled Calculator; it later became widely and more conveniently known as the Harvard Mark I. On its completion in 1944, IBM donated the machine to Harvard, where it ran for many years, initially under the control of the US Navy.

Every aspect of Mark I was determined by its role in mathematical calculation. Like Zuse's machines, it was equipped with a number of general purpose registers, or counters, which stored results and allowed them to be retrieved when needed. Aiken explained the need for storage registers as a consequence of the pragmatics of conventional mathematical notation:

> The use of parentheses and brackets in writing a formula requires that the computation must proceed piecewise. [...] This means that a calculating machine must be equipped with means of temporarily storing numbers until they are required for further use. Such means are available in counters. (Aiken 1937, 198)

The counters stored incoming numbers by adding them to their existing contents, thus enabling Mark I to carry out addition in general. Subtraction was carried out using complements. There were specialized units for multiplication and division, to compute the values of selected exponential and trigonometric functions, and to interpolate between values read from a paper tape. But the heart of the machine was the sequence mechanism. This read a list of coded instructions that had been punched onto a paper tape and invoked the corresponding operations. By simply changing the tape, Mark I could be instructed to carry out any desired computation.

6.4 The Structures of Computation

The sequence of operations performed by the Z3 or Mark I was determined by the sequence of instructions read from the machines' tapes. To evaluate a simple formula, the tape would simply contain one instruction for each operation that the machine was to execute, but this approach did not scale up well to more complex problems. Many calculations have an iterative structure in which a small sequence of operations is repeatedly performed. It would be wasteful to punch a tape with the same instructions over and over again, and in many cases this would not even be possible. In general, a mathematician cannot tell in advance how many iterations of the operations will be required and instead has to rely some property of the results obtained so far to determine when the calculation should stop.

The conditional branch instructions of modern programming languages address these issues by allowing computations to diverge when necessary from the default sequence of instructions. The earliest computers did not have branch instructions, however, and various ad hoc approaches were adopted instead. Babbage proposed mechanisms to "back up" the Analytical Engine's cards so that instructions could be repeated, while Mark I's tapes were made "endless" by gluing one end to the other. An endless tape would loop indefinitely, carrying out the instructions on it over and over again. To interrupt a loop, Mark I had a conditional instruction that stopped the machine when, say, the results obtained so far reached certain limits of tolerance, but in order to continue with the next stage of the computation a new tape had to be mounted by the operators and the machine restarted. The first machine to fully automate computation, allowing loops and conditional branches to be freely utilized, was ENIAC.

ENIAC was the brainchild of a physicist, John Mauchly, who had taken up wartime employment at the Moore School.[4] The school had a long-standing collaboration with the Army Ordnance Bureau's proving ground in Aberdeen, in nearby Maryland, and in particular with its Ballistics Research Laboratory (BRL), an important centre of calculation in the interwar years. BRL had supported the Moore School's acquisition of a differential analyzer, with the understanding that in the event of war it would be made available for BRL's use. Developed by Vannevar Bush (1931) at MIT, the analyzer was a cutting-edge analogue machine which used mechanical integrators to solve differential equations.

When war broke out, BRL faced the challenging task of compiling firing tables for a vast range of new ordnance and ammunition. These tables integrated large amounts of experimental data and ballistic calculation, and told gunners how to aim their weapons to hit a specific target. To compile a table, many trajectories—the predicted paths of projectiles fired from the gun—had to be calculated, each requiring the solution of a set of differential equations that could take a human computer several hours. Invoking the terms of the earlier agreement, BRL set up a satellite computing centre at the Moore School overseen by Herman Goldstine, a mathematician whose wartime commission had seen him posted to BRL. Goldstine and his wife Adele were responsible for training and supervising teams of computers calculating trajectories. The Moore School's differential analyzer was extensively used in these calculations.

Mauchly was not directly involved in the firing table work, but he supervised a group carrying out manual computation and was familiar with the design and use of the analyzer. He had a long-standing interest in the use of electronics for calculation, and in August 1942 brought these interests and experience together in the form of a brief proposal for an electronic analogue of the differential analyzer. He estimated that the use of "high-speed vacuum tubes" would allow trajectories to be calculated in a fraction of the time taken by the mechanical analyzer, let alone by manual calculation. The proposal eventually came to the attention of Herman Goldstine, who saw great potential in it. Mauchly and Presper Eckert, a talented electronic engineer who had trained Mauchly when he first arrived at the Moore School, wrote a more detailed outline and a collaboration was soon agreed whereby the Moore School would build an electronic machine for BRL.

Although it was envisaged that the machine would spend a lot of its time calculating trajectories, its design was not limited to that particular application. As Mauchly had explained:

> There are many sorts of mathematical problems which require calculation by formulas which can readily be put in the form of iterative equations. [...] Since a sufficiently approximate solution of many differential equations can be had simply by solving an associated difference equation, it is to be expected that one of the chief fields of usefulness

[4]The following two sections draw extensively on the material in Haigh et al. (2016).

> for an electronic computor [sic] would be found in the solution of differential equations. (Mauchly 1942)

Mauchly's use of the phrase "electronic computer" seems very natural to modern-day readers, but would have been quite unfamiliar in 1942. Mauchly had described the new machine as an "electronic difference analyzer", but "computer" was soon added to the machine's name to reflect its potential generality, as Grist Brainerd, the Moore School academic in charge of the project, explained:

> The electronic difference analyzer and computer is a proposed device never previously built, which would perform all the operations of the present differential analyzers and would in addition carry out numerous other processes for which no provision is made on present analyzers. It is called a "difference" analyzer rather than a "differential" analyzer for technical reasons. (Brainerd 1943)

The new machine soon became terminologically independent of its predecessor, being dubbed the "Electronic Numerical Integrator and Computer", or ENIAC.[5] The numerical solution of differential equations by iterative means became ENIAC's signature application, but over the course of its working life it was applied to a much wider range of calculations than simply trajectories. Nevertheless, as late as the early 1950s, "artillery and bomb ballistics computation" made up a quarter of its workload (Reed 1952).

Mauchly may have used the term "computer" to emphasize that ENIAC, unlike a simple calculator, would be automatically sequenced and, like a human computer, able to work independently. In this respect, electronic speed was problematic, as it meant that the familiar technique of reading coded instructions from paper tape was simply too slow. Instead, the team adopted what they later described as a stop-gap solution in response to the urgency of a wartime project and designed ENIAC as a collection of specialized calculating units. They shared with Zuse and Aiken the view that calculations could be specified as sequences of instructions, but they adopted a different technological approach to realizing the instruction sequences. Instructions were set up on "program controls" on each unit, and computations were sequenced by cabling these controls together in problem-specific configurations. As it turned out, this gave ENIAC a flexibility that allowed a greater degree of automation than was possible on the tape-controlled machines.

The ENIAC team delivered their first progress report at the end of 1943, 6 months after the start of the contract funding the project. After extensive research into the existing state of the art, a new design for the machine's basic electronic counters had been decided on, but nothing had been constructed apart from a few test circuits. Plans for some units were fairly well advanced, but others had barely been started. There were many open questions about the design of the machine, and it had not yet been demonstrated that large numbers of unruly electronic valves could be persuaded to collaborate reliably and work as required.

---

[5]In a 1962 affidavit, Brainerd recalled that Paul Gillon of the Ordnance Bureau, an enthusiastic supporter of the project, came up with the new name.

Despite the provisional and incomplete state of the hardware design, however, the report was accompanied by detailed plans showing how ENIAC could be set up to calculate a ballistic trajectory.[6] Acting as a kind of feasibility test, this exercise enabled the team to settle many aspects of ENIAC's design. Different algorithms were investigated, the choice between them being governed by a variety of practical considerations. Would the numerical properties of the equations allow a reasonable degree of accuracy to be preserved throughout the calculation? Would the number of operations to be carried out and intermediate values to be stored physically fit onto ENIAC? Analysis by the Moore School mathematician Hans Rademacher showed that the relatively unfamiliar Heun method would be suitable, and the problem was reduced to a set of 24 simple difference equations. This analysis also enabled the size of ENIAC's accumulators to be fixed: numbers had to be stored to a precision of ten decimal digits to enable the computed results to be sufficiently accurate for BRL's purposes.

The analysis of the structure of the computation was just as significant as the numerical work. The trajectory calculation was split into four basic sequences of instructions: setting up the initial conditions, performing an integration step, printing a set of results, and carrying out a check procedure. These sequences were combined in a complex structure which included two nested loops: after the initial sequence was complete, a loop would print a set of results and carry out the check procedure 40 times; each set of results was calculated by performing the integration step 50 times.[7] In 1943, the team had little idea how a computation of such complexity would be controlled, and proposed a unit called the "master programmer" which would control the repetition of instruction sequences and move from sequence to sequence when required.

In the following months, the team set to work on the master programmer. Central to its design was a multi-functional device known as a "stepper" which controlled the initiation of up to six program sequences, one after the other. Each stepper had a counter associated with it to keep track of how many times the current sequence had been executed. Once a sequence had been repeated a specified number of times, the stepper would move the machine on to the next sequence. Conditional control was enabled by routing pulses derived from the results already calculated into a special "direct input" socket which advanced the stepper independently of the number of repetitions that had been counted.

ENIAC, then, was designed to solve a specific type of mathematical problem, but it had to be able to do so completely automatically: if its operators had to change

---

[6]Haigh et al. (2016) attributed these plans to Arthur Burks. Subsequent archival research suggests that the work was in fact split between Burks and Adele Goldstine, with Goldstine taking the lead on the mathematical analysis of the problem, expressed in a "setup form", and Burks mapping it onto ENIAC's distributed programming system in the form of a "panel diagram".

[7]ENIAC would therefore carry out 2000 integration steps to calculate a single trajectory, many more than was feasible in a hand calculation. This was one reason why the numerical properties of the method to be used had been studied so closely: with many more arithmetical operations being carried out, errors could be expected to accumulate more rapidly.

instruction tapes, as on Mark I, the advantages of its electronic speed would be lost. The analysis of the trajectory calculation revealed the level of control flexibility required and led to the design of the master programmer, a device capable of controlling highly complex computations built up using the fundamental structures of loops and conditional branches. As a result, ENIAC was capable of tackling a wide range of problems, although in practice physical constraints such as its small amount of high-speed storage limited its scope of application (Reed 1952).

6.5 The Computer as Mathematical Instrument

By the summer of 1944, ENIAC's design was virtually complete and the team were beginning to think about the future. Anxious to secure a new contract before the generous wartime funding arrangements dried up, they proposed to BRL's director, Colonel Leslie Simon, a new research and development project for a machine which would address two perceived weaknesses of ENIAC's design: its paucity of high-speed storage, and the time-consuming way in which problems were set up.

At around the same time, John von Neumann discovered ENIAC. Despite the fact that he had been a member of BRL's Scientific Advisory Committee since 1940, he only found out about the machine, according to Herman Goldstine, thanks to a chance meeting at Aberdeen railway station. Within a month of this meeting, however, an agreement had been reached for a contract to develop a new machine. Historians have speculated about von Neumann's role in helping to bring about this decision, but one likely consequence of his involvement was to establish just what the new machine was for. In 1943, BRL had a very clear idea of why they needed ENIAC: they faced a bottleneck in the calculation of firing-table trajectories, and the need to address this requirement shaped ENIAC's design in many ways. By contrast, in a context where it was cutting back on long-term research projects, the Bureau of Ordnance might not have been so keen to support a proposal framed in terms of the need to address shortcomings in a machine it was still in the process of paying for. As Babbage had discovered a century earlier, this is not a great strategy for winning a funding body's support.

Matters moved quickly. On August 29, at a meeting attended by both Goldstine and von Neumann, BRL's Firing Table Reviewing Board decided to support a new contract with the Moore School, to develop "a new electronic computing device". The Board minuted that the new machine would be "cheaper and more practical to maintain" than ENIAC, would be able to store large quantities of numerical data, and would be easy to set up for new problems. The Board also noted that the new machine would be "capable of handling many types of problems not easily adaptable to the present ENIAC" (see Haigh et al. 2016, 134).

Von Neumann brought to the meeting the perspective of a user, not a computer builder. Although he proved more than capable of engaging with the gritty details of vacuum tubes, he was also engaged in a continent-wide search for raw computing power for a variety of projects, including the Manhattan Project at Los Alamos. In

March he had used the IBM punched card machines at BRL to carry out some test calculations on hydrodynamical shock problems, noting that:

> The actual computations on each problem required 6-12 working hours net, and the entire program (setting up, etc), insofar as these three problems were concerned, took less than ten days. [...] In the truly many-dimensional cases the possibility of using other types of machines will also have to be investigated. (von Neumann 1944b, 375, 379)

The possibility of using ENIAC for similar work was quickly investigated. By August 21, as Goldstine reported:

> Von Neumann is displaying great interest in the ENIAC [...] He is working on the aerodynamical problems of blast and runs into partial differential equations of a very complex character. By greatly simplifying his equations he is able to get a one dimensional equation that is solvable in four hours on the IBM's. We calculate that it will take ten seconds on the ENIAC counting the printing time. (Goldstine 1944b)

But not even ENIAC was powerful enough. The day after deciding to support the new contract, the Firing Table Reviewing Board sent a detailed memo to Simon outlining the rationale for their decision. Since the new machine would be more flexible and capable of storing large amounts of numerical data:

> It would make possible the solution of the complete system of differential equations of exterior ballistics [...] these equations are too complicated in character to be handled by the differential analyzer, the Bell Telephone machines, the IBM machines, or the present ENIAC in a reasonable length of time. (BRL 1944)

The Board also noted the application of the new machine to the "extensive and unusual computations" needed to make use of the data produced by BRL's new wind tunnel. Existing machines, including ENIAC, would be "most useful in extensive but less complicated routine calculations". The wind tunnel played a prominent role in selling the new proposal to BRL and its paymasters. In mid-September Brainerd wrote to Colonel Paul Gillon of the Bureau of Ordnance referring to:

> some rather extensive discussions concerning the solution of problems of a type for which the ENIAC was not designed. [...] Dr. Von Neumann is particularly interested in mathematical analyses which are the logical accompaniment of the experimental work which will be carried out in the supersonic wind tunnels. Unfortunately practically all of these problems are tied up in non-linear partial differential equations, the solutions of which is is impractical to obtain with any known equipment now existing or being built. (Brainerd 1944a)

Brainerd was now careful to suggest that ENIAC's perceived shortcomings were not defects, but rather adaptations to the particular type of problem it was designed to solve. These representations evidently had the required effect: towards the end of October, a supplement to the ENIAC contract was signed authorizing a 9-month contract on "an Electronic Discrete Variable Calculator", starting on January 1,

1945.[8] Von Neumann's existing role at BRL was expanded, allowing him to act as a consultant to the Moore School for the new project (Goldstine 1944a).

Simon received yet another memo on the subject, this time from von Neumann himself, in January 1945. Von Neumann emphasized the importance of "general aerodynamical and shock-wave problems" and the need to make "full and efficient use of the Supersonic Windtunnel", and he pointed out that ENIAC and the Bell Labs machines were not really suited to the kind of calculations required:

> The differential equations are usually partial and 2 or 3 dimensional, and they are therefore in the simplest cases just on the margin of what the present equipment can handle, and in all other cases far outside its compass. [...] The EDVC [sic] is being designed with just this type of problem in view. (von Neumann 1945a)

In the latter part of 1944, then, a rather vague aspiration to build a machine that would address some of ENIAC's shortcomings was refined into a proposal for a computer optimized to solve a class of problems of pressing concern to BRL, multi-dimensional, non-linear, partial differential equations. Brainerd was quick to spell out the connections between this application and the team's technical ambitions:

> If a two-dimensional problem is to be solved [...] many thousands of values of quantities must be stored while the process is being carried on. It is on this point of the great amount of storage capacity required that existing and contemplated machines fall down. There is also a further point that the programming of the carrying out of the solutions is far more complicated than permitted by existing or contemplated machines. (Brainerd 1944a)

EDVAC, then, needed a large high-speed store because the calculations it was being built to carry out generated large amounts of numerical data. But this also suggested a solution to the problem of setting up the machine quickly:

> To evaluate seven terms of a power series took 15 minutes on the Harvard machine of which 3 minutes was set up time, whereas it will take at least 15 minutes to set up ENIAC and about 1 second to do the computing. To remedy this disparity we propose a centralized programming device in which the program routine is stored in a coded form in the same type storage devices [sic] suggested above [to hold numerical data]. (Goldstine 1944c)

All previous automatic computers had used different storage media for numbers and program instructions: numbers were stored in counters of various kinds, while instructions were read from paper tape or, in the case of ENIAC, set up on dedicated pieces of hardware. If instructions were to be available at electronic speed, they could not be read when needed from an external medium, but had to be placed on the machine before the computation began. As Goldstine noted, a new device—mercury delay lines—had been proposed for the cost-effective storage of large amounts of numerical data. If instructions were coded as numbers, as on the Harvard and Bell Labs machines, it would obviously be possible to use the same kind of device to hold the instructions.

---

[8]Brainerd (1944b) described the machine thus in a memo to the Bureau of Ordnance. By the time the project's first progress report was issued, at the end of March 1945, it had firmly acquired the acronym EDVAC, in which the C stood for "computer".

At this point, Goldstine's proposal was that, instead of using different media to store numbers and instructions, they could be held in storage devices of the same type. What is often taken to be a defining characteristic of the modern computer, storing data and instructions in a *single* device, was adopted rather cautiously. In the *First Draft*, after carefully listing all the different types of information that EDVAC would have to store, von Neumann commented:

> While it appeared, that various parts of this memory have to perform functions which differ somewhat in their nature and considerably in their purpose, it is nevertheless tempting to treat the entire memory as one organ, and to have its parts even as interchangeable as possible for the various functions enumerated above. (von Neumann 1945b, 6)

Some problems needed lots of programming instructions but used little numerical data, while others were exactly the reverse. As Eckert explained the following year, a single store would give EDVAC valuable flexibility:

> Aside from simplifying the construction of the machine, this move eliminates for the designer the problem of attempting to find the proper balance between the various types of memory [. . .] The proper subdivision of the memory, even for a restricted set of problems, such as the ENIAC is designed to handle, is too variable from problem to problem to permit an economical compromise. (Eckert 1946, 112)

However, the code proposed in the *First Draft* clearly distinguished numbers and instructions, and treated the two kinds of data rather differently. EDVAC's memory would still be explicitly partitioned, recreating on a problem-by-problem basis the separate storage devices that Goldstine envisaged.

The tape of Alan Turing's universal machine of 1936 also held both data and coded instructions, a fact that has led some writers to suppose that there is a simple "stored program concept", invented by Turing and subsequently implemented by the new machines of the mid-1940s. The complexities and confusions surrounding the term "stored program" have been analysed by Haigh et al. (2014), and it is sufficient here to note that EDVAC's unitary memory was not the result of the application of an insight drawn from mathematical logic, but of a series of pragmatic engineering decisions taken during the design of a machine requiring an unprecedentedly large store in order to address a particular class of mathematical problem.

A more significant innovation of the *First Draft* was to give programs the ability to modify their own instructions in certain ways as computations progressed. This had profound consequences for EDVAC's mathematical capabilities, making it feasible to write programs that operated on large vectors and matrices, not just on a small number of individual variables. Without this, the machine would not have been able to solve the partial differential equations of interest to von Neumann. There is nothing like this in Turing's earlier logical work.

Like ENIAC, then, EDVAC was designed and sold to its sponsor as a mathematical instrument with a rather specific purpose. Designing a machine capable of carrying out the required calculations led to a number of features that are now considered definitional of the modern computer, such a large unitary memory and code that allows programs to modify their own instructions. There is no need to postulate an injection of mathematical logic in order to explain the origins of the computer and little, if any, evidence in the archival record of such an injection.

6.6 Planning and Coding

The last two sections have examined some of the ways in which the computer's mathematical origins shaped its technological design. The influence of mathematics was not limited to hardware, however: the following sections will explore how the work practices within which automatic computers were situated profoundly affected early conceptions of computer programming.

The ENIAC progress report issued at the end of 1943 positioned the plans for the trajectory calculation in the context of a "general setup procedure" consisting of three phases. The first phase was mathematical, involving "the reduction of the given set of equations or relations to such a form that they can be solved by the ENIAC" (Moore School 1943, XIV (1)). This involved transforming the equations so that they only used the basic operations provided by ENIAC and ensuring that the computation would fit within the limits of its hardware, both in terms of the number of operations involved and the accuracy of the results that would be obtained. For the trajectory calculation, this phase resulted in a set of difference equations allowing a numerical solution of the equations of exterior ballistics to be calculated.

The second phase involved mapping these difference equations onto ENIAC's hardware. Variables were assigned to accumulators, and decisions were taken about numerical matters such as the number of decimal places and the position of the decimal point. Once this was done:

> this phase of setup reduces to the somewhat routine task of scheduling the operations and the corresponding connections. There are many possible arrangements for each problem, however, so that some skill is involved in chasing a suitable and preferred one. (Moore School 1943, XIV (2))

The results of this phase were given in a "setup form" describing the sequencing of the operations and the numerical details, and a "panel diagram" giving details of exactly how switches would be set and connections plugged so that ENIAC would perform the operations in the required order. Plans for the trajectory calculation were attached to the report. Based on the information in the panel diagram, the third phase of the procedure was rather more routine, involving "the manual plugging in of the various conductor cables and the manual setting of the various program switches" (Moore School 1943, XIV (3)).

The report claimed no originality for this three-phase procedure, pointing out its similarity to the way equations were set up on the differential analyzer. But its roots go back much further than that: the three phases correspond quite clearly to the basic division of labour devised by de Prony in the eighteenth century. The first phase, putting "the given equations [...] in a form suitable for the machine", corresponds to the work of the mathematicians in de Prony's first section, and Adele Goldstine and Arthur Burks, developing detailed computational plans that used only basic arithmetical operations, would have been natural members of the second section. The most significant change is a direct consequence of automation: the arithmetical hairdressers of de Prony's third section have been replaced by ENIAC, and human labour is only required to set up ENIAC to perform the calculation and supervise it while in operation.

The similarity extends even to the checking of calculations. De Prony had called for calculations to be carried out in two different ways, with the workers of the second section checking that the results were consistent. Every time ENIAC printed a set of results, a single integration step would be carried out with known data and the results printed. These would be checked by the operators, and any discrepancies with the expected results would indicate that a hardware fault had occurred.

Howard Aiken's group at Harvard employed a similar division of labour when preparing computations for Mark I. The first step was taken by "the mathematician who chooses the numerical method best adapted to computation by the calculator" (Harvard 1946, 50). Relevant factors considered at this stage included the accuracy and the speed of the calculation, and also the ease with which it could be checked.

The chosen method was then expressed in terms of Mark I's basic operations. A variety of notations were used at this stage. Coding sheets (Harvard 1946, 49) were used to define the basic sequence of operations to be punched onto instruction tapes, and diagrams were prepared showing how to wire the plugboards that some of Mark I's more complex units possessed.

Mark I was not fully automatic, however, and its operators were a integral part of computations, being required, for example, to change tapes when necessary. As well as coded instructions for Mark I, therefore, detailed operating instructions had to be drawn up for each calculation. In this context, the equivalent of de Prony's third section was the cyborg assemblage of Mark I and its operators. The Harvard group preserved the traditional status distinctions between the sections: operators were "enlisted Navy personnel" (Bloch 1999, 87), whereas the mathematicians were civilians or commissioned officers.

Historians have sometimes described the origins of programming as a secondary process that followed the development of the computing hardware. For example, Nathan Ensmenger (2010, 34) writes that programming was "little more than an afterthought in most of the pioneering wartime computing projects". At least in the case of ENIAC and EDVAC, this is not true: detailed plans, or programs, were prepared as part of the design process in both projects and directly influenced central aspects of the machines' design, such as ENIAC's master programmer.

It would be more accurate to say that the participants in these wartime projects did not view programming as being something particularly novel or problematic. Machines were built to carry out specific mathematical tasks and their designers assumed that existing well-understood procedures for planning and organizing large-scale calculations could be straightforwardly applied to the new situation. Moving from human to automatic computation led to changes in the way that the accuracy of calculations was estimated and their results checked, but the overall workflow of the planning process was unchanged. The biggest difference was that instead of handing a computation sheet to a human, the instructions it contained had to be translated into machine-readable form, but once the sequence of low-level operations had been decided on, this was thought to be a straightforward procedure. The changes brought about by automation were localized at a late stage in the overall planning process, as von Neumann pointed out when preparing a "tentative computing sheet" for a Monte Carlo simulation. It was, he said,

> neither an actual "computing sheet" for a (human) computer group, nor a set-up for the ENIAC, but I think it is well suited to serve as a basis for either. (von Neumann 1947, 152)

In the first of an influential series of reports on *Planning and Coding of Problems for an Electronic Computing Instrument*, Goldstine and von Neumann (1947) gave a detailed account of how existing practices of large-scale calculation could be adapted for use with automatic computers. Although the word "programming" was being used in its modern sense as early as 1944,[9] Goldstine and von Neumann chose not to use it. Instead, they split the overall workflow into the two major phases of "planning" and "coding". The division between the two phases marked the point at which techniques specific to automatic computers became important.

Goldstine and von Neumann described planning as a "mathematical stage of preparations". Echoing the approach taken by the ENIAC and Mark I designers, they explained that planning involved developing equations to model the problem at hand, reducing these to "arithmetical and explicit procedures", and estimating the "precision of the approximation process". They emphasized that all three steps in the planning stage were "necessary because of the computational character of the problem, rather than because of the use of a machine" (Goldstine and von Neumann 1947, 19).

The coding phase was less familiar and so discussed in much more detail. It was divided into two stages. A "macroscopic" stage corresponded to the second phase of the ENIAC setup procedure. It began by expressing the structure of the program in diagrammatic form, using the new flow diagram notation that Goldstine and von Neumann had developed, and drawing "storage tables" summarizing the data used by the program. The subsequent "microscopic" stage corresponded more closely to what in understood by "coding" today, and involved expressing the contents of the various boxes in the flow diagram in machine code. Some routine manipulations of the code were then carried out to turn it into its final machine-readable form.

By 1948, two further *Planning and Coding* reports, containing a number of worked examples, had been issued. The reports were highly influential and the flow diagram notation was widely adopted. Ensmenger (2016) has pointed out that as programming industrialized, flow diagrams came to function as boundary objects, notations inhabiting "multiple intersecting social and technical worlds" and flexible enough to enable communication between groups as disparate as managers, system analysts and programmers. Initially, however, they sat on the boundary between

---

[9]Goldstine was an early adopter of the terminology; see, for example, the uses of "programming" and "program routine" by Brainerd (1944a) and Goldstine (1944c) quoted in the previous section.

the planning and coding stages of the program preparation process. As computers came to be used for tasks that were not exclusively mathematical, or where a "mathematical stage of preparation" became less applicable, development began with a stage of "analysis" whose results, documented as a flow diagram, became the input for the more machine-oriented aspects of the workflow.

As experience with the new machines was gained, it quickly became apparent that planning and coding was not quite as straightforward as expected. The exercise of preparing instructions for a machine revealed the extent to which planners had relied on the humans of the third section to display intuition and common sense, even when they were supposedly acting "mechanically". The English mathematician Douglas Hartree, one of ENIAC's first users, commented on a typical breakdown in an automated calculation:

> A human computor, faced with this unforeseen situation, would have exercised intelligence, almost automatically and unconsciously, and made the small extrapolation of the operating instructions required to deal with it. The machine without operating instructions for dealing with negative values of z could not make this extrapolation. (Hartree 1949, 92)

The moral that Hartree drew from this experience was that programmers needed to take a "machine's-eye view" of the instructions being written, and this blurring of the boundaries between human and machinic agency is nicely captured in the image of the human "automatically" exercising intelligence. However, it was more common to call for a more exhaustive and rigorous planning process. In what is often described as the first programming textbook, the Cambridge-based team of Maurice Wilkes, David Wheeler, and Stanley Gill explained that:

> A sequence of orders [...] must contain everything necessary to cause the machine to perform the required calculations and every contingency must be foreseen. A human computer is capable of reasonable extension of his instructions when faced with a situation which has not been fully envisaged in advance, and he will have past experience to guide him. This is not the case with a machine. (Wilkes et al. 1951, 1)

As this indicates, Goldstine and von Neumann's view of computer programming as a form of planning quickly became standard. The first challenge to the perceived limitations of this approach would not emerge until the mid-1950s, a development outlined in the final section of this chapter.

6.7 From Tables to Subroutines

The influence of mathematical practice on the use of automatic computers is visible not only in the organization of complete computations, but also in the details of specific programming techniques. An interesting example of this is the relationship between the use of tables in manual computation and the development of the idea of the subroutine.

The use of tables was so engrained in mathematical practice that the Harvard Mark I's designers put it on a par with the familiar operations of addition, subtraction, multiplication and division, writing that the machine was designed to carry out computations involving "the five fundamental operations of arithmetic": the fifth operation was described as "reference to tables of previously computed results" (Harvard 1946, 10). Tables were a ubiquitous feature of manual computation. A typical table would hold the precomputed values of a function, and when a value was required the (human) computer would interrupt work on the main calculation, take the appropriate volume of tables down off the shelf, look up the required value, and copy it into the appropriate place on the worksheet. Interpolation was used to obtain values for arguments that fell between those printed in the table.

Mark I contained dedicated hardware to support each arithmetic operation. Table look-up was implemented by three "interpolation units". These units read numerical data from tapes containing equally spaced values of the function argument, each followed by the coefficients to be used in the interpolation routine (Harvard 1946, 38, 47). When a function value was required, the argument was sent to an interpolation unit. The unit would then search the tape for the appropriate value of the argument, read the interpolation coefficients, and carry out a hardwired routine to calculate the required value.

Mark I also had special-purpose units to compute logarithms and values of the exponential and sine functions. Unlike the interpolators, these units did not read a tape, but executed a built-in algorithm to compute the required values directly. Nevertheless, the units were described as "electro-mechanical tables" (Harvard 1946, 11), a terminological choice that makes clear that Mark I's designers were not only transferring the use of mathematical tables in manual computation into the world of automatic machinery, but also using the experience of the past as a way of making sense of the new machine.

ENIAC's designers also considered table look-up to be one of their machine's basic capabilities (Moore School 1943, XIV (1)), and took a similarly explicit approach to supporting the use of tables. Numerical information was stored on three "portable function tables", large arrays of switches on which a table of around 100 values could be set up, indexed by a two-digit argument. This data was read by a "function table" unit, the whole arrangement being optimized to make it convenient to read the five values required for a biquadratic interpolation. Unlike Mark I, however, ENIAC had no dedicated interpolation unit. It was left to the user to set up an interpolation routine suitable for the problem at hand, and many examples of such routines are presented in Adele Goldstine's 1946 manual and other reports.

There is a tension apparent in Mark I and ENIAC between the alternatives of looking up tabular data and computing values when needed. While mathematical functions could be computed on demand, some applications, such as calculating a trajectory, made use of empirical data for which no formula was available. There was no alternative to storing such tables explicitly. The volume of tabular data to be stored was one of the issues that the EDVAC team considered when estimating the size of memory the machine would need, and von Neumann summarized the situation as follows:

> In many problems specific functions play an essential role. They are usually given in form of a table. Indeed in some cases this is the way in which they are given by experience [...], in other cases they may be given by analytical expressions, but it may nevertheless be simpler and quicker to obtain their values from a fixed tabulation, than to compute them anew (on the basis of the analytical definition) whenever a value is required. (von Neumann 1945b, 4–5)

He suggested that common functions such as log, sin and their inverses could be treated by table look-up rather than calculation. Interestingly, Mark I's designers had made precisely the opposite choice, providing the dedicated electromechanical tables to compute the values of these elementary functions on demand.

Large computations would typically have to look up many values, and so perform multiple interpolations. On Mark I, this would simply require repeated calls to the interpolation units, but the situation was a bit more complicated on ENIAC where the interpolation routine was set up by the programmer. Clearly, setting up the instructions repeatedly would be a wasteful and ultimately infeasible approach. To perform multiple interpolations, the designers had to find a way to return to a different place in the main instruction sequence each time the interpolation routine was carried out. This capability was provided by the versatile steppers, the key components of ENIAC's master programmer. The mid-1944 progress report explained how this could be done, making the connection with interpolation explicit:

> Thus within a given step of integration a certain interpolation process may be used several times. This sequence need be set up only once; by means of a stepper the same sequence can be used whenever needed. (Moore School 1944, IV-40)

This idea of "computation on demand" was naturally soon generalized, and it was recognized that it would be useful to be able to easily reuse any sequence of instructions, not only those computing familiar mathematical functions. In August, 1944, von Neumann reported to Robert Oppenheimer on the progress of the Bell Labs machine. Like Mark I, this machine would read instructions from paper tape, but unlike the Harvard machine, it would have more than one sequence unit. As von Neumann (1944a) noted, it would employ "auxiliary routine tapes [...] used for frequently recurring sub-cycles". There is no suggestion that these auxiliary tapes would be limited to the purpose of interpolation or table look-up.

This turned out to be an issue even on Mark I: its electromechanical tables took a long time to calculate a value, as they used the full numerical precision of the machine. Programmers Richard Bloch and Grace Hopper soon found it necessary to develop more efficient routines for specific problems. As Mark I only had one sequence mechanism, however, they had no alternative to recording and reusing these routines by hand, as Hopper recalled:

> And if I needed a sine subroutine, angle less than $\pi/4$, I'd whistle at Dick and say, "Can I have your sine subroutine?" and I'd copy it out of his notebook. (Hopper 1981)

It quickly became clear that it would be useful to plan in advance, and to make routines that were likely to be generally useful available for reuse. In 1945, to test the usability of the EDVAC code he had designed, von Neumann wrote a program to merge two sequences of data. After completing the code, he noted the potential generality of the procedure and commented that it could

> be stored permanently outside the machine, and it may be fed into the machine as a "sub routine", as a part of the instructions of any more extensive problem, which contains one or more [merge] operations. (von Neumann 1945c, 25-6)[10]

Subroutines were extensively discussed by the EDVAC group in the summer of 1945, and in September Eckert and Mauchly provided the following account in a progress report:

> It is by the use of "subsidiary chains" of orders, to be called into use from time to time, as they are needed, by a "higher" set of orders, that a computational routine can be compactly represented. What is more, this corresponds to the way in which mathematical processes are most easily and naturally thought about. The rule for interpolation is not written down anew each time it must be used, but is regarded as a "subsidiary routine" already known to the computer, to be used when needed. (Eckert and Mauchly 1945, 40)

Eckert and Mauchly made here the familiar connection between subroutines and interpolation, and hence the use of tables, but it is striking that the direction of the metaphor is now reversed and the terminology of automatic computing is used to characterize a familiar and long-established mathematical practice.

The idea that subroutines would be recorded in a notebook already seemed outdated, and the benefits of more systematic ways of storing and sharing code were becoming recognized. Herman Goldstine (1945) commented that "[e]vidently one would collect in his library tapes for handling standard types of problems such as integrations and interpolations", and even in Harvard sequence tapes of "general interest" were "preserved in the tape library" (Harvard 1946, 292).

The idea of a subroutine library soon caught on and the developers and users of various machines began to plan standard libraries. As well as convenience, the promise of greater reuse made it economic to analyze the library routines to ensure that they were efficiently coded and would work correctly in a range of contexts. In a January 1947 report on EDVAC programming, Samuel Lubkin (1947, 20, 28) gave an example of a "standard subroutine" to compute square roots "in the form it would take in a library of subroutines", while at around the same time ENIAC operator Jean Bartik was contracted by BRL to run a programming group charged with developing "the technique of programming the production of trigonometric and exponential functions" along with a number of other routines of interest to ballisticians.[11] Some years later, the library concept and techniques for writing and using subroutines

[10] It is not clear whether the term "subroutine" originated with von Neumann or whether he took it over from the Mark I programmers. Assuming that Hopper in 1981 was not providing a verbatim report of her 1944 conversation with Bloch, von Neumann's manuscript is the earliest documented usage that I know of, and it is perhaps significant that the term does not appear in Harvard (1946). In fact, the more general term "routine" seems to appear only once in that volume (on page 98), suggesting that it was not in common use in Harvard.

[11] See Anonymous (1947) for the complete list of problems assigned to the group. As Bartik (2013, 115–120) described, however, much of the group's effort was diverted to developing EDVAC-style codes in advance of ENIAC's conversion to central control.

were more widely disseminated in the textbook by Wilkes et al. (1951) which made, as its subtitle promised, "special reference to the EDSAC and the use of a library of subroutines".[12]

The metaphor of the "library" is telling. Authors working in libraries consult reference books, but the texts they are writing do not form part of the library. At best, they will be added to the shelves only after being completed, published and found worthy of preservation. Similarly, a subroutine thought to be generally useful might, after extensive checking, be placed in a library, but the main routines written to solve specific problems were treated quite separately and were less likely to be permanently stored. Work practices reinforced the distinction between the two types of code. Wilkes et al. (1951, 43) described how EDSAC subroutines and master routines were punched on separate tapes and only combined at the last minute to form a program tape for an actual computation. The subroutines themselves were punched on coloured tape and stored in a steel cabinet, while the master copies were kept under lock and key. In contrast to these complex and bureaucratized procedures, the master routine tapes could be treated very casually, as the story of Wilkes' Airy program reveals (Campbell-Kelly 1992). At Harvard (1946, 292), there was also a contrast between the care that would go into the preparation of a library tape for Mark I and one intended to be run but once.

Subroutines, then, are a technique with roots in the mathematical practice of table use that allowed programs to be efficiently structured and written. However, while a mathematician carrying out a complex calculation would not normally develop a new interpolation routine, say, programmers did identify new and unanticipated subroutines while writing new programs. Among the first to notice this were BRL mathematicians Haskell Curry and Willa Wyatt who in 1946 planned an interpolation routine for ENIAC. They divided the program into a number of "stages" and, noting that some stages could be reused to avoid having to recode them, went on to make the methodological recommendation that programmers identify reusable stages by looking for repeated code: "the more frequently recurring elements can be grouped into a stage by themselves" (Curry and Wyatt 1946, 30).

However, other writers did not follow this lead. Subroutines were not explicitly represented in the flow diagram notation, and in the *Planning and Coding* reports Goldstine and von Neumann offered no guidance on how to identify useful new subroutines. Some of the library subroutines described by Wilkes et al. (1951), such as those carrying out integration, made use of "auxiliary subroutines" which defined the function being integrated, but more general uses of user-defined subroutines were not considered.

As a result, perhaps, ad hoc subroutines were rather uncommon in practice. Of the 30 stages in Curry and Wyatt's interpolation program, only four were identified as being reusable. In the Monte Carlo programs run on ENIAC in 1948, there was only one subroutine (to compute a pseudo-random number) in approximately 800

[12]Not to be confused with EDVAC, EDSAC was an electronic computer developed in Cambridge by a team led by Maurice Wilkes. It came into operation in 1949.

program instructions (Haigh et al. 2016, 183–6). Programming guidelines for the Harvard Mark II even suggested that in general "the method which involves the fewest routines [. . .] is the logical choice" (Harvard 1949, 266).

The emphatic distinction between master routines and subroutines had another consequence, namely that calling hierarchies were rather flat. Typically, a master routine would call a small number of subroutines, but it was rather rare for one subroutine to call another. The techniques used for subroutine call and return further meant that recursive calls, where a subroutine calls itself, were not possible.

The practices of subroutine use that emerged in the early years of automatic computing, then, reflected the ways in which tables were used in manual calculation. Like a set of tables, a subroutine library is a resource that is available in advance of a computation, and subroutine use was largely restricted to calling routines from a library. Looking up a table is an exceptional task that takes the computer away from the normal process of working through a computation sheet and, similarly, calling a subroutine is an exceptional occurrence. Looking up a table is a self-contained and non-recursive operation: when looking up a value in a table, you rarely have to look up a second table in order to complete the operation. Similarly, complex structures of calling relationships between subroutines appear to be uncommon.

These assumptions were still in evidence 10 years later in the first widely-used programming language, Fortran. Like the computer itself, Fortran was intended for mathematical application. The source code was described as "closely resembling the ordinary language of mathematics" and "intended to be capable of expressing any problem of mathematical computation" (IBM 1956, 2). Subroutines were understood by analogy with mathematical functions. A formula containing a function, such as $a - \sin(b - c)$, could be translated directly into Fortran as `A-SINF(B-C)` (IBM 1956, 12). Fourteen functions were provided as "built-in subroutines" of the language, but these were for rather simple operations such as returning the absolute value of a number. Functions that would typically have been tabulated, such as the trigonometric and exponential functions, were not built in and were left for users to define.

However, new subroutines could not be defined in the Fortran language itself, but had to be written in machine code, and then added to the library in rather a complex and labour-intensive process.

> Library subroutines exist on the master FORTRAN tape in relocatable binary form. Placing a new subroutine on that tape involves (1) producing the routine in the form of relocatable binary cards, and (2) transferring these cards on to the master tape by means of a program furnished for that purpose. (IBM 1956, 40)

Only with the arrival of Fortran II in 1958 did the language provide more general support for the definition and use of functions and subroutines.

> The FORTRAN II subprogram facilities are completely general; subroutines can in turn use other subroutines to whatever degree is required. These subroutines may be written in source program language. For example, subprograms may be written in FORTRAN II language such that matrices may be processed as units by a main program. (IBM 1958, 1)

6.8 Conclusions

This chapter began by considering the view expressed by Davis and Mahoney that since EDVAC the computer has been intrinsically a universal logic machine, and hence that its subsequent application to a host of application areas was, if not always straightforward in practice, at least unproblematic in theory. A consequence of this view is that the computer's origins as a technological innovation to automate specific mathematical processes are reduced to the level of an incidental detail.

In contrast, this chapter has shown that EDVAC, like its predecessors, was planned, promoted, designed and built for very specific mathematical purposes. This perspective dominated much computer development throughout the rest of the 1940s, and I have argued elsewhere (Priestley 2011, 147–153) that the identification of machines based on the EDVAC design with Turing's idea of a universal machine was not widely made until the early 1950s. As Mahoney might have pointed out, the story of the adoption of the computer by non-mathematical communities is often the story of how the mathematical orientation of the early machines was overcome. As Christopher Strachey, one of the first people to write substantial programs for non-mathematical applications, commented:

> the machines have been designed principally to perform mathematical operations. This means that while it is perfectly possible to make them do logic, it is necessarily a rather cumbersome process. (Strachey 1952)

What was invented in the 1940s was not just the automatic computer, however, but modern computing. The machines were conceived as replacements for human computers engaged in mathematical calculation. As Stibitz made clear, this is why they are called computers. The computers' job was to carry out, in ways specified by an explicit plan, a sequence of operations, and the central innovation of modern computing was to automate the task of instruction following. Rather than describing the take-up of a uniquely capable technology, Mahoney's "histories of computing" were to be the stories of how different communities came to reformulate their existing work practices in the form of computer programs.

The task of preparing instructions for the new machines to execute, the activity that we now call programming, naturally became of central importance. Sections 6.6 and 6.7 showed how early thinking about programming was profoundly shaped by the mathematical context in which the new computers were built. At the organizational level, existing techniques for managing large-scale calculation were preserved as far as possible. Goldstine and von Neumann's *Planning and Coding* reports dealt largely with mathematical applications and were rooted in a division of labour dating back to the late eighteenth century. Machine-specific techniques were categorized as coding issues, and it was assumed that the overall planning of a computation could proceed along familiar lines. At a more detailed level, the particular ways in which subroutines were used to make programming more efficient reflected aspects of the use of mathematical tables in manual computation. This is not to say, of course, that the use of subroutines was limited to mathematical functions—the EDSAC library also included crucially important input and output subroutines. The point is, rather, that the role of subroutines within programs and the ways in which they were used were constrained by their association with existing practices of using mathematical tables.

These two aspects are characteristic of a general approach to programming that was widely accepted in the late 1940s and early 1950s. Many of the developments of the 1950s, such as the move to automate coding that led to the development of high-level programming languages such as Fortran, were aimed at making technical improvements within this framework but did not break away from the overall model or the mathematically-oriented thinking that underlay it.

The first explicit reflection on and challenge to this approach emerged, perhaps unsurprisingly, in a non-mathematical context. In 1955, Allen Newell and Herbert Simon began to consider the prospects of writing programs to solve what they called "ultracomplicated problems" such as chess playing and theorem proving. They chose the latter as a testbed, and by 1956 had developed the Logic Theorist (LT), a program capable of finding proofs in the propositional calculus. They found existing programming technique inadequate for developing LT, developing instead a notion of "heuristic programming".[13]

Newell and Simon's critique of current approaches to programming focused on precisely the two issues that I have taken as being emblematic of the mathematical approach to programming. They first addressed the belief that computations had to be planned in advance in exhaustive detail.

> But one of the sober facts about current computers is that, for all their power, they must be instructed in minute detail on everything they do. To many, this has seemed to be harsh reality and an irremovable limitation of automatic computing. It seems worthwhile to examine the necessity of the limitation of computers to easily specified tasks. (Newell and Simon 1956, 1)

Secondly, they noted that the design of LT made extensive use of subroutines. Recognizing that "most current computing programs [. . .] call for the systematic use of a small number of relatively simple subroutines that are only slightly dependent on conditions", they argued for a view of program structure that was quite different from the prevailing view of a program as a sequence of statements. Whereas "[a] FORTRAN source program consists of a sequence of FORTRAN statements" (IBM 1956, 7), Newell and Simon held that:

> a program [. . .] is a system of subroutines [. . .] organized in a roughly hierarchical fashion. [. . .] The number of levels in the main part of LT is about 10, ignoring some of the recursions which sometimes add another four or five levels. (Newell and Shaw 1957, 234–8)

This vision of the use of subroutines is quite different from the prevailing model discussed in Sect. 6.7 of this chapter. Rather than corralling subroutines in libraries that enforced limited and rather stereotypical patterns of use, Newell and Simon viewed them as being fundamental programming structures on a par with loops

[13] See Priestley (2017) for a more detailed account of Newell and Simon's critique and the take-up of their work by the nascent AI community.

and conditional branching. Their programming work was highly influential in the late 1950s in the emerging field of artificial intelligence (Feigenbaum and Feldman 1963), and it is very striking that in applying automatic computers to this new area of application they rejected two aspects of the traditional approach that directly reflected the specific practices of mathematical computation.

Certain aspects of Newell and Simon's approach can be found in the personal styles of earlier writers. In his proposal for the ACE, Turing gave some examples of the "paper technique of using the machine", culminating in the definition of a routine CALPOL to calculate the value of a polynomial. The program for CALPOL, or "instruction table" in Turing's terminology, made use of eight subsidiary routines, and its code bore out Turing's general comment that:

> The majority of instruction tables will consist almost entirely of the initiation of subsidiary operations and transfers of material. (Turing 1946, 28)

Turing was exceptional among the computer developers of the early 1940s in having no significant experience of large-scale manual computing. The intellectual roots of his famous 1936 paper on computable numbers were in the logical theory of recursive functions, which proceeds by building up complex definitions from simpler ones. Turing adapted this approach for his machine table notation, and the table defining the universal machine is built up largely by combining many simpler tables (Priestley 2011, 77–92). It is precisely this style of thought that is reflected in his practical programming examples such as the table for CALPOL.

Curry and Wyatt's 1946 interpolation program for ENIAC was constructed by combining a large number of small program fragments. Although he spent the war as a BRL mathematician, Curry's background and interests were, like Turing's, in mathematical logic rather than practical computation. In two later reports Curry developed this approach into a general theory of program construction, one that he explicitly opposed to the Goldstine/von Neumann model of subroutines and that bore more than a passing resemblance to his work in combinatory logic (De Mol et al. 2013).

Neither Turing's example nor Curry's theory made an immediate impact, however. Rather than developments in logical theory, it was the stimulus to develop programs for a new class of essentially non-mathematical problems that led, in the mid-1950s, to the establishment of an alternative to the prevailing approach to programming.

Acknowledgements I would like to thank Martin Campbell-Kelly, Tom Haigh, and Sven Ove Hansson for their helpful comments on earlier versions of this chapter.

References

AWB-IUPUI Arthur W. Burks Papers, Institute for American Thought, Indiana University–Purdue University Indianapolis.

ETE-UP ENIAC Trial Exhibits, Master Collection, 1964–1973, University of Pennsylvania Archives and Records Center, Philadelphia.

MSOD Moore School of Electrical Engineering, Office of the Director Records, 1931–1948, UPD 8.4, University Archives and Records, University of Pennsylvania, Philadelphia.

Aiken, H. H. (1937). Proposed automatic calculating machine. In Randell (1982) (pp. 195–201).

Anonymous (1947). Problems 1947–1948. MSOD box 13.

Babbage, C. (1822). On the application of machinery to the purpose of calculating and printing mathematical tables. In M. Campbell-Kelly (Ed.), *1989, The works of Charles Babbage* (Vol. 2, pp. 6–14). London: Pickering.

Bartik, J. J. (2013). *Pioneer programmer*. Kirksville: Truman State University Press.

Bloch, R. M. (1999). Programming Mark I. In I. B. Cohen & G. W. Welch (Eds.), *Makin' numbers: Howard Aiken and the computer*. Cambridge: The MIT Press.

Brainerd, J. G. (1943). Letter to Dean Harold Pender. MSOD box 51.

Brainerd, J. G. (1944a). Letter to Lt. Col. Paul N. Gillon. MSOD box 55a.

Brainerd, J. G. (1944b). Letter to Lt. Thomas H. Bogert. AWB-IUPUI.

BRL. (1944). Ballistics Research Laboratory firing table reviewing board, memorandum to Colonel Leslie E. Simon. AWB-IUPUI.

Bush, V. (1931). The differential analyzer: A new machine for solving differential equations. *Journal of the Franklin Institute, 212*(4), 447–488.

Campbell-Kelly, M. (1992). The Airy tape: An early chapter in the history of debugging. *IEEE Annals of the History of Computing, 14*(4), 16–26.

Curry, H. B., & Wyatt, W. A. (1946). A study of inverse interpolation of the Eniac (BRL report No. 615). Aberdeen Proving Ground.

Davis, M. (2000). *The universal computer*. New York: W. W. Norton & Company.

Daylight, E. G. (2015). Towards a historical notion of 'Turing—the father of Computer Science'. *History and Philosophy of Logic, 36*(3), 205–228.

De Mol, L., Carlé, M., & Bullynck, M. (2013). Haskell before Haskell: An alternative lesson in practical logics of the ENIAC. *Journal of Logic and Computation, 25*(4), 1011–1046.

Eckert, J. P., Jr. (1946). A preview of a digital computing machine. In M. Campbell-Kelly & M. R. Williams (Eds.), *The Moore school lectures: Theory and techniques for design of electronic digital computers* (Volume 9 of Charles Babbage Institute reprint series for the history of computing, pp 109–126). The MIT Press.

Eckert, J. P., & Mauchly, J. W. (1945). Automatic high speed computing: A progress report on the EDVAC. Moore School of Electrical Engineering.

Ensmenger, N. (2010). *The computer boys take over*. Cambridge: MIT Press.

Ensmenger, N. (2016). The multiple meanings of a flowchart. *Information & Culture, 51*(3), 321–351.

Feigenbaum, E. A., & Feldman, J. (1963). *Computers and thought*. New York: McGraw-Hill.

Goldstine, H. H. (1944a). Letter to John von Neumann. ETE-UP, Plaintiff's Trial Exhibit Number 2512.

Goldstine, H. H. (1944b). Letter to Lt. Col. P. N. Gillon. ETE-UP, Plaintiff's Trial Exhibit Number 2343.

Goldstine, H. H. (1944c). Letter to Lt. Col. P. N. Gillon. ETE-UP, Plaintiff's Trial Exhibit Number 2395.

Goldstine, H. H. (1945). Letter to Haskell B. Curry. ETE-UP, Plaintiff's Trial Exhibit Number 3556.

Goldstine, H. H., & von Neumann, J. (1947). *Planning and coding of problems for an electronic computing instrument* (Vol. 1). Princeton: Institute for Advanced Study.
Grattan-Guinness, I. (1990). Work for the hairdressers: The production of de Prony's logarithmic and trigonometric tables. *Annals of the History of Computing, 12*(3), 177–185.
Grier, D. A. (1998). The math tables project of the Work Projects Administration: The reluctant start of the computing era. *IEEE Annals of the History of Computing, 20*(3), 33–50.
Haigh, T. (2014). Actually, Turing did not invent the computer. *Communications of the ACM, 57*(1), 36–41.
Haigh, T., Priestley, M., & Rope, C. (2014). Reconsidering the stored-program concept. *IEEE Annals of the History of Computing, 36*(1), 4–17.
Haigh, T., Priestley, M., & Rope, C. (2016). *ENIAC in action: Making and remaking the modern computer*. Cambridge: MIT Press.
Halsey, F. A. (1896). Some special forms of computers. *Transactions of the American Society of Mechanical Engineers, 18*, 70–74.
Hartree, D. R. (1949). *Calculating instruments and machines*. Urbana: The University of Illinois Press.
Harvard. (1946). *A manual of operation for the Automatic Sequence Controlled Calculator*. Vol. I of the Annals of the Computation Laboratory of Harvard University, Harvard University Press.
Harvard. (1949). *Description of a relay calculator*. Vol. XXIV of the Annals of the Computation Laboratory of Harvard University, Harvard University Press.
Hering, C. (1891). *Universal wiring computer for determining the size of wires for incandescent electric light leads*. New York: W. J. Johnston.
Hopper, G. M. (1981). Keynote address. In R. L. Wexelblat (Ed.), *History of programming languages* (pp. 7–20). New York: Academic.
Hurd, C. C. (1985). A note on early Monte Carlo computations and scientific meetings. *Annals of the History of Computing, 7*(2), 141–155.
IBM. (1956). *The FORTRAN automatic coding system for the IBM 704 EDPM: Programmer's reference manual*. New York: Applied Science Division and Programming Research Department, IBM.
IBM. (1958). *Reference manual: FORTRAN II for the IBM 704 data processing system*. New York: IBM Corporation.
Knuth, D. E. (1973). *The art of computer programming*. (Vol. 3: Sorting and searching). Reading: Addison-Wesley.
Lubkin, S. (1947). *Proposed programmning for the EDVAC*. MSOD, box 8.
Mahoney, M. S. (2005). The histories of computing(s). *Interdisciplinary Science Reviews, 30*(2), 119–135.
Mauchly, J. W. (1942). The use of high speed vacuum tubes for calculating. In Randell (1982) (pp. 355–358).
Mindell, D. A. (2002). *Between human and machine: Feedback, control, and computing before cybernetics*. Baltimore: John Hopkins University Press.
Moore School. (1943). *The ENIAC: A report covering work until December 31, 1943*. Moore School of Electrical Engineering, University of Pennsylvania.
Moore School. (1944). *The ENIAC: Progress report covering work from January 1 to June 30, 1944*. Moore School of Electrical Engineering, University of Pennsylvania.
Newell, A., & Shaw, J. C. (1957). Programming the logic theory machine. In *Proceedings of the Western Joint Computer Conference* (pp. 230–240).
Newell, A., & Simon, H. A. (1956). *Current developments in complex information processing* RAND corporation report P-850.
Priestley, M. (2011). *A science of operations: Machines, logic and the invention of programming*. London: Springer.
Priestley, M. (2017). AI and the origins of the functional programming language style. *Minds and Machines, 27*(3), 449–472.
Randell, B. (1982). *The origin of digital computers: Selected papers* (3rd ed.). Springer.

Reed, H. L. (1952). Firing table computations on the ENIAC. In *Proceedings of the 1952 ACM National Meeting*, Pittsburgh (pp. 103–106).

Stibitz, G. R. (1945). *Relay computers* (AMP Report 171.1R). Applied Mathematics Panel, National Defense Research Committee.

Strachey, C. S. (1952). Logical or non-mathematical programmes. In *Proceedings of the 1952 ACM National Meeting*, Toronto (pp. 46–49).

Turing, A. (1946). *Proposed electronic calculator* (DSIR 10/385). London: The National Archives.

von Neumann, J. (1944a). *Letter to Robert Oppenheimer* (LA-UR-12-24686). Los Alamos National Laboratories.

von Neumann, J. (1944b). Proposal and analysis of a new numerical method for the treatment of hydrodynamical shock problems. In A. H. Taub (Ed.) (1963) *John von Neumann, collected works, volume VI: Theory of games, astrophysics, hydrodynamics and meteorology* (pp. 361–379). Princeton: Pergamon Press.

von Neumann, J. (1945a). *Memorandum on mechanical computing devices, to Col. L. E. Simon*. Philadelphia: Herman Heine Goldstine Papers, box 21, American Philosophical Society.

von Neumann, J. (1945b). *First draft of a report on the EDVAC*. Moore School of Electrical Engineering, University of Pennsylvania.

von Neumann, J. (1945c). Untitled typescript containing discussion of EDVAC program. American Philosophical Society, Mss. 510.78.V89p.

von Neumann, J. (1947). Letter to Robert Richtmyer, March 11, 1947. In Hurd (1985) (pp. 149–153).

Wilkes, M. V., Wheeler, D. J., & Gill, S. (1951). *The preparation of programs for an electronic digital computer*. Cambridge: Addison-Wesley Press, Inc.

Zuse, K. (1936). Method for automatic execution of calculations with the aid of computers. In Randell (1982). (pp 163–170).

Chapter 7
Cryptology, Mathematics, and Technology

Sandy Zabell

Abstract Cryptology furnishes an ideal example of the synergy between mathematics and technology. This is illustrated by events before, during, and after World War II: manual methods of encryption were replaced by faster and more secure methods of machine encryption; these methods were then attacked during the war by mathematicians using a combination of mathematics and machines; and after the war machine encryption was in turn eventually supplanted by computers and computer-based encryption algorithms. Random number generation illustrates one aspect of this: physical randomization has been completely replaced by the use of pseudo-random number generators. A particularly striking example of the impact of mathematics on cryptography is the development of public key encryption.

Tracing developments in cryptology can pose interesting challenges for the historian because of a desire for secrecy, but it is occasionally possible to see behind the veil; the last section of this chapter discusses some interesting instances of this.

7.1 Introduction

Cryptology is the science of secret communication. It has two branches: *cryptography*, designing secret methods of communication; and *cryptanalysis*, developing ways and means of attacking cryptographic systems. These two branches are archrivals: cryptographers attempt to design their systems to be resistant to even the most imaginative attacks; cryptanalysts attempt to circumvent such defenses by all possible means.

Cryptology is an ideal case study of the synergy between mathematics and technology: the cryptographer develops new methods of encryption, based on advances in either technology or mathematics, to combat vulnerabilities in current methods; the cryptanalyst in turn develops new technology and mathematics to

---

S. Zabell (✉)
Department of Mathematics and Statistics, Northwestern University, Evanston, IL, USA
e-mail: zabell@math.northwestern.edu

S. O. Hansson (ed.), *Technology and Mathematics*, Philosophy of Engineering and Technology 30, https://doi.org/10.1007/978-3-319-93779-3_7

attack such systems. Sometimes a new technology is found to be vulnerable as a result of careful mathematical analysis; sometimes new mathematical methods of encryption are attacked by developing new technologies (as was the case at Bletchley Park during World War II in their attack on German methods of encryption).

7.1.1 The Four Ages of Cryptology

Subdividing history into discrete periods necessarily involves an element of oversimplification; but, suitably qualified, it does aid organizing material. In that spirit, there are the *four ages of cryptology*:

1. *The classical period*: manual methods (up to the end of World War I).

Up until the end of the First World War, almost all cryptographic systems were manual ("paper and pencil"). Herbert Yardley's 1931 *The American Black Chamber*, and Fletcher Pratt's 1939 *Secret and Urgent* convey a vivid picture of the subject as it existed at that time. Mathematics (for example, in the guise of statistical attacks) and technology (for example, secret inks) were both employed, but with few exceptions this was only at the most basic level (see, e.g., the book by Abraham Sinkov 1968).

2. *The middle ages*: the rise of the machines (1918–1973).

The First World War made clear the limitations of hand methods. There were two basic problems: *speed* and *security*. The rise of modern warfare and commerce led to an unprecedented increase in wireless communication, and this in turn raised the issue of secure communication using a method subject to interception by third parties. The sheer volume of messages called for mechanization. Furthermore, after the war it soon became clear just how vulnerable the traditional methods of manual encryption were: the Germans learned that their naval codes had been compromised; the Japanese that the US government had been reading their diplomatic messages during sensitive negotiations after the war.

This led to the development of a variety of mechanical devices designed to efficiently encrypt a large volume of messages while at the same time being immune to classical cryptanalytical attacks. These included the commercial Hagelin machines of the commercial firm Crypto A. G., the German Enigma, Japanese "Red" and "Purple" machines, the British Typex, the US Sigaba, and so on.

3. *The modern era*: the advent of the computer (1973–present).

Although representing a great leap forward in sophistication, speed, and security, these machines suffered from a number of disadvantages. Foremost of these were the constraints arising from the use of a mechanical device to perform encryption. Obvious theoretical improvements might be ruled out on the basis of practical engineering considerations. And change was necessarily slow: replacing one system by another meant recalling potentially thousands of devices world-wide. Using a computer as the basis for encryption freed one from the limitations of a machine, replacing hardware by software, and had the advantage that software updates could be accomplished in days, not months or years.

Dating the start of this period is even more arbitrary than the preceding one. But developments such as computer networking, the design of the Unix operating system, the Unix utility *crypt*, and Feistel's *Scientific American* article (the last two in 1973), suggest the year 1973 as a reasonable point to date this change.

4. *The postmodern era*: public key encryption (1976–present).

All methods of cryptography – classical, mechanical, and computer – up to 1976 required a shared secret: a *private key(s)*, shared by sender and receiver, that enabled one to encrypt a message and the other to decrypt it. And this in turn required some secure channel by which at least one party could communicate this secret to the other. But in 1976 and 1977 a remarkable discovery was made: it was possible to securely communicate between two parties *without* a prior secure key exchange: a key could be sent from one party to the other *over a public channel* without compromising any subsequent encrypted communication. This astounding discovery – breaking with more than two millennia of past cryptographic theory – we refer to as the postmodern era in cryptography.

7.2 Classical Cryptography

The role of both mathematics and technology in classical cryptology was relatively limited; the existing literature on it is vast. Nevertheless, some brief discussion of it, in order to set the stage for later developments, is necessary.

The need for and use of methods of secret communication is as old as man himself. For example, during the Persian siege of the Greek city of Potidaea in the winter of 480–79 BC, the Persian commander Artabazus exchanged messages with Timoxenus, a military officer inside the city.

> Whenever Timoxenus and Artabazus wished to communicate with one another, they wrote the message on a strip of paper, which they rolled round the grooved end of an arrow, and the arrow was then shot to some predetermined place. Timoxenus's treachery was finally discovered when Artabazus, on one occasion, missed his aim, and the arrow, instead of falling in the spot agreed upon, struck a Potidaean in the shoulder. As usually happens in war, a crowd collected round the wounded man; the arrow was pulled out, the paper discovered, and taken to the commanding officers. [Herodotus 8.128, Aubrey de Sélincourt translation.]

Strictly speaking, this is an instance of *steganography*: the message is hidden rather then encrypted. The *Caesar cipher* instead is a method of enciphering messages that goes back to Gaius Julius Caesar (100–44 BC). The Roman historian Suetonius tells us:

> There exist letters from Caesar to Cicero and acquaintances on topics in which Caesar, when he wished to transmit them confidentially, wrote in cipher. That is, he changed the order of letters in such a way that no word could be made out. If somebody wanted to decipher it and understand the content, then he had to insert the fourth letter of the alphabet, that is D, for A, and so on. [*Lives of the Caesars*, 56; translation modified from that of Beutelspacher.]

That is, one *enciphers* the message by substituting three letters back:

$$A \to X, \quad B \to Y, \quad C \to Z, \quad \ldots, \quad Z \to W;$$

and one *deciphers* the enciphered message by reversing this process:

$$A \to D, \quad B \to E, \quad C \to F, \quad \ldots, \quad Z \to C.$$

Thus if the message were

VENI VEDI VICI,

("I came, I saw, I conquered"), then Caesar would have enciphered this as:

SBKF SBAF SFZF.

(Strictly speaking, the Roman alphabet of Caesar's time was smaller than the 26 letter alphabet of today.)

The Caesar cipher is a special instance of what is termed a *monoalphabetic substitution cipher*: one replaces each letter of the alphabet by another letter (its cipher equivalent), each letter being used as a cipher equivalent precisely once. The result is a permutation of the 26 letters of the alphabet. The Caesar cipher is a very special permutation of the alphabet, a *shift permutation*. There are a total of 26 such permutations: if σ_k represents a shift by k, then there are 26 shifts σ_k $(0 \leq k \leq 25)$. (Note a shift back by k is equivalent to a shift forward by $26 - k$.)

Other simple permutation methods are known; for example, one can step forward by a multiple k of the position. For example, if $k = 3$, then A in position 1 is replaced by C (the third letter in the alphabet), B in position 2 is replaced by F (the sixth letter in the alphabet), C in position 3 is replaced by I (the ninth letter in the alphabet), and so on. (The letter H is replaced by X, and then one cycles around, so that I is replaced by A.) It can be shown that the result is a permutation of the alphabet provided the multiplier k is not divisible by either 2 or 13 (the factors of 26). The process is sometimes described as one of *decimation*, and the result a decimated substitution alphabet.

Of course, if one knows a monoalphabetic shift or decimation cipher is being used, such a system affords little security: one can use brute force to try out the 26 possible shifts or 12 possible decimations and (provided the message is long enough) only one shift or decimation will produce a meaningful message. But if one does not confine oneself to a shift or decimation, and chooses an arbitrary permutation of the alphabet, the system becomes much more secure: there are a total of 26! or

$$403, 291, 461, 126, 605, 635, 584, 000, 000$$

monoalphabetic substitutions. The brute force method is no longer feasible.

Monoalphabetic substitution, however, is still not a very secure method; one can exploit, given a message of sufficient length, the statistical regularities present in a language to determine the particular permutation being used. There are a number of famous stories in literature illustrating the method; for example, Edgar Allan Poe's short story "The Goldbug", and Sir Arthur Conan Doyle's "The Adventure of the Dancing Men". For reference, the approximate order of occurrence of the most common letters in ordinary English is:

ETAOIN SHRDLU

(Of course, frequency of occurrence depends on both context and language. Thus, for example, one would not particularly expect this order to hold for military German.)

Polyalphabetic substitution ciphers, in contrast, use a sequence of different permutations for several successive letters. For example, in the so-called *Vigenère cipher*, one uses a *key word* (such as ISP), and each letter in the key word indicates the number of letters to shift forward in a Caesar cipher type substitution. (So if the key word is ISP, then I = 9, S = 19, P = 16 shifts are employed, followed by another set of three such shifts, and so on.) Even these ciphers have their weaknesses, however: if the length of the keyword is guessed, say k, then one can divide the message into k groups (each corresponding to a letter in the keyword) and subject each group to the classical attack used in the case of a monoalphabetic cipher. What one would need would be an encryption key as long as the message itself, and even here there are vulnerabilities if the key were itself some form of plaintext (say a passage from the Bible). In the end security would depend on a key consisting of a random string of letters as long as the message itself (a "one-time pad"). The exigencies of commerce, diplomacy, or defense seldom permit one such a luxury; what is needed is a compromise between the total security afforded by the one-time pad (or tape in the case of a mechanical implementation), and the essentially total insecurity of the monoalphabetic cipher.

The desire to avoid such vulnerabilities motivated the design of encryption devices (such as the German military Enigma) in the years leading up to World War II.

7.3 The Rise of the Machines

The advent of mechanical or electromechanical methods of encryption posed novel problems, and this in turn led to the need for novel personnel.

7.3.1 Mathematics Comes to the Fore

Apart from such basic statistical analyses such as frequencies of letters (or digraphs or trigraphs, and so on), classical attacks on encryption systems were basically linguistic in nature, and gifted cryptanalysts often came from the humanities. For example, during the First World War, Room 40, the cryptanalytic section of the British Admiralty's Naval Intelligence Division, employed a number of classical scholars such as Frank Adcock (1886–1968, Professor of Ancient History at the University of Cambridge from 1925 to 1951), Alfred "Dilly" Knox (1884–1943, Fellow of King's College from 1909), and John Beazley (1885–1970, Professor of Classical Archaeology and Art at the University of Oxford from 1925 to 1956), as well as Frank Birch (1889–1956, who after the war was a Fellow of King's College and Lecturer in History until he turned to the stage in the 1930s), and Walter Bruford (1894–1988, Professor of German at Edinburgh and the University of Cambridge). Adcock and Birch made sufficiently important contributions to the war effort that afterwards both were awarded the OBE (Order of the British Empire).

Although such skills remained valuable even during the Second World War (for example, Knox continued on in British Intelligence until 1943, and Adcock, Birch, and Bruford worked at Bletchley Park after the outbreak of war in 1939), it eventually became clear that mathematicians (or at least individuals of mathematical bent) were also needed. The intelligence services of various countries came to realize this sooner or later. The US and Poland were among the first.

7.3.1.1 US Mathematical Cryptologists

In 1930 the US Army established the Signal Intelligence Service, headed by William Frederick Friedman (1891–1969). Although not a mathematician himself, Friedman had done graduate work in genetics and made extensive use of mathematical techniques in cryptology while at the Riverbank Institute from 1914 to 1921. In 1921 he was hired by the Army as a cryptographer and later became the Army's chief cryptanalyst. It was during this period (1923) that he wrote his *Elements of Cryptanalysis*, later expanded into the four-volume classic *Military Cryptanalysis* (Friedman 1938–1941).

In April 1930 the first three individuals Friedman hired for his fledgling organization were all mathematics teachers: Frank B. Rowlett (1908–1998), Solomon Kullback (1907–1994) and Abraham Sinkov (1907–1998). Both Kullback and Sinkov were sufficiently advanced in their studies that they received doctorates in mathematics shortly afterwards (in 1934 and 1933, respectively) from the nearby George Washington University. All three were subsequently to play an important part in US cryptology.

7.3.1.2 Polish Mathematical Cryptologists

During the Polish-Soviet war of 1919–1921, Polish military intelligence employed several outstanding research mathematicians (Stanisław Leśniewski, Stefan Mazurkiewicz, and Wacław Sierpiński), who succeeded in breaking a number of Soviet ciphers. Presumably because of this positive experience, after the German military began to use the Enigma, an electromechanical device, to encrypt their messages starting in the late 1920s, Poland hired three mathematicians in 1932 – Marian Rejewski (1905–1980), Jerzy Różycki (1909–1942), and Henryk Zygalski (1908–1978) – to work on attacking the device. Using an approach grounded in group theory developed by Rejewski (and aided by information provided by French Intelligence), in 1933 the three were able to begin reading Enigma traffic. In doing so they were aided by a number of mechanical devices that were developed especially for the purpose. These included the *cyclometer* (c. 1934) and the *bomba* (1938), as well as other aids such as the *Zygalski sheets*; see Rejewski (1981).

The Polish contribution to the ability of the Allies to read the Enigma during the Second World War was considerable. Up to July 1939 the UK had no success in attacking the military Enigma. But then, sensing the impending outbreak of war, the Poles convened a special meeting outside of Warsaw where they revealed to their British and French counterparts their success, even providing each with a copy of the machine, including the internal wiring of its wheels. This, together with their extensive knowledge of intercepts and how the machine was used, was to prove invaluable; see Welchman (1986).

7.3.1.3 British Mathematical Cryptologists

The British were somewhat slower to exploit the skills and talents of mathematicians. But when it became clear in 1938 that war was coming soon, GC & CS (the Government Code and Cypher School) began to recruit "men of the professor class", including the *phenom* Alan Turing, who took training courses in cryptology prior to the outbreak of war and reported to Bletchley Park on September 4, 1939 (the day after war was declared). By the end of the war dozens of research mathematicians had been hired, including J. W. S. Cassels, I. J. ("Jack") Good, Philip Hall, Peter Hilton, M. H. A. ("Max") Newman, David Rees, Derek Taunt, William Tutte, Gordon Welchman, J. H. C. Whitehead, and Shaun Wylie. Many of these performed outstanding feats of cryptanalysis during the war.

7.3.1.4 German Mathematical Cryptologists

The Germans, although they were a towering presence in world mathematics (at least until the Nazis came to power), were curiously late in coming to the game. There were essentially no mathematicians in German signals intelligence prior to 1937. (Dr. Ludwig Föppl, 1887–1976, was a notable exception, serving during World War I; see Brückner 2005 and Samuels 2016.) When Dr. Erich Hüttenhain (1905–1990) was hired that year, it was essentially by pure chance. Dr. Hüttenhain, who was a mathematical astronomer, had become interested in cryptography because of an interest in Mayan astronomical chronology (the Mayan language then being largely unknown). Hüttenhain subsequently submitted a design for a cryptographic system to the German military, and on the basis of this was offered a job in 1937 in their cipher section, housed in the Reichskriegsministerium. In 1938 this cipher section was transferred to the newly formed OKW (*Oberkommando der Wehrmacht*), and thenceforth called OKW/Chi (*Oberkommando der Wehrmacht/Chiffrierabteilung*).

Hüttenhain rose rapidly in the organization and was soon tasked with hiring more mathematicians for it. The first of these was Wolfgang Franz (1905–1996), who joined OKW/Chi on July 17, 1940. Other subsequent hires included Ernst Witt (1911–1991), Otto Teichmueller (1913–1943), and Karl Stein (1993–2000). Teichmueller was killed in action after rejoining his unit and appears to have accomplished little, but the others all survived the war and in some cases went on to careers of considerable distinction. Both Franz and Stein later became Presidents of the DMV (*Deutsche Mathematiker Vereinigung*, or German Mathematical Society). Indeed it appears that one of Hüttenhain's motivations in hiring the mathematicians he did was to ensure their survival. But it is striking and telling that OKW/Chi only began hiring new mathematicians in addition to Hüttenhain nearly a year after the outbreak of war.

The one other branch of the German military that also began to hire mathematicians in considerable numbers for wartime cryptologic purposes was the German Army proper (the *Heer*, as opposed to the *Kriegsmarine* or *Luftwaffe*) housed in OKH (*Oberkommando der Heeres*). At least 14 Ph.D.s in mathematics, from Berlin, Göttingen, and Dresden, among other universities, were eventually hired, along with others working in statistics, economics, and actuarial science. (The most distinguished of these was Willy Rinow, later to become yet another President of the DMV.) But here too these individuals were only brought in after the outbreak of war, and this was to cost the Germans heavily on the defensive side of ensuring the security of their encryption devices. For further information on German mathematical cryptologists during World War II, see Weierud and Zabell (2018).

7.3.2 The Enigma

During World War II, the German military used an encryption device called the *Enigma* for sending enciphered messages. After one of 26 keys on a typewriter keyboard was pressed, an electric current entered the machine from the right, passed through a series of three moveable wheels (termed the right, middle, and left wheels) while traveling from right to left, and entered a fourth, fixed wheel which reversed the direction of current. The current then passed in the opposite direction through the left, middle, and right wheels, exited the machine, and one of 26 small lamps lit up indicating the enciphered version of the input letter.

Each of the moveable wheels had 26 contacts on each side, so that current could enter either side and exit on the other. Because of the internal wiring of the wheel, current entering a contact on the right, say, would exit at a different contact on the left. The result was a permutation of the alphabet. During the 1930s Polish cryptanalysts were able to determine the internal wiring of the moveable and fixed wheels of the German military Enigma. This was an impressive feat given that the total number of possible wheel wirings (that is to say, possible permutations of the 26 letters of the alphabet) is, as previously noted, on the order of 4×10^{27}.

In cryptography, *Kerckhoffs's principle* (named after Auguste Kerckhoffs, 1835–1903, a Dutch linguist and cryptologist) states that a cryptographic system should be secure (immune to attack) even if all aspects of the design of the system are known except the key (a specific item of information needed to decipher a message, preferably varying from message to message or day to day, or some relatively short period of time). A system that relies for its security on a lack of knowledge of the system by an opponent ceases to be secure when a copy of the device is obtained or a spy provides its specifications or (as in the case of the Poles) a cryptanalyst deduces it.

Thus the Germans did not rely on a lack of knowledge of the wiring of the wheels of their Enigma to ensure its security. Instead they relied on the daily setting. When the machine was set for sending messages on a given day, the three moveable wheels were chosen from a set of 5. The wheel order (or *Walzenlage*) specified which wheels were to be selected and how they were to be placed in the machine. (For example, 2 5 3 means wheel 2 on the left, wheel 5 in the middle, wheel 3 on the right.) There were therefore a total of ${}_5P_3 = 5 \cdot 4 \cdot 3 = 60$ possible wheel orders on any given day. For extra security, the German Naval Enigma selected its 3 wheels from an enlarged set of 8, rather than just 5. This increased the number of possible wheel orders from 60 to ${}_8P_3 = 8 \cdot 7 \cdot 6 = 336$.

In order to set the wheels on a given day, each wheel had a lettered ring attached to its left side, the 26 letters of the alphabet appearing on the rim of the ring. If we think of the 26 contacts on the wheel as numbered from 1 to 26, the ring could be rotated so that the letter A on its rim was next to contact 1, or next to contact 2, and so on, up to contact 26. Specifying how each of the three rings were set relative to each of the three moveable wheels was called the ring setting (or *Ringstellung*). Since each ring could be set in any of 26 different ways, the total number of ring settings was

$$26^3 = 17,576.$$

Before the current entered the wheels, it passed through a plugboard which subjected the letters of the alphabet to an initial permutation by interchanging selected pairs of letters. (For example, the letters a and b, and c and d might be interchanged, and the remaining 22 letters left unchanged.) On a given day, the usual practice was to cross-plug (or "stecker") 10 pairs of letters (for a total of 20), and thus leave the remaining 6 letters unsteckered. If n pairs of letters were steckered, then the number of possible steckerings (or *Steckerverbindungen*) was

$$\frac{26!}{2^n \cdot n! \cdot (26 - 2n)!}$$

The table below gives the number of *Steckerverbindungen* for anywhere from $n = 1$ to $n = 13$ letter pairs. (Curiously, the choice of 10 letter pairs does not maximize the number of possible steckerings: the maximum is reached for $n = 11$.)

n =	Number of *Steckerverbindungen*
1	325
2	44,850
3	3,453,450
4	164,038,875
5	5,019,589,575
6	100,391,791,500
7	1,305,093,289,500
8	10,767,019,638,375
9	53,835,098,191,875
10	150,738,274,937,250
11	205,552,193,096,250
12	102,776,096,548,125
13	7,905,853,580,625

Thus, the total number of possible daily settings for the Army Enigma (*Walzenlagen, Ringstellungen, Steckerverbindungen*) was

$$60 \cdot 17,576 \cdot 150,738,274,937,250 = 158,962,555,217,826,360,000.$$

Presumably for this reason the German authorities considered the Enigma to be a highly secure encryption device. In reality the Allies were able to decipher a substantial fraction of the Enigma messages that they intercepted. They were able to do this in part because of a variety of errors on the part of the German operators (insecure practices), but also because of the "Bombe", a special purpose mechanical device devised under the leadership of Alan Turing (1912–1954) and Gordon Welchman (1906–1985).

The Bombe consisted of 36 replicas of the three-wheel Enigma. Each replica consisted of 3 drums (one for each wheel), these would collectively spin through the $26^3 = 17,576$ different settings of the three wheels in approximately 18 minutes. Before a run, a *menu* was prepared: a *crib* (conjectured plaintext) was identified, and on the basis of it a graph was constructed summarizing relationships implied by the crib between different letters being encrypted at different stages. (This process was assisted by the fact that in the Enigma a letter could never encrypt to itself.) On the basis of this graph, the Bombe was wired accordingly and a run begun. Whenever the drums came to a setting consistent with the menu, then current would flow and this setting would be noted; this was a *stop*. Although there could be "false stops" corresponding to an incorrect setting, Turing calculated early on that these would be few enough given a menu of sufficient complexity. Further, his initial design was considerably improved in 1940 by Welchman's invention of the *diagonal board*, which exploited the reciprocal nature of the Enigma (the same setting on the Enigma was used to both encrypt and decrypt the same message).

Although some of the German cryptologists had some appreciation of the potential weakness of the Enigma, they viewed these as largely theoretical in nature, requiring rooms full of mechanical equipment to effect an attack. The willingness of the British to make precisely such an outlay was a key element in their success.

Some of the credit for the success of Bletchley Park is due to the then Prime Minister, Winston Churchill, who had a keen understanding of the value of science and technology in pursuit of Britain's war aims. On September 6, 1941, Churchill had visited Bletchley Park, and expressed appreciation for their efforts. But, frustrated with then inadequate resources, 6 weeks later, on October 21, 1941, the heads and deputy heads of Hut 6 (cryptanalysis of the Army and Luftwaffe Enigma) and Hut 8 (cryptanalysis of the Naval Enigma) wrote a letter directly to Churchill noting with frustration impediments to the cryptanalysts's work, such as the absurdity that some messages were not being decrypted due solely to a "shortage of trained typists". The letter was hand-delivered to Churchill's private secretary at 10 Downing, who promised it would be given directly to Churchill. This was done and Churchill promptly wrote a memo (headed "Action This Day") directing his principal staff officer: "Make sure they have all they want on extreme priority and report to me that this has been done". Not surprisingly, there were no problems after that.

7.3.3 Tunny

Tunny was the codename the British gave to another important German encryption device, the SZ40/42 (the "Schlusselzusatz", or Cipher Attachment, manufactured by Lorenz). The Enigma was an off-line device (that is, encrypting or decrypting a message was performed separately from sending and receiving the message), intended for short communications (say 200 characters or less). The SZ40/42, in contrast, was an online teleprinter encryption device intended for much longer messages, containing thousands of characters. (So, for example, it is possible that Hitler's infamous message to the Commandant of Paris shortly before the city fell to the Allies in August 1944, instructing that Paris be destroyed, was sent in encrypted form using this device.)

The Poles were familiar with the basic structure of the Enigma, in part because it was a modified version of a commercially available device (and in part thanks to a spy, Hans Thilo Schmidt). The SZ40, in contrast, had been designed by the German military, and so there was no corresponding model to work from. In a tour-de-force of cryptological skill, however, members of Bletchley Park were eventually able to deduce the entire structure of the machine thanks to a single operational slip-up on the part of the German operators, who had once sent two long and virtually identical messages at the same setting (such messages are said to be "in depth").

This was no mean feat, given the complexity of the device. Letters were represented in it using the then standard five impulse *Baudot code* (so, for example, $A = 00011$, $B = 11001, \ldots$). The encryption used

- five "chi" wheels (employing regular motion)
- five "psi" wheels (employing irregular motion)
- two "mu" wheels (determining when irregular motion occurs)

(Here "irregular" means that sometimes the wheels moved, and sometimes did not.)

Despite its apparent complexity, the process of encryption may be simply and schematically represented as:

$$P \quad \rightarrow \quad P + \psi \quad \rightarrow \quad P + \psi + \chi \quad = \quad C$$

(P denoting "plaintext", C "ciphertext").

Despite its impressive appearance, Tunny suffered from a serious design flaw: when the five psi (irregularly moving) wheels did move, they did so simultaneously. As a result, a crafty combination of the output of a pair of wheels (in the initial stage of the attack, the ψ_1 and ψ_2 wheels) resulted in a biased stream of 0-1 bits. This could be used as a test for the correct setting of the chi wheels for the given message. Because there were 1271 ($= 41 \cdot 31$) possible settings for the χ_1 and χ_2 wheels, respectively, if the correct setting was used to decrypt this test stream, this would strip off the chi layer of encryption and the resulting 0–1 stream would be a *biased* sequence of 0s and 1s; whereas if one of the other 1270 incorrect settings were used to decrypt the test stream, the resulting 0–1 stream would remain and appear as *unbiased*. Thus the task of setting the first two chi wheels was converted into the purely statistical task of finding the one biased stream among the 1271.

This required a vast amount of computing, and for this the *Colossus* was constructed (see Copeland 2006, for a detailed discussion of this device from a variety of viewpoints). It has been argued that in many ways the Colossus was the first programmable computer, not because it could store a program in memory, but because it could be (relatively) easily rewired to perform different tasks. (This was in contrast with the Bombe, which was a special purpose device, designed and constructed for the sole task of attacking the Enigma.)

Once one pair of chi wheels had been set, then by a similar process other pairs of chi wheels could be set, eventually resulting in setting all five wheels. This work was performed in the *Newmanry*, named after its head, M. H. A. ("Max") Newman, who although a pure mathematician had initially proposed the feasibility of such an attack. After the chi wheels had been set by primarily statistical means, the message and settings were sent to the *Testery* (named after Major Ralph Tester, who headed it), where the psi wheel layer of encryption was then stripped off by primarily linguistic means of attack. Thus the attack on Tunny combined technology, mathematics, and linguistics, and rooms of equipment, but was most certainly worth it. See Reeds et al. (2015) and Zabell (2015).

Of course both the Allies and the Axis powers used a wide variety of devices for encryption and decryption; see Pröse (2006) for a detailed scholarly discussion. For a general overview of the cryptologic war, 1939–1945, see Budiansky (2000). For a discussion of Turing's Bayesian viewpoint in his own words, see Turing (2012) and a commentary on it, Zabell (2012).

7.4 The Modern Era: Computers

One of the impediments the Germans encountered was that their methods of encryption were limited by purely mechanical considerations (as well as the difficulty in replacing old equipment by new if a new method of encryption were thought to improve on an old one). This changed with the advent of the computer: now there was no purely physical limitation on the length (number of bits that could be set) of a wheel, or the number of wheels, or the algorithm used to combine different inputs from different components at any stage in the process of encryption. Eventually highly secure, publicly available algorithms such as DES (the Data Encryption Standard, first published in 1975) and AES (the Advanced Encryption Standard, first published 1998) became available. Obviously the rise of computer networks was a factor in this development.

This subject could easily be the subject of a book, so in this chapter we focus on one particular aspect of the use of computer algorithms.

7.4.1 Generating Random Numbers

We have already seen the importance of generating random numbers in cryptography, in its role in producing one-time pads.

The resort to random selection was already widespread in the ancient world. Aristotle, for example, in his *Athenian Constitution*, describes an elaborate two-stage procedure that the Athenians used for selecting members of a jury (Moore 1975, pp. 303–307); and during the Roman Republic lots were commonly used to assign provinces to the consuls and other major state officials. The use of randomization for scientific and mathematical purposes is of course much more recent. Stigler (1999, Chapter 7) discusses a number of nineteenth century examples.

True physical randomization, however, is often difficult to achieve (and in the hands of the unwary is often not achieved). One celebrated example is the famous 1970 draft lottery debacle (Fienberg 1971). One remedy for this is the construction of tables of random numbers that researchers can use with confidence. The earliest of these was L. H. C. Tippett's table of random numbers published in 1927, prepared at the suggestion of Karl Pearson, to facilitate carrying out random sampling experiments for simulation studies; see Pearson (1990, pp. 88–95). The culmination of such efforts was the construction of the Rand Corporation's "A Million Random Digits With 100,000 Normal Deviates" (Rand 1955).

In the last several decades such physically generated tables have been entirely supplanted by algorithmic random number generators executed by computer code. Strictly speaking, of course, sequences of numbers generated this way are only "pseudo-random", although this typically suffices for almost all practical purposes. Nevertheless, there are challenges here too; see Knuth (1997, Chapter 3). To appreciate just what a change this represents, consider that the following two lines of R (a statistical programming language) code will generate the entire contents of the Rand Tables in only a fraction of a second:

```
rand.rd <- sample(0:9, 10^6, replace = TRUE)
rand.nd <- rnorm(10^5)
```

The failure to enforce randomness in a cryptographic protocol can have serious consequences. For example, in the Naval Enigma, a trigram "message indicator" was encrypted using one of 9 *bigram tables*. The sender chose a pair of trigrams, say LQR and CPY, from a *Kenngruppenbuch*, added a pair of "haphazard" letters, say G and O, and then encrypted each column of the resulting two-by-four array using the bigram table:

$$\begin{array}{cccc} G & L & Q & R \\ C & P & Y & O \end{array} \qquad \begin{array}{cccc} T & A & L & I \\ U & H & S & U \end{array}$$

Then $TALI\ UHSU$ was sent.

The receiver of $TALI\ \ UHSU$, who knew the bigram table in use that day, reversed this process, to find the message indicator CPY. (Strictly speaking, CPY was not the actual message indicator: using the *Grundstellung* or general daily setting, CPY was in turn encrypted and the resulting trigram was the final setting used to encrypt the message.)

This apparently impressive procedure had, however, two fatal weaknesses. The first was that the trigrams were not selected randomly by the operators from the *Kenngruppenbuch*: there was a tendency to pick trigrams from the tops of pages. Turing devised an attack that exploited this (using a "sampling of species" approach). The other weakness was that humans are also very poor at randomly selecting individual letters. As Good later related:

> I noticed on one night shift that about 20 messages were enough to identify which digraph table was in use, because the 'haphazard' letters (G and O in the example) were not 'flat-random'. This discovery then provided the routine method for identifying the table. [Good 2000, p. 109]

For such reasons experienced cryptographers go to great lengths to ensure genuinely random selection is employed. For example, in a code book one replaces letters, words, and phrases by, say, groups of five digits. If the purpose of the encoding is not just data compression (as was sometimes the case for commercial code books) but also secure communication, then two books must be used: one listing in alphabetic order the letters, words, and phrases next to randomly assigned five-digit numbers; and another listing in numerical order the five-digit numbers next to the corresponding letters, words, and phrases.

In the 1930s the Signal Intelligence Service of the US Army attempted to construct such a code book, and encountered great difficulty in randomly assigning the code equivalents to the letters, words, and phrases. They were faced with the challenge of "scrambling" 60,000 cards. At first they dumped drawers of the cards onto the floor and attempted to mix them by hand, but found it did not mix the cards enough. Then they started throwing handfuls of cards into the air, and even turned on the wall fans to maximum speed; "the results were still far from satisfactory". It was only after in addition to all this when they began placing cards on cleared desktops in an irregular way and repeated the entire process several times that they were "at last able to achieve an acceptable randomization of all the plaintext cards" (Rowlett 1998, pp. 53–54).

Here is an instructive illustration of the importance of randomness in cryptology, a topic briefly mentioned earlier.

7.4.2 The One-Time Pad

In many cryptographic systems the goal is to transform a given plaintext into a ciphertext that is indistinguishable from a "flat random" (that is, uniformly distributed) sequence. For example, suppose a plaintext $P = (P_1, P_2, \dots P_n)$ is written in a t-letter alphabet (for instance $t = 2$ for bits, $t = 26$ for letters, and so on). Suppose that an additive key sequence $K = (K_1, K_2, \dots, K_n)$ is flat random (in the sense that every n-long sequence in the t-letter alphabet has a probability of t^{-n} of occurring). Then it is not hard to see that (addition being mod t) that the cipher text sequence $C = (C_1, C_2, \dots, C_n)$ defined by

$$C_j = P_j + K_j$$

is itself flat random. That is, *whatever statistical regularities may have been present in the plaintext P have been entirely obliterated by addition of the flat random key sequence K*. This is the theoretical basis for the use of the "one-time pad" (a pad containing such a key sequence that is then added once – and only once – to a plaintext).

In principle the use of the one-time pad is the basis of a theoretically unbreakable encipherment if carried out in a correct and secure manner.

One famous example of its misuse was the subject of the NSA's *Venona Project*. During World War II, some Soviet agents in the US used one-time pads to communicate with their masters in Moscow. But for reasons that are not understood (but presumably reflected wartime conditions in the Soviet Union) pages from the pads were reprinted and the same sequences of numbers reused. The upshot was that as a result of a monumental effort on the part of US cryptanalysts a substantial fraction (more than 10%) of the content of these highly secret messages were in fact read, giving insight for example into the Soviet atomic spy ring. See generally Haynes and Klehr (2000).

7.5 Postmodern Era: Public Key Encryption

Three may keep a secret if two are dead – Benjamin Franklin.

In classical, private-key crypto-systems, A and B securely communicate over a channel in which a third party (C) may be eavesdropping.

$$\begin{array}{ccc} & C & \\ & \downarrow & \\ A \longrightarrow & \text{public channel} & \longrightarrow B \end{array}$$

They do this by means of a *private key* that has been previously sent via a secure channel. Classical examples of this include DES (the *data encryption standard*) and the more recent AES (*advanced encryption standard*).

In the 1970s cryptography was revolutionized by the introduction of *public key systems*, where no prior exchange of a private key over a secure channel is necessary. This possibility is the basis of the *https* protocol, which enables you and Amazon (say) to securely exchange information about credit cards even if someone is “listening in”.

How is this possible? The key lies in the use of “trap–door” functions: a function, say E, which is easy to compute, but whose inverse $D = E^{-1}$ is hard to compute (unless additional, private information is available). For example, think of computing x^2 vs. $\sqrt{x}$.

7.5.1 RSA Encryption

RSA (for Rivest et al. 1978) is an early and still very important example. In the following, $\phi(n)$ is the *Euler phi function*, the number of integers k, $1 \leq k \leq n$, relatively prime to n, that is, $(k, n) = 1$. (A good reference for the number theory that appears in RSA encryption is Kraft and Washington 2014.)
Encryption method:

1. Choose n and e: here $n = pq$ (p, q two large primes, *private*), and e is an exponent such that $(e, \phi(n)) = 1$. Here both n and e are *public*.

2. Translate letters into their numerical equivalents, forming blocks P of the largest permissible size; for example, MATH $\to$ 12001907. (In this example, the encoding is A $\to$ 00, B $\to$ 01, ..., Z $\to$ 25.)
3. Encrypt: $C := E(P) \equiv P^e \pmod{n}, 0 \le C < n$.

Now comes the clever part, which appeals to Euler's theorem. Recall that Euler's theorem tells us that $a^{\phi(n)} \equiv 1 \pmod{n}$ provided $(a, n) = 1$. Suppose $(P, n) = 1$. Because $(e, \phi(n)) = 1$,

$$d := e^{-1} \pmod{\phi(n)}$$

exists, hence $ed \equiv 1 \pmod{\phi(n)}$, hence

$$ed = k\phi(n) + 1.$$

So to decrypt, *if you know* d, you just compute

$$C^d = (P^e)^d = P^{ed} = P^{k\phi(n)+1} = (P^{\phi(n)})^k P = P \pmod{n}.$$

In order for this to be secure, one needs d to be difficult to find given just n and e (which are public), but easy to compute given p and q (which are to be private).

Now if we know $\phi(n)$, then solving $ex \equiv 1 \pmod{\phi(n)}$ (to invert e and find d) is easy, because it is equivalent to solving the first order Diophantine equation $ex + \phi(n)y = 1$, for which the (extended) Euclidean algorithm is available. The relevance to RSA is this: in order to find d, we need to know $\phi(n)$. Now if we know the factorization $n = pq$, then it is easy to find $\phi(n)$, since $\phi(n) = (p-1)(q-1)$. But

"multiplication is easy, factoring is hard",

so it is easy to go in one direction (use the private p and q to find n and $\phi(n)$), but hard to go in the other (factor the public n to find p and q): we are in the trap-door function situation.

Objection: maybe there is some other way of finding $\phi(n)$ without factoring $n = pq$.

Response: *no*, given $n = pq$, factoring n is equivalent to computing $\phi(n)$.

In one direction this is immediate, given the factorization $n = pq$, just use the formula $\phi(n) = (p-1)(q-1)$. For the other direction, since $n = pq$ and $\phi(n) = (p-1)(q-1)$, observe that

$$p + q = n - \phi(n) + 1.$$

Thus we know the sum $p + q$ as well as the product pq if we know both n and $\phi(n)$. Thus the question reduces to the

Problem: given pq and $p + q$, find p and q. In general, given the sum and product of two numbers, find the two numbers.

Solution: If the numbers are a and b, consider

$$f(x) = (x - a)(x - b) = x^2 - (a + b)x + ab.$$

We know the sum and product, $a + b$ and ab, so we are given a quadratic equation, and our mission is to find a, b, *the roots* of $f(x)$! This is easy: just use the quadratic formula. (Computing square roots is easy for a computer.)

The bottom line: given n and $\phi(n)$, it is easy to find p, q.

Note: This does *not* prove that factoring is hard; only that it is equivalent to computing $\phi(n)$. Note also that "hard" is a function of current technology. (Some things that were hard 50 years ago are easy today; and some things that are hard today may be easy 50 years from now.)

7.5.2 Key Exchange Protocols

In *key exchange* or *key establishment* protocols, the goal is for two parties to arrive at a common, secret key for use in a cryptosystem, doing this while communicating over an insecure channel. The original idea for this goes back to back to Whitfield Diffie and Martin Hellman in 1976, and is an attractive application of primitive roots.

7.5.2.1 Diffie-Hellman Key Exchange

In Diffie-Hellman key exchange, Alice and Bob communicate over a public (and potentially insecure) channel. The two agree on a large prime number p, and a fixed number $q < p$. (Technically, q is a *primitive root* mod p.) Alice has private key a,

Bob has private key b. Alice sends $q^a \pmod{p}$ to Bob, and Bob sends $q^b \pmod{p}$ to Alice. Then Bob computes $(q^a)^b \pmod{p}$ and Alice computes $(q^b)^a \pmod{p}$. The two numbers agree because

$$(q^a)^b = q^{ab} = q^{ba} = (q^b)^a \pmod{p}.$$

The security of the method resides in the fact that even if a third party (Carol) intercepts q^a or q^b, she cannot find the values of a or b even if she knows q and p; this involves the computationally challenging task of finding the *discrete logarithms*

$$a = \log_q(q^a), \qquad b = \log_q(q^b).$$

Note the two clever ingredients of the Diffie-Hellman method. First, Alice and Bob exchange information that enables each to construct a common key: Alice gives Bob $q^a \pmod{p}$; Bob gives Alice $q^b \pmod{p}$.

The potential insecurity in the key exchange arises from the fact that the information Alice sends Bob obviously has to bear some relation to the use Alice makes of the information Bob sends her. The common element is her private key a: Alice uses it both to compute q^a and q^{ba}. If Carol could learn the value of a and intercept q^b, she could figure out q^{ba}. But the only public glimpse of Alice's private key a is when Alice sends q^a to Bob. Thus it is essential that this step not compromise the security of the private key a. This is the second clever element of the method: the use of a mathematical procedure that is readily computable in one direction (otherwise the method would be impractical), but computationally intractable in the other.

7.5.2.2 Massey-Omura Key Exchange

In this scheme just a single large prime p is public; Alice has private keys e_a and d_a (such that $e_a d_a \equiv 1 \pmod{p-1}$), Bob has private keys e_b and d_b (such that $e_b d_b \equiv 1 \pmod{p-1}$). The following exchange from Alice to Bob then takes place:

$$q^{e_a} \to q^{e_a e_b} \to q^{e_a e_b d_a} \to q^{e_a e_b d_a d_b} = (q^{e_a d_a})^{e_b d_b} \equiv q \pmod{p}$$

Think of this as follows: Alice puts a lock on a box (e_a) and sends it to Bob; Bob puts a second lock on the box (e_b) and sends it back to Alice. Alice then removes her lock using her key (d_a), and sends the box back to Bob. Finally, Bob removes his lock using his key (d_b), and opens the box, revealing the shared secret q.

7.6 Through a Glass Darkly

Signals intelligence organizations are often quite chary of revealing their secrets or successes. This presents challenges – but also opportunities – for the historian.

One might expect former employees would want to publicize their past exploits, but there can be serious disincentives here. The history of signals intelligence contains a number of celebrated instances of old hands feeling free to publicize their cryptologic exploits, only to suffer serious consequences. Two cautionary tales here are those of Herbert Yardley (1889–1958) and Gordon Welchman (1906–1985).

Yardley had been the head of the "American Black Chamber", a highly-successful code-breaking organization having its origins in the World War I Cipher Bureau MI-8, and which later became a joint operation run by the US Army and Department of State. But when, in 1929, administrations changed and the new Secretary of State Henry L. Stimson met with Yardley and was briefed on its operations, the American Black Chamber was promptly shut down. "Gentlemen", Stimson declared, "do not read each other's mail". (Twelve years later, in the aftermath of the attack on Pearl Harbor, Stimson, now Secretary of War, presumably came to feel differently.) Now out of work, in the middle of the Great Depression, and informed that his operation was no longer worth while, Yardley went on to write his fantastic book *The American Black Chamber* (1931), narrating with gusto the many exploits of that organization. But although the Department of State many have looked down on Yardley's reading of gentlemen's mail, the Department of the Army did not: they were furious with Yardley for revealing so many of their secrets and he became persona non grata for the rest of his life. (When the Canadians hired Yardley several years later to help run their own fledging signals intelligence organization, he was dismissed after less than a year at the insistence of the US and UK.) For an outstanding account of Yardley's life, see Kahn (2004).

Gordon Welchman provides another cautionary tale. Welchman had headed Hut 6 (Army and Luftwaffe cryptanalysis) at Bletchley, and was responsible for many important advances during the war. (He was also the moving force behind the letter to Churchill in late 1941 that resulted in Bletchley Park being given virtual carte blanche in obtaining personnel and materiel.) Welchman emigrated to the US in 1948, and spent the rest of his life working primarily for the US defense establishment. He kept scrupulously quiet for more than 35 years about his outstanding contributions to the Allied war effort. But in the late 1970s, as more and more revelations about Bletchley Park came out, Welchman concluded that total silence was no longer required. And so he came to write his highly informative *The Hut Six Story*, which detailed the many successes in the attack on the Enigma, the devices (such as the Bombe and diagonal board used in its attack), and Alan Turing's crucial role in all this. But he made a fatal error: he failed to submit his book for prepublication review. He was promptly stripped of his security clearance, forbidden to speak to the press, and remained under a cloud for the (sadly short) remainder of his life. For a recent biography of Welchman, see Greenberg (2014).

Silence about technical achievements would seem particularly important. Nevertheless it is possible from time to time to identify individuals who were able to publish technical results – stripped of their cryptographic origins – before their involvement on the dark side became known. Here are two case studies.

7.6.1 Case Study 1: I. J. Good

I. J. (Irving John, "Jack") Good (1916–2009) was an undergraduate at Jesus College, Cambridge (1934–1938) before receiving his Ph. D. at Cambridge in 1941, under the supervision of G. H. Hardy and A. S. Besicovitch. Shortly after, he reported to Bletchley Park on May 27, 1941 (the day the *Bismarck* was sunk). He was fortunate in this: he spent his first 2 years (1941–1943) at Bletchley working in Hut 8 (Naval cryptanalysis) under Alan Turing, from whom he learned the Bayesian approach to statistics; and his last 2 years (1943–1945) working in the Newmanry (recall this was one of two sections devoted to cryptanalysis of the SZ40/42, an online teleprinter system) under M. H. A. ("Max") Newman, using an attack centered on the use of the "Colossus".

After the war Good spent a few years at the University of Manchester (1945–1948), and then returned to GCHQ (Government Communications Headquarters, the postwar successor to GC & CS), where he remained for 11 years (1948–1959). After visiting several institutions for 2–3 year stints (Admiralty Research Laboratory, 1959–1962; Institute for Defense Analyses, 1962–1964; Trinity College, Cambridge, 1964–1967), he became a Professor at Virginia Polytechnic Institute, where he remained for the rest of his life.

After the war Good wrote a classic book, *Probability and the Weighing of Evidence* (1950), espousing the subjective, Bayesian viewpoint (but with a strong pragmatic streak running throughout). But even though the book appeared in 1950, a first draft had been written in 1946, immediately after Good left Bletchley. In retrospect it is clear that the book advances a view of the subject that Good had acquired directly from Turing. In its preface, Good thanks Turing, Newman, and Donald Michie (that is, his two bosses at Bletchley and his closest collaborator in the Newmanry) for reading the first draft.

But Good's Bletchley Park-inspired contributions to statistics in the years immediately after the war were not confined to just a general advocacy of the Bayesian viewpoint. He proceeded to publish (always scrupulously crediting Turing) developments and refinements of a number of technical advances Turing had developed during the war. As Good later explained:

> Turing did not publish these wartime statistical ideas because, after the war, he was too busy working on the ground floor of computer science and artificial intelligence. I was impressed by the importance of his statistical ideas, for other applications, and developed and published some of them in various places. Much of my delay was caused by the wartime attitude that everything was classified, from Hollerith cards to sequential statistics,

to empirical Bayes, to Markov chains, to decision theory, to electronic computers. These extreme standards of secrecy only gradually abated after the war [Good 2001, p. 211].

Up until 1976, however, Good remained entirely silent about his actual wartime work. At this point in time it is hard to appreciate the total silence up to then regarding Allied successes in attacking German encryption devices; one illustration among many is provided by David Kahn's pathbreaking book *The Codebreakers* (1967): although it contains an entire chapter about the US success in reading the Japanese "Purple" cipher, and several chapters on German signals intelligence, it is entirely silent about Bletchley and Ultra.

All this changed in 1973, when General Gustave Bertrand (1896–1976) wrote *Enigma, ou la plus grande énigme de la guerre 1939–1945* ("Enigma, or the Greatest Enigma of the War of 1939–1945"). This revealed that since 1932 the Poles had been reading the Enigma, as well as the Polish-French collaboration. This apparently served as an inducement to the British to lift a year later (1974) their total embargo on any discussion of their cryptologic successes during the war; the first beneficiary of this change in policy was F. W. Winterbotham's *The Ultra Secret* (1974). After this the floodgates opened, and an ever-increasing succession of books and papers appeared; a small (but significant) sampling of these include Hinsley (1979–1990), Rejewski (1981), Welchman (1982), Hinsley and Stripp (1993), and Reeds et al. (2015). I. J. Good has returned to this subject many times (in what might be termed the "dance of the seven veils"); see Good (1976, 1979, 1993, 2000, 2001, 2006).

For further information on Good's life, see the outstanding interview by David Banks (1996).

7.6.2 Case Study 2: Aleksandr Alekseevich Borovkov

A. A. Borovkov (1931–) is a prominent Russian mathematician, working in the areas of probability and statistics. He did his undergraduate work at the University of Moscow, graduating in 1954. After completing his graduate studies under the great A. N. Kolmogorov, he then moved to Novosibirsk in 1960 to become "head of the recently created probability theory and mathematical statistics section of the Institute of Mathematics of the Siberian Branch of the Academy of Sciences of the USSR" (Borisov et al. 2001, p. 1009). He has remained there since. He is perhaps best known for his work in the field of large deviations, for example boundary crossing probabilities for random walks, the subject of his thesis. This was an interesting (if risky) choice of topic: although large deviation theory is currently one of the most active areas of research in mathematical probability, it was a relatively unexplored area at the time and virtually nothing had been done in Borovkov's particular area of study. What led him to his interest in this field?

Borisov et al. (2001) discretely tell us that after the completion of his undergraduate studies in 1954, Borovkov "worked for several years in an organization doing applied research" (p. 1009). Borovkov himself was willing just a few years later (2004) to be much more explicit:

> I began to study this problem under the following circumstances. After graduating from Moscow State University in 1954 I was assigned to a covert organization (despite the recommendation of Kolmogorov for graduate work) that solved cryptography problems by using computers. For this purpose, one of the most powerful computers of the time was created. The approach was based on exhaustion of diverse versions of decoding, which produces a great number N of variants of decoding on the output, that is, sequences of letters $a_1, a_2, \ldots, a_n$. Among them one should recognize the true 'decoded' text (that is, a text in English corresponding to the correct version of decoding). Since the number N was very large (say, of order $10^8 - 10^{10}$), it was impossible to perform this work 'manually', and a 'computer algorithm' for recognition of the decoded text was used. This algorithm was based on the statistical criterion of sequential analysis that was to distinguish between two hypotheses: $H_1 = \{$the text is chaotic, that is, the a_i are independent, $P(a_i = k) = q(k) = 1/26, k = 1, \ldots, 26\}$ and $H_0 = \{$the text is decoded$\}$. In the latter case, diverse simplest models were used, for example, $\{a_i\}$ was assumed to be a sequence of independent variables with the known probabilities $p(k) = P(a_i = k)$ or a Markov chain with the known probabilities $p_{jk} = P(a_i = k/a_{i-1} = j)$.

So Borovkov's public work in large deviations was a direct consequence of his working for a "covert organization" interested in cryptography!

What is particularly interesting (and impressive) about Borovkov's work on this subject is that it was *not* encouraged by Kolmogorov – quite the contrary:

> At that time I was successful in enrolling in the correspondence graduate programme with the support of Kolmogorov, and I decided to take the problem as a thesis project. This was quite risky, because nothing was known at the time about the problem, and Kolmogorov told me at once that he had no ideas about it. (He even suggested that I choose another problem, but I declined.) The risk turned out to be serious, because I could not get anything for almost three years. The solution in the case of bounded lattice variables ξ_i (this was the very case we needed) was found in 1958 in a purely analytic way.

7.7 Discussion

Modern methods of communication involve the transmission of massive amounts of information over channels that are either insecure or potentially insecure (subject to interception). The early part of the twentieth century saw this in the case of wireless transmissions over long distances; the last several decades with the rise of computer networks, LANs and WANs, and the internet. This gave rise in turn to the need for rapid methods of ensuring privacy, authentication of sender, and guaranteeing integrity of message.

In the era before computers this was accomplished by the constructing of machines, often impressive by the standards of their day, used to encrypt an increasing volume of military, diplomatic, and commercial information; these were in turn often attacked by methods devised by mathematicians and implemented by the construction of new and sophisticated machines. In the modern computer era encryption has, not surprisingly, become the output of computers rather than special purpose mechanical, electronic, or electromechanical devices. These modern methods of encryption now almost exclusively call upon the resources of modern mathematics, as do the efforts of cryptanalysts to defeat them.

Documenting the evolution of modern encryption is a challenging one for the historian: it is in the nature of the subject that the more successful one is, the less one wants others to know about it. The career of I. J. Good illustrates this: it took decades before his part in Allied successes during World War II became known even in outline; and a number of statistical advances that arose out of his war work became public knowledge only after their cryptanalytic origins were hidden.

References

Banks, D. L. (1996). A conversation with I. J. Good. *Statistical Science, 11*, 1–19.

Borisov, I. S., et al. (2001). Aleksandr Alekseevich Borovkov (on his 70th birthday). *Russian Mathematical Surveys, 56*, 1009–1012.

Borovkov, A. A. (2004) Kolmogorov and boundary problems of probability theory. *Russian Mathematical Surveys, 59*, 91–102.

Brückner, H.-D. (2005). Germany's first cryptanalysis on the Western Front: Decrypting British and French naval ciphers in World War I. *Cryptologia, 29*, 1–22.

Budiansky, S. (2000). *Battle of wits: The complete story of codebreaking in World War II*. New York: Free Press.

Copeland, B. J. (Ed.). (2006). *Colossus: The secrets of Bletchley Park's codebreaking computers*. Oxford: Oxford University Press.

Feistel, H. (1973). Cryptography and computer privacy. *Scientific American, 228*(5), 15–23.

Fienberg, S. E. (1971). Randomization and social affairs: The 1970 draft lottery. *Science, 171*, 255–261.

Friedman, W. F. (1938–1941). *Military cryptanalysis*. Washington, DC: US Government Printing Office. [Several subsequent editions, later expanded in collaboration with Lambros D. Callimahos, under the title *Military Cryptanalytics*.].

Good, I. J. (1950). *Probability and the weighing of evidence*. New York: Hafner Publishing Company.

Good, I. J. (1976). Early work on computers at Bletchley. *Cryptologia, 3*, 65–77. Reprinted with an addendum in *Annals of the History of Computing, 1*, (1979), 38–48.

Good, I. J. (1979). Studies in the history of probability and statistics. XXXVII A. M. Turing's statistical work in World War II. *Biometrika, 66*, 393–296.

Good, I. J. (1993). Enigma and Fish. In Hinsley and Stripp (1993) (pp. 149–166).

Good, I. J. (2000). Turing's anticipation of Empirical Bayes in connection with the cryptanalysis of the Naval Enigma. *Journal of Statistical Computing and Simulation, 66*, 101–111.

Good, I. J. (2001). Introductory remarks for the article in *Biometrika, 66* (1979). A. M. Turing's Statistical Work in World War II. In J. L. Britton (Ed.), *Collected works of A. M. Turing* (Vol. 2, pp. 211–223). Elsevier.

Good, I. J. (2006). From Hut 8 to the Newmanry. In Copeland (2006) (pp. 204–222).

Greenberg, J. (2014). *Gordon Welchman: Bletchley Park's architect of Ultra intelligence*. London: Frontline Books.

Haynes, J. E., & Klehr, H. (2000). *Venona: Decoding Soviet espionage in America*. New Haven: Yale University Press.

Hinsley, F. H. (1979–1990). *British intelligence in the Second World War* (5 Vols.). London: Her Majesty's Stationery Office.

Hinsley, F. H., & Stripp, A. (1993). *Codebreakers: The inside story of Bletchley Park*. Oxford: Oxford University Press.

Kahn, D. (1967). *The codebreakers: The comprehensive history of secret communication from ancient times to the Internet*. New York: Macmillan.

Kahn, D. (2004). *The reader of gentlemen's mail: Herbert O. Yardley and the birth of American codebreaking*. New Haven: Yale University Press.

Knuth, D. E. (1997). *The art of computer programming* (Volume 1: Fundamental algorithms, 3rd ed.). Reading: Addison-Wesley.

Kraft, J. S., & Washington, L. C. (2014). *An introduction to number theory with cryptography*. Boca Raton: CRC Press.

Moore, J. M. (1975). *Aristotle and Xenophon on democracy and oligarchy*. University of California Press.

Pearson, E. S. (1990). *'Student': A statistical biography of William Sealey Gosset*. Oxford: Oxford University Press/Clarendon Press.

Pratt, F. (1939). *Secret and urgent: The story of codes and ciphers*. New York: Bobbs-Merrill.

Pröse, M. (2006). *Chiffriermaschinen und Entzifferungsgeräte im Zweiten Weltkrieg: Technikgeschichte und informatikhistorische Aspekte*. Munich: Martin Meidenbauer Verlagsbuchhandlung.

Rand. (1955). *A million random digits with 100,000 normal deviates*. Illinois: Rand Corporation. Reissued 2001.

Reeds, J., et al. (2015). *Breaking teleprinter ciphers at Bletchley Park: An edition of I.J. Good, D. Michie and G. Timms: General report on Tunny with emphasis on statistical methods (1945)*. Piscataway: IEEE Press.

Rejewski, M. (1981). How Polish mathematicians deciphered the Enigma. *Annals of the History of Computing, 3*, 213–234.

Rivest, R., Shamir, A., & Adelman, L. (1978). A method for obtaining digital signatures and public-key cryptosystems. *Communications of the ACM, 21*, 120–126.

Rowlett, F. B. (1998). *The story of Magic: Memoirs of an American cryptologic pioneer*. Laguna Hills: Aegean Park Press.

Samuels, M. (2016). Ludwig Föppl: A Bavarian cryptanalyst on the Western front. *Cryptologia, 40*, 355–373.

Sinkov, A. (1968). *Elementary cryptanalysis: A mathematical approach*. New York: New Mathematical Library, Random House.

Stigler, S. M. (1999). *The statistics on the table: The history of statistical concepts and methods*. Cambridge: Harvard University Press.

Turing, A. M. (2015). The applications of probability to cryptography. unpublished paper, c. 1941. UK National Archives, HW 25/37, 2012. For a typeset version, see Ian Taylor, arXiv:1505.04714v2 [math.HO].

Weierud, F., & Zabell, S. (2018). *German mathematicians and cryptology during World War II*. Submitted to *Cryptologia*.

Welchman, G. (1982). *The Hut six story: Breaking the Enigma codes*. New York: McGraw-Hill.

Welchman, G. (1986). From Polish bomba to British bombe: The birth of Ultra. *Intelligence and National Security, 1*, 71–110.

Winterbotham, F. W. (1974). *The Ultra secret*. London: Weidenfeld and Nicolson.

Yardley, H. O. (1931). *The American Black Chamber*. Indianapolis: Bobbs-Merrill.

Zabell, S. L. (2012). Commentary on Alan M. Turing: The applications of probability to cryptography. *Cryptologia, 36*, 191–214.

Zabell, S. L. (2015). Statistics at Bletchley Park. In Reeds et al. (2015) (pp. lxxv–ci).

Part III
Technology in Mathematics

Chapter 8
The Epistemology of Computer-Mediated Proofs

Selmer Bringsjord and Naveen Sundar Govindarajulu

Abstract Epistemology includes in large part investigation of the conditions by which rational human knowledge and belief, of the propositional variety, can be secured. Our particular instance of this investigation arises from the stipulation that a human (a) receives a partial or complete formal argument/proof ($\mathcal{A}$) for/of a conclusion ϕ, where some computing machine $\mathcal{M}$ "stands between" or *mediates* a's receiving $\mathcal{A}$ and ϕ. The mediation can take any number of forms, ranging from the simple and mundane (e.g., a is a teacher who types in to a text-editing system a proof of some easy theorem for a math class, and then prints out the proof for subsequent study and presentation to the class) to the exotic and famous (e.g., a receives a too-big-to-survey printout of a computer-generated proof of the four-color theorem). Under what conditions is it rational for a to believe ϕ? Once we have erected at least a reasonably precise framework for understanding the structure of arguments and proofs, classifying computing machines, ranking strength of knowledge and belief, and distinguishing at least roughly between types of computer mediation, this result, as we indicated, is a framework in which this pair of questions (and other, related ones) can eventually be answered.

8.1 Introduction

Epistemology includes in large part investigation of the conditions by which rational human knowledge and belief, of the propositional variety (a.k.a. learning of declarative content), can be secured. Our particular instance of this investigation

We are deeply grateful for the vision (and patience!) of, and guidance provided by, editor Sven Ove Hansson. We are also indebted to both ONR and AFOSR for support that has made the investigation of forms of advanced logicist learning (of conclusions of proofs and arguments) possible.

S. Bringsjord (✉) · N. S. Govindarajulu
Rensselaer AI & Reasoning (RAIR) Lab, Department of Computer Science, Department of Cognitive Science, Lally School of Management, Rensselaer Polytechnic Institute (RPI), Troy, NY, USA
e-mail: selmer.Bringsjord@gmail.com

S. O. Hansson (ed.), *Technology and Mathematics*, Philosophy of Engineering and Technology 30, https://doi.org/10.1007/978-3-319-93779-3_8

arises from the stipulation that a human (a) receives a partial or complete formal argument/proof ($\mathcal{A}$) for/of a conclusion ϕ, where some computing machine $\mathcal{M}$ "stands between" or *mediates* a's receiving $\mathcal{A}$ and ϕ. The mediation can take any number of forms, ranging from the simple and mundane (e.g., a is a high-school math teacher who types in to a text-editing system a proof of Euclid's Theorem for a math class, and then prints out or displays the proof for subsequent study and presentation to the class) to the exotic and famous (e.g., a receives a too-big-to-survey printout of a computer-generated proof of the four-color theorem). In this context, here is the most-general form of the question that drives our investigation:

(QB) Where $\mathcal{M}$ mediates as provisionally described above, under what conditions is it rational for a to believe ϕ?[1]

Once we have erected at least a reasonably precise framework for understanding the structure of arguments and proofs, classifying computing machines, ranking strength of knowledge and belief, and distinguishing between some types of computer mediation, the result is a framework in which (QB) can be answered.[2] We try herein to provide some evidence for this optimism, by applying the framework in somewhat concrete ways, and by pointing toward next-steps concretization in connection with proof systems more exotic and powerful than standard extensional ones associated with first- and second-order logic.

The sequel unfolds in accordance with this plan: In the next section (Sect. 8.2), we provide a brief but serviceable clarification of the mediating machine $\mathcal{M}$ in our overarching framework. This section also includes a rapid discussion of what we take proofs (and also, for reasons to be given, arguments) to be. We next (Sect. 8.3) present a "high-altitude" view of the overarching process with which we are concerned, one going from the ingredients being given to $\mathcal{M}$ by a human, eventually to a final epistemic attitude (specifically, as we have said, belief) on

[1]We are sorry to disappoint those readers who will wish to have this different question addressed as well or instead:

(QK) Where $\mathcal{M}$ mediates as provisionally described above, under what conditions does a in our instance really *know* that ϕ?

We leave (QK) aside in favor of (QB) and its variants because the conditions under which rational belief becomes knowledge have been notoriously difficult to set out to the satisfaction of most, let alone nearly all, thinkers. The most efficient way to confirm this is to read any decent overview of the "Gettier Problem" (GP) a problem generated by consideration of ingenious thought-experiments from Gettier (1963) in which an agent *seems* to know some proposition, but by any of the traditional accounts of knowledge as justified (= rational) true belief going back to Plato, doesn't. E.g. see this cogent overview: (Ichikawa and Steup 2012). Plato's original defense can be found in the *Theaetetus*, which can in turn be found in (Hamilton and Cairns 1961). A final word related to the GP conundrum: We encourage readers to join us in resisting the affirmation of any such principle as that if an agent a believes but doesn't know ϕ, the agent can't have learned ϕ—this being resistance that protects the position that a learns ϕ in the computer-mediated arrangement considered in the present paper, at least in cases in which the strength of the belief that ϕ on the part of a is high.

[2]And eventually (QK) as well.

the part of that human. Next, in Sect. 8.4, we infuse our overarching framework with *gradations* of belief, which allows us to refine the framework in such a way that nuanced sub-questions under the umbrella of question (QB) can be sensibly addressed. For example, this sub-question becomes expressible:

(QB_6) Where $\mathcal{M}$ mediates as provisionally described above, under what conditions is it rational for a to believe ϕ with *certainty*?

In Sect. 8.5, we spend some time exploring some concrete instantiation of our framework, using some proof-oriented technology in our laboratory (for standard, extensional logic), and featuring some real proofs. We wrap the paper up by spending a bit of time explaining the next steps that can be taken in order to achieve further concretization of our framework (Sect. 8.6); we include discussion here of situations rather more exotic than those arising from use of straightforward extensional logic: viz., the cases of infinitary logic, and intensional logic.

8.2 Computing Machines, Arguments/Proofs

Computing machines in the present discussion shall range across pretty much everything one might consider to be a candidate, from a device or process that simply prints out the input it receives, to a computer program for discovering and checking a proof, to all the abstractions at and below a standard Turing machine (e.g., abaci, register machines, etc.), to so-called "hypercomputers" (which are formally specified machines capable of computing functions beyond Turing-level machines).[3] In all cases, we refer simply to the computing machine in our analysis as $\mathcal{M}$ (and we use subscripts and superscripts to refer to more than a single such machine; e.g. we can ask the reader to consider two machines $\mathcal{M}_1$ and $\mathcal{M}_2$).

Next, note that an argument or proof $\mathcal{A}$ is what we can harmlessly call an *abstract type*. The basis for this generic terminology is the same as what allows various thinkers to refer to, say, "Henkin's completeness proof" without any relevant physical object in play. The same thing is going on when people refer (successfully) to, say, Gödel's proof of the completeness theorem, rather that any particular inscriptions of it upon paper, or a computer display, etc. (It's the *tokens* of Gödel's proof that vary considerably across textbooks, classrooms, notebooks, etc.) We denote a physical token of this type as $\widehat{\mathcal{A}}$. Argument/proof tokens are physical instantiations of the corresponding abstract arguments/proof; they can be written down on paper and other media, read, inspected, erased, copied, transmitted, and so on. We follow the same notation and simple ontology for machines as well, and hence distinguish between $\mathcal{M}$ and $\widehat{\mathcal{M}}$.

[3] An elegant example is infinite-time Turing machines; see (Hamkins and Lewis 2000). For a list of hypercomputing devices (in the context of a case, entirely separate from purposes driving the present chapter, for the proposition that human persons can hypercompute), see (Bringsjord and Arkoudas 2004).

As to what an argument for our investigation is, we let $\mathcal{A} = (\Gamma, \vdash^*, \phi)$ denote an arbitrary argument, which usually[4] has finitely many premises Γ and proceeds by inferential steps $\vdash_1, \vdash_2, \ldots, \vdash_n$ to the conclusion, ϕ ($\vdash^*$ refers simply to all the inferences collected together). But why do we refer to *arguments*, not *proofs*, given that the title of the present chapter of course revolves around the phrase 'Computer-Mediated *Proofs*'? The explanation is simply that the arguments with which we concerned are *formal* arguments, and the only difference between them and what are customarily classified as proofs is that the latter are usually distinguished by appearing in a particular context (e.g., one in which the community uses 'proof' instead of 'argument'), whereas that isn't necessary for arguments. We recognize that some will wish to count our formal arguments as proofs. That is fine; nothing we say hinges on the absence of this conflation. And a final point: Just as some computer programs can be invalid, so some proofs can be invalid—despite the fact that they are still proofs. It's an odd fact, but a fact nonetheless, that some harbor the notion that a "proof is a proof"; that is, that by definition a proof is a valid progression of reasoning. This is an odd notion, because undeniably we can and often do speak of defective computer programs, and we are perfectly entitled to likewise speak sensibly of defective proofs. In light of this situation, computer-mediated proofs certainly can suffer from the mediation in question. Indeed, computer-mediation can turn a valid proof into an invalid one. This possibility, and related ones, is what makes the epistemology of computer-mediated proofs interesting, important, and "real-world."

8.3 The Compu-Mediated Epistemological Framework

In our opening paragraph we provided a *provisional* account of mediation, to which the questions (QB) and (QB_6) referred. Now we get a bit more precise. Our framework for systematizing the epistemology of computer-mediated arguments/proofs is a generalization and expansion of what is presented in (Arkoudas and Bringsjord 2007) for analysis of proof-checking (in connection with e.g. proofs of the Four-Color Theorem) and what is presented in (Bringsjord 2015) in connection with a defense of a particular approach to program verification.[5] The diagram shown in Fig. 8.1 sums up in end-to-end fashion the entire framework that anchors the present chapter, and we explain this framework now.

First, a quick point regarding notation: $\rightarrow$ is the material conditional, while $\rightsquigarrow$ denotes the causal production of what is to the right from what is to the left. Next, note that there is an important temporal dimension to the framework: time is

[4]Not invariably. See our coverage of infinitary logic in Sect. 8.6.1.

[5]The second of these papers uses a scheme that generalizes, expands, and relaxes the scheme set out and employed in the first.

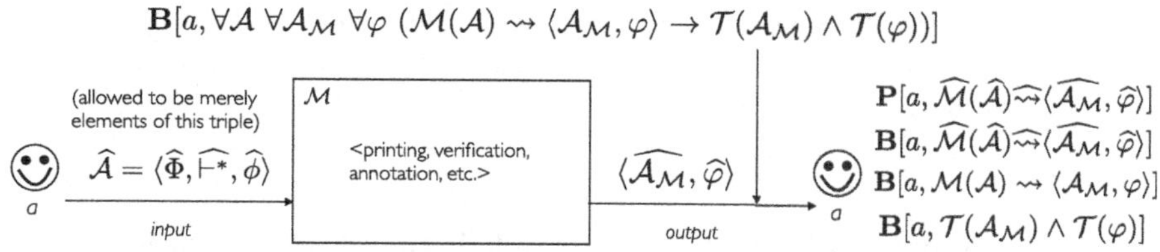

Fig. 8.1 Overarching diagrammatic depiction of the proof-mediation flow

basically flowing from the left to the right, and also from the top down to the bottom when we see the column of formulae on the right side of the picture in question:

- $\mathbf{P}[a, \widehat{\mathcal{M}}(\widehat{\mathcal{A}}) \widehat{\rightsquigarrow} \langle \widehat{\mathcal{A}_{\mathcal{M}}}, \widehat{\varphi} \rangle]$
- $\mathbf{B}[a, \widehat{\mathcal{M}}(\widehat{\mathcal{A}}) \widehat{\rightsquigarrow} \langle \widehat{\mathcal{A}_{\mathcal{M}}}, \widehat{\varphi} \rangle]$
- $\mathbf{B}[a, \mathcal{M}(\mathcal{A}) \rightsquigarrow \langle \mathcal{A}_{\mathcal{M}}, \varphi \rangle]$
- $\mathbf{B}[a, \mathcal{T}(\mathcal{A}_{\mathcal{M}}) \wedge \mathcal{T}(\varphi)]$

Hence, at the final timepoint in the progression, our agent a believes that the machine-mediated argument $\mathcal{A}_{\mathcal{M}}$ is true (= is valid), and that its conclusion φ is so as well. Notice that at this concluding moment the agent's belief is directed not at a particular token, but at the abstract types in question. The topmost formula in Fig. 8.1 is a crucial one and is "applied" to the output produced by $\mathcal{M}$; it says that the agent a believes that what the machine gives as output is worthy of assent. Specifically, the $\mathcal{M}$-mediated argument is believed true/valid, and the conclusion of this argument is believed true as well. We view the situation as one in which our agent has *learned* ϕ.

We now explain what happens as time flows on. The overall progression starts with an argument token as input given to the mediating machine for processing of some kind. The output token consists of a pair composed, first, of an argument $\widehat{\mathcal{A}_{\mathcal{M}}}$ token that is an argument causally related to the one in the original input, and, second, the conclusion $\widehat{\varphi}$ corresponding in turn to the original, earlier input conclusion $\widehat{\phi}$. This output pair from $\mathcal{M}$, given the agent's belief in the general principle that is the topmost formula (discussed above), is combined with the fact that the agent a perceives (**P** is a perception operator) that pair is so produced (i.e., combined with the fact that $\mathbf{P}[a, \widehat{\mathcal{M}}(\widehat{\mathcal{A}}) \widehat{\rightsquigarrow} \langle \widehat{\mathcal{A}_{\mathcal{M}}}, \widehat{\varphi} \rangle]$), leads to the state-of-affairs expressed by the next formula in the column on the right, which simply reflects the move from perception to belief. Belief targeting tokens then move to belief targeting propositions (types), and then finally we come to the concluding formula in the column, which we have already explained.

With our framework now in hand, the questions we have isolated thus far can be refined. For example, the general question driving our inquiry now becomes this one:

(QB′) Where $\mathcal{M}$ mediates as laid out in Fig. 8.1, under what conditions is it rational for a to believe φ?

And the sub-question (QB_6) is now supplanted with this more precise version:

(QB'_6) Where $\mathcal{M}$ mediates as laid out in Fig. 8.1, under what conditions is it rational for a to believe φ with *certainty*?

8.4 The Epistemological Framework Infused with Graded Belief

Despite the foregoing, we do not assume belief to be simply an "on versus off" matter. On the contrary, belief, at least of the human variety, is modulated by at least some version of strength, confidence, likelihood, or probability. Therefore the framework we have presented above, and diagramed in Fig. 8.1, is objectionably simplistic, and hence must be refined. We have already hinted at this via question (QB_6), which refers to "believing with certainty."

Let's consider some simple examples to start. You surely believe that in ordinary base-10 arithmetic (τ) $2+2=4$. You also believe that (π) the objects you currently perceive in front of you (the characters composing this parenthetical in this sentence, e.g.) are actually in front of you. And you also believe that (μ) some humans have in the past landed on the Moon. But these are very different beliefs, strength-wise. In the case of τ, you are, we can safely say, **certain** that this proposition holds. What about π, that those characters really exist in the physical world? Here things aren't certain. You might be a brain in a vat, or you might be dreaming, or Descartes' "evil demon" might be deceiving you. Nonetheless, unless you have evidence that your senses are compromised by such exotica, we can say that while your belief that π holds isn't at the level of certainty, it's as close as a cognizer of our kind can get to certainty without getting all the way there. We shall say that π is at the level of the **evident**.

From here, we can continue to work down to a point where a proposition that is the target of belief is a "toss up"; in this case, we say that the belief that ϕ (where ϕ expresses the proposition in question) is **counterbalanced**. In between evident and counterbalanced are four strength factors; the gist of what they mean should be clear from the words selected to express them. These words, and the entire spectrum of six (positive) values, are shown in Fig. 8.2. For what it's worth, we suspect that in most real-world cases in which relevant professionals in the formal sciences consider computer-mediated arguments/proofs with the aim of accepting or rejecting some statement φ, if they do accept, their corresponding belief that φ holds is at the level of **overwhelmingly likely**.

Please note that we intentionally dodge having to deal in the present chapter with probability and inductive logic, and hence employ the minimalist scheme of Fig. 8.2 in order to articulate a basic foundation for the epistemology of computer-mediated proof, upon which our successors can perhaps build. Trying to use probability with epistemic operators would make for very heavy and controversial going, and the situation would be made all the worse by the brute fact that we haven't here the

Fig. 8.2 Positive epistemic values

certain (6)
evident (5)
overwhelmingly likely (4)
beyond reasonable doubt (3)
likely (2)
more likely than not (1)
counterbalanced (0)

space of a monograph at hand. In this regard, nothing much, alas, has changed since the early days of modern inductive logic, 50 odd years ago. To take in the situation then, one can start with the seminal (Hintikka and Hilpinen 1966). In addition, Kyburg (1970) sums up the "chaos of non-consensus" regarding what probability is, and the implications of this state-of-affairs for inductive logic, as of a half-century back. That things haven't changed all that much in the half-century since Kyburg's survey is why we believe the circumspect move to make, in the present chapter, is to employ a scheme that is extremely general, and skirts the still-buzzing hornet's nest of probability and inductive logic.[6]

Of course, ultimately the truly mature epistemology of computer-mediated proof will need to include, minimally, some determinate stand regarding probability and its role (or possibly its lack thereof) in this epistemology.[7] Given this, we volunteer only that presumably the most natural interpretation of probability to

[6]Sanguine, skeptical readers can see some very recent publications which reveal that even today the role of probability in supporting rational belief, whether or not that belief is about arguments, proofs, and the conclusions therefrom, is highly controversial. E.g., the recently released *Argument & Inference: An Introduction to Inductive Logic* (Johnson 2016) divides the non-deductive basis for rational belief and decision-making into one part that leaves probability (in any guise) aside, and then another side that embraces and employs probability—and yet on the other hand, other overviews of this non-deductive basis assume that it *must* be probabilistic in nature (e.g. see Hawthorne 2004/2012).

[7]There are purely technical reasons for opting herein to use a streamlined multi-valued continuum for graded belief, given the current state of inductive logic, and for that matter of contemporary fields that also draw directly from probability, such as artificial intelligence (AI). (Bayesian reasoning in the AI of today is central to the field. For an overview, see Russell and Norvig 2009.) One reason is that today inductive logic, AI, and other probability-infused fields invariably make use of formal languages that are too inexpressive in the context of real-world proofs. Real-world proofs routinely make use of constructions that are infinitary in nature, and hence, taken at face value, these proofs, in the context of computer-mediation and epistemology, explode the bounds of formal languages that are the bases for probabilistic processing today. This is true because these languages are rooted in the space running from the propositional calculus to fragments of simple extensional logics like first-order logic (FOL) to FOL itself. This is despite seminal work long ago in the assignment of probabilities to formulae in *infinitary* formal languages of logics; (e.g. see Scott and Krauss 1966).

investigate in connection with the epistemology of computer-mediated proof would presumably be the *epistemological* interpretation of probability. An early but lucid and still-informative summary is the chapter "The Epistemological Interpretation of Probability" in Kyburg (1970).[8]

It's worth noting that while the multi-valued scheme we employ here, as we have cheerfully confessed, is intended to avoid explicit alliances with persistently murky and controversial concepts and interpretations of probability, it can nonetheless be shown that this scheme is in conformity with at least the vast majority of the generic frameworks for so-called *plausibility relations* (Friedman and Halpern 1995), and hence is a scheme far from idiosyncratic. More specifically, the binary relation in question must satisfy four axioms (with additional axioms serving to regiment further, more-specific constraints), and it can be proved that these axioms, suitably instantiated, hold for our scheme. For instance, where '$\phi \preceq \psi$' is read as 'ϕ is of less or equal plausibility to ψ,' the first axiom proposed by Friedman and Halpern (1995) says that

> if ϕ is a tautology and ψ includes a contradiction, then $\phi \not\preceq \psi$,

and it's an easily obtained theorem that this axiom, appropriately rendered more precise,[9] holds for our ordering (which is of course generated by letting $\leq$ for the natural numbers be $\preceq$).

We are now in position to appreciate that every occurrence of a belief operator **B** in Fig. 8.1 carries a strength parameter in its superscript. We don't present a new overarching picture with these parameters to supplant the one given in Fig. 8.1, but rest content with a single example to fix the situation for the reader: Consider again the final moment in the progression, corresponding to this formula:

$$\mathbf{B}[a, \mathcal{T}(\mathcal{A}_{\mathcal{M}}) \wedge \mathcal{T}(\varphi)].$$

The question here is: What is the level of confidence in this belief? Parameterized, we have:

$$\mathbf{B}^k[a, \mathcal{T}(\mathcal{A}_{\mathcal{M}}) \wedge \mathcal{T}(\varphi)],$$

where k is the parameter; and of course the parameter is instantiated with some value from the gradations given above. For instance, if our human a's belief is at the level of 'beyond reasonable doubt,' we would have:

$$\mathbf{B}^3[a, \mathcal{T}(\mathcal{A}_{\mathcal{M}}) \wedge \mathcal{T}(\varphi)].$$

---

[8]Terminology used to denote competing interpretations of probability has evolved. In the modern survey (Hájek 2002/2011), 'logical probability' is essentially used instead. Kyburg's (1970) terminology is sustained, and modernized, in the excellent (Galavotti 2011).

[9]Specifically, for the proof, take ϕ to be a theorem established by some valid proof known to be valid by an arbitrary agent a, and stipulate that the second conjunct of the antecedent in this axiom is cashed out as both that $\{\psi\} \vdash \bot$, and that a knows this.

8.5 Some Concretization of the Framework

In order to present the promised concretization, we start by looking at standard *first-order logic* (FOL), which as is well-known is purely extensional.[10] FOL is in the literature associated with many different proof calculi. Among these, one family is *resolution*-based, and another is *natural-deduction*-based; these are the two most commonly used kinds of proofs systems in question. Resolution-based proof calculi are made up of a few simple inference schemata, and require formulae to be in a clausal form. Such calculi are primarily used by automated theorem provers (e.g. by SNARK (Stickel et al. 1994), one of our personal favorites), as it is quite easy to optimize, relative to other automated proof systems for FOL. On the other hand, natural deduction has more inference schemata/rules than resolution. Natural deduction's schemata are more complex, and they correspond to the kind of supposition-based reasoning that humans use in the formal, deductive sciences. At any point in a natural-deduction proof in progress, there are a large number of choices. Natural-deduction systems are primarily used in pedagogy, in proofs that have to be authored manually, and in proofs that are supposed to be read by humans. That said, natural-deduction provers, though fewer in number, have been built (Pelletier 1998).[11]

Before we go further, we first announce two definitions of what it means in our framework for a human to understand a proof in first-order logic:

> *a* **strongly understands** $\widehat{A}$
> Understanding a proof, $\widehat{\mathcal{A}} \equiv \Gamma \vdash \phi$, requires checking it for its accuracy firsthand and verifying that for <u>all</u> models m and interpretations $\mathcal{I}$, if $m \models_{\mathcal{I}} \Gamma$ then $m \models_{\mathcal{I}} \phi$.

The above definition requires checking through possibly an infinite number of models. The definition below is more lenient but less general, but we feel fits more closely what non-logicians (mathematicians, students, etc.) do when they encounter a formal proof or argument.

---

[10] A readable overview can be found in (Boolos et al. 2003). An overview more suitable for consumption by those with some mathematical maturity, and wonderfully economical, is provided in (Ebbinghaus et al. 1994).

[11] One still in existence and available is OSCAR, created by John Pollock, and revived after his passing by our laboratory's Kevin O'Neill. Resurrected (and improved) OSCAR can be obtained here: http://rair.cogsci.rpi.edu/projects/automated-reasoners/oscar.

***a* weakly understands $\widehat{A}$**
Understanding a proof, $\widehat{\mathcal{A}} \equiv \Gamma \vdash \phi$, requires checking it for its accuracy firsthand, and verifying that in the intended model m and interpretation $\mathcal{I}$, if $m \models_{\mathcal{I}} \Gamma$ then $m \models_{\mathcal{I}} \phi$.

These definitions are supposed to reflect the fact that humans have a semantic picture of what a proof says before saying that they understand the proof. Now, given a human a and a proof $\widehat{\mathcal{A}}$ from a resolution or natural-deduction system $\mathcal{M}$, there are a few different possibilities that can ensue; they are shown below. Assume that a believes with strength $c_1 \in \mathcal{E}$ that $\mathcal{M}$ is implemented correctly as $\widehat{\mathcal{M}}$.

Possibilities

- P_1 a (strongly/weakly) understands the proof.
- P_2 a does not (strongly or weakly) understand the proof, but can check the proof first-hand without deploying a proof checker.
- P_3 a neither understands the proof nor can check it manually, but can deploy a proof-checker or proof-verifier ν that a knows is built correctly.
- P_4 a neither understands the proof nor can check it manually, but can deploy a proof-checker or proof-verifier ν. a has belief with some confidence $c_2 \in \mathcal{E}$ that ν is built correctly, but does not know that it is built correctly.
- P_5 a can neither understand nor check the proof.

Given the above five overarching cases, we now walk through what could happen.

In the first case, the proof is fully understood and fully checked by the human, firsthand. In this case, the machine's role could be to "merely" discover an unknown proof or present a proof that the human has not seen before (this would e.g. be natural in a class). One instance of the former case is the theorem in algebra *"All Robbins algebras are Boolean algebras."* This statement was conjectured in the 1930s and the proof was finally (and, for theorem-proving aficionado, famously) completed by a machine in 1996 (Wos 2013). The proof in question was simple and understandable.[12]

[12]The proof can be obtained from http://www.cs.unm.edu/~mccune/papers/robbins/

Case P_1
There are two sub-possibilites:

1. If a strongly understands the proof $\widehat{\mathcal{A}}$, a's strength of belief in the proof is *certain*.
2. If a only weakly understands the proof $\widehat{\mathcal{A}}$, a's strength of belief in the proof is *evident*.

In the second case, a does not understand the proof, but can check the proof manually. For example, a could be a student getting acquainted with the process of formal theorem-proving. For instance, in the Slate (Bringsjord et al. 2008; Bringsjord and Taylor 2017) proof-engineering system used in Bringsjord's formal-logic classes, in one standard setup students are asked to prove a conclusion from a

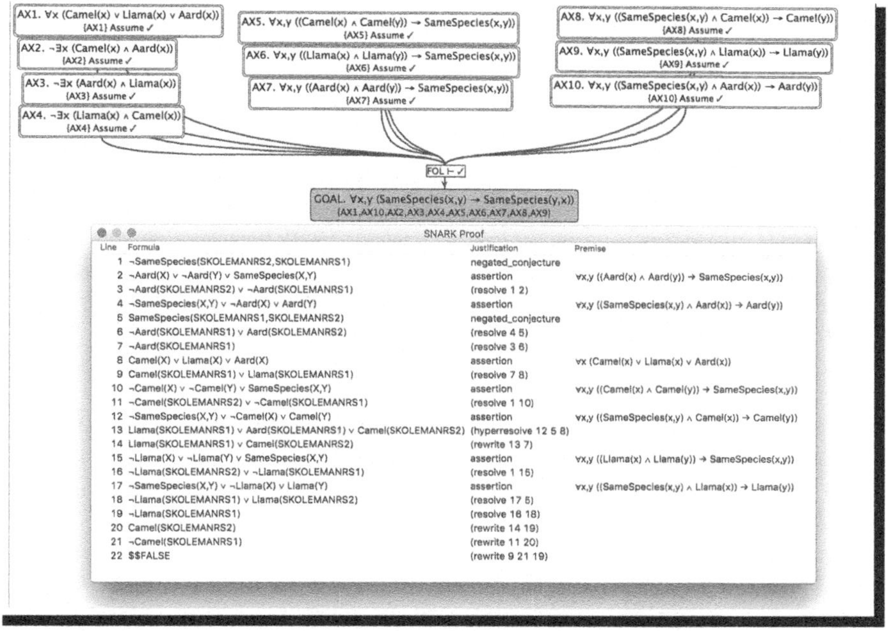

Fig. 8.3 A complex resolution proof. (The resolution proof is obtained by SNARK after a call from higher-level, manually crafted, hypergraphical natural deduction that is unique to Slate)

set of given premises using natural deduction. Before they embark on this process, students can summon a resolution prover as an oracle, and examine the fruit of its effort. The resolution prover's individual steps are usually quite simple, but it can be rather hard to understand the proof generated by the prover. For example, see the proof from Slate shown in Fig. 8.3. The proof simply shows that the relation SameSpecies is symmetric.

Case P_2
Agent a's belief is at the level of *overwhelmingly likely.*

In the third case, a is given a proof $\widehat{\mathcal{A}}$ that is quite difficult to understand and check manually. Fortunately, a has at hand a proof verifier ν that can check whether $\widehat{\mathcal{A}}$ is correct or not. Agent a also fully understands and knows that the proof-verifier is correct. The proof enterprise of the Four-Color Theorem by Appel and Haken falls into this category (for details, see Arkoudas and Bringsjord 2007). To get a sense of the scale of the proof, here are Appel and Haken describing it:

"This leaves the reader to face 50 pages containing text and diagrams, 85 pages filled with almost 2500 additional diagrams, and 400 microfiche pages that contain further diagrams and thousands of individual verifications of claims made in the 24 lemmas in the main sections of text. In addition, the reader is told that certain facts have been verified with the use of about twelve hundred hours of computer time and would be extremely time-consuming to verify by hand. The papers are somewhat intimidating due to their style and length and few mathematicians have read them in any detail."

It is obviously not possible for a human to *fully* understand such a proof in any fashion. One simply has to rely on proof verifiers. This fact is discussed in some detail in (Arkoudas and Bringsjord 2007), where it is explained that the verifier can itself be verified.

Case P_3
a's belief is at the level of *beyond reasonable doubt.*

Now onto the fourth case. When might we have a proof in first-order logic that we don't have full confidence in? Actually, there are numerous possibilities. One arises from the use of what are called *procedural attachments* in theorem provers. Given a theorem prover ρ, if it supports procedural attachments, we can supply it with a set of oracles $\{o_{R_1}, o_{R_2}, \ldots, o_{R_n}\}$ for the set of predicates $\{R_1, R_2, \ldots, R_n\}$. For example, if we are seeking a proof about ancestry and lineages, a computer program `motherOf` could be supplied as an oracle for the `motherOf` predicate symbol.[13] Then any proof will contain, in addition to the standard set of proof steps, calls to an/the oracle/s. In such situations, the strength c_2 of our belief in the correctness of the proof checker is limited to how effective the proof checker is in validating the oracles' behavior. Even if we have full confidence in the verifier, we should not be more confident of the overall proof than we are in the third case above. Hence the assignment below:

Case P_4
a's belief is at the level of the lower of *beyond reasonable doubt* and c_2.

In the final case, we are looking at proofs that are neither fully understandable nor fully verifiable. What could give rise to such a situation? While this situation may at first thought seem exotic, ordinary mathematics in fact frequently gives rise to such states-of-affairs. For example, when faced with a challenging conjecture, mathematicians commonly use analogies with other simpler situations and mathematical domains to construct a hope-filled proof-sketch rather than a full, conclusive proof. A well-known example of this is Gödel's first incompleteness theorem (GI),[14] the proof of which is usually presented in analogy with the much simpler liar paradox (= The Liar).[15] In a formalized version of this approach in which analogical inference is itself allowed, the machine would present, rather than a full proof, an analogy and a corresponding partial proof-sketch. For instance, see Licato et al. (2013), in which an intelligent machine generates just such a formal argument for GI. In this case, we have the following:

---

[13]Many theorem provers also support what are called *rewrite codes*. These are computer functions that rewrite complex function expressions to simpler forms. Since function expressions can be written using just relation symbols, our discussion covers this too. See SNARK's documentation for examples of procedural attachments and rewrite codes in action: http://www.ai.sri.com/~stickel/snark.html.

[14]Wiles' proof of Fermat's Last Theorem is a more-recent case in point; see (Wiles 1995) and (Wiles and Taylor 1995).

[15]A fully technical, elegant version of GI based explicitly on The Liar can be found in (Ebbinghaus et al. 1994).

Case P_5
a's belief is at the *likely* level.

Given our coverage of the above possibilities, we have provided at least preliminary answers to a number of questions that are instances of (QB_k).

8.6 Next Steps

All serious readers will have realized early in their study of the present essay that we have contributed but a prolegomenon for sustained investigation of the epistemology of computer-mediated proofs. We do maintain, however, that the framework we have erected will serve as a firm foundation on which to build future analyses of scenarios beyond those we entertained above. Even so, there is much work to be done; here, in particular, are the next two steps we will be taking.

8.6.1 *Infinitary Proof Systems*

First, given some of our earlier work devoted to systematic consideration of infinitary formal reasoning and infinitary computation (e.g. Govindarajulu et al. 2013a; Bringsjord and Govindarajulu 2011), it will be necessary to expand our framework so that it includes not just finite proofs, but also infinite ones.[16]

One particularly interesting non-finitary inference schema is the **ω-rule**. This rule has historically been a candidate for use with theories of arithmetic. Any theory of arithmetic has the following standard terms $\{\overline{0}, \overline{1}, \overline{2}, \ldots, \}$, whose interpretations in the standard model $\langle \mathbb{N}; +, *, \ldots \rangle$ are the usual suspects $\{0, 1, 2, \ldots, \}$. The ω-rule is then:

ω-rule

$$\frac{\phi(\overline{0}), \phi(\overline{1}), \phi(\overline{2}), \ldots, \phi(\overline{n}), \ldots}{\forall x : \phi(x)} \; \omega\text{-rule}$$

Readers familiar with Gödel's first incompleteness theorem will know that any *nice* theory of arithmetic, Γ, is *negation-incomplete*, that is, there are one or more

[16] See Footnote 7 if you haven't done so already.

troublesome ϕ such that $\Gamma \nvdash \phi$ and $\Gamma \nvdash \neg\phi$.[17] These ϕ can be tamed one way or another if we add the ω-rule to the proof calculus in question (that is $\Gamma \vdash_\omega \phi$; or $\Gamma \vdash_\omega \neg\phi$).[18] Unfortunately, and this can be obviously seen, the ω-rule is not that useful in standard practice as it has an infinite number of premises. This is where the **restricted ω-rule** steps in. In this rule, instead of an infinite number of formulae, we provide a computer program m_ϕ which takes in as input $n \in \mathbb{N}$ and provides a proof of $\phi(\overline{n})$ from Γ. We use this finite computer program as a premise.

Restricted ω-rule

$$\frac{m_\phi}{\forall x : \phi(x)} \text{ restricted } \omega\text{-rule}$$

Obviously, checking a proof that has the restricted ω-rule can be a quite difficult task. Such checking involves verifying whether a computer program always behaves according to a certain set of requirements. This is not always possible, but can be possible in some set of cases. It should be noted that in such a case, we could have a finite proof that is correct but still uncheckable, not due to any practical circumstances, but rather due to strong fundamental limits. The resultant strength of the belief in the proof would be then strongly tied to the strength of the belief in the computer program m_ϕ. (For a rigorous proof that any system that uses the restricted ω-rule is not even semi-decidable, see our Govindarajulu and Bringsjord 2014.)

8.6.2 *Intensional Systems*

Plain first-order logic (and indeed, for that matter, n-order logic) is not capable of correctly modeling knowledge, beliefs, and other internal states of information-processing agents. This can be most easily demonstrated by simply trying one's level best to model knowledge. For example, consider an agent a investigating a murder. The agent does not know that *jack* is the murderer, when in fact *jack* is the murderer. The agent does trivially know that $(jack = jack)$. The agent's knowledge could be derived from the agent knowing a stronger statement such $\forall x.(x = x)$, which is a simple theorem in first-order logic. Straightforward modeling in first-order logic quickly leads to a contradiction, as can be seen below. In fact, other sophisticated

[17] A nice theory Γ is one that allows *representations* (it can prove facts about the primitive-recursive relations and functions), is *decidable* (for any ϕ, it is decidable whether $\Gamma \vdash \phi$) and is *consistent*. See Smith (2013) for a good introduction to the two Gödelian incompleteness theorems.

[18] For a more in-depth discussion of the ω-rule and its uses, see Baker et al. (1992) and Franzén (2004). For a proof that deploys it "brazenly" (i.e., in a manner that simply takes it to be wholly legitimate) see Bringsjord and van Heuveln (2003).

schemes also quickly disintegrate, as we show in (Bringsjord and Govindarajulu 2012).

Modeling Knowlege in First-order Logic

$$\dfrac{\dfrac{\textbf{Knows}(a, jack = jack) \quad jack = murderer}{\textbf{Knows}(a, jack = murderer)} \quad \neg\textbf{Knows}(a, jack = murderer)}{\phi \land \neg\phi}$$

It is well-known that any sophisticated cognitive modeling requires at the least a quantified modal logic. For example, see our modeling of the false-belief task in (Arkoudas and Bringsjord 2008), modeling of *akrasia* in Bringsjord et al. (2014), and self-consciousness in Bringsjord et al. (2015). In addition, rigorous semantics of natural language calls for modal logic (Govindarajulu et al. 2013b). Common to all these investigations is a family of systems that we have termed *cognitive calculi*.[19] Unlike traditional logics that deal with intensional operators, cognitive calculi eschew the use of possible-worlds semantics and instead opt for proof-theoretic semantics. That is, $\Gamma \models \phi$ is defined via a function $\mu(\Gamma, \phi, \rho_1, \ldots, \rho_n)$, where ρ_i are proofs in the system. We feel that proof-theoretic semantics is not only more cognitively plausible but also more feasible computationally, if we have to build agents that understand proofs and arguments. For example, see (Francez and Dyckhoff 2010) for a proof-theoretic semantics of natural language that is also trivially mechanizable. There are two major questions in this domain. The first question, of the same structure as the questions isolated and presented above, is: Given a computer-mediated proof ρ, what ought to be the strength of our belief in the proof? The second question is this: What ought to be the strength of belief in a computer-mediated proof ρ for an agent a, given we have at hand what the agent knows, believes, etc.?

Wrapping up with a concession, we do admit that our framework is currently somewhat limited by the system of "graded belief" we have employed, even though that system is substantially more nuanced than Chisholm's (1966).[20] We are in the process of generalizing our multi-valued epistemic logic so that it incorporates the epistemological interpretation of probability (recall our remarks above regarding probability and inductive logic).

[19] One specimen in the family is the *Deontic Cognitive Event Calculus*. See http://www.cs.rpi.edu/~govinn/dcec.pdf for an overview.

[20] Including frameworks presented in subsequent, revamped-and-expanded editions of his epistemological theory, given in (Chisholm 1977, 1987).

References

Arkoudas, K., & Bringsjord, S. (2007). Computers, justification, and mathematical knowledge. *Minds and Machines, 17*(2), 185–202. http://kryten.mm.rpi.edu/ka_sb_proofs_offprint.pdf

Arkoudas, K., & Bringsjord, S. (2008). Toward formalizing common-sense psychology: An analysis of the false-belief task. In T.-B. Ho & Z.-H. Zhou (Eds.), *Proceedings of the Tenth Pacific Rim International Conference on Artificial Intelligence (PRICAI 2008)* (Lecture notes in artificial intelligence (LNAI), Vol. 5351, pp. 17–29). Springer. http://kryten.mm.rpi.edu/KA_SB_PRICAI08_AI_off.pdf

Baker, S., Ireland, A., & Smaill, A. (1992). On the use of the constructive omega-rule within automated deduction. In A. Voronkov (Ed.), *Logic programming and automated reasoning* (Lecture notes in computer science (LNCS), Vol. 624, pp. 214–225). Berlin: Springer.

Boolos, G. S., Burgess, J. P., & Jeffrey, R. C. (2003). *Computability and logic* (5th ed.). Cambridge: Cambridge University Press.

Bringsjord, S. (2015). A vindication of program verification. *History and Philosophy of Logic, 36*(3), 262–277. This URL goes to a preprint. http://kryten.mm.rpi.edu/SB_progver_selfref_driver_final2_060215.pdf

Bringsjord, S., & Arkoudas, K. (2004). The modal argument for hypercomputing minds. *Theoretical Computer Science, 317*, 167–190.

Bringsjord, S., & Govindarajulu, N. S. (2011). In defense of the unprovability of the church-turing thesis. *Journal of Unconventional Computing, 6*, 353–373. Preprint available at the URL given here. http://kryten.mm.rpi.edu/SB_NSG_CTTnotprovable_091510.pdf

Bringsjord, S., & Govindarajulu, N. S. (2012). Given the web, what is intelligence, really? *Metaphilosophy, 43*(4), 361–532. This URL is to a preprint of the paper. http://kryten.mm.rpi.edu/SB~NSG~Real~Intelligence~040912.pdf

Bringsjord, S., & Taylor, J. (2017). *Logic: A modern approach: Beginning deductive logic, advanced*. Troy, Motalen. This is an e-book edition of 23 Jan 2017. The book is accompanied by the Slate software system, ISBN of 978-0-692-60734-3, and version of 25 Jan 2016.

Bringsjord, S., & van Heuveln, B. (2003). The mental eye defense of an infinitized version of Yablo's paradox. *Analysis, 63*(1), 61–70.

Bringsjord, S., Taylor, J., Shilliday, A., Clark, M., & Arkoudas, K. (2008). Slate: An argument-centered intelligent assistant to human reasoners. In F. Grasso, N. Green, R. Kibble, & C. Reed (Eds.), *Proceedings of the 8th International Workshop on Computational Models of Natural Argument (CMNA 8)* (pp. 1–10). Patras, University of Patras. http://kryten.mm.rpi.edu/Bringsjord_etal_Slate_cmna_crc_061708.pdf

Bringsjord, S., Govindarajulu, N., Thero, D., & Si, M. (2014). Akratic Robots and the computational logic thereof. In *Proceedings of ETHICS 2014 (2014 IEEE Symposium on Ethics in Engineering, Science, and Technology)*, Chicago (pp. 22–29). IEEE Catalog Number: CFP14ETI-POD. Papers from the Proceedings can be downloaded from IEEE at URL provided here. http://ieeexplore.ieee.org/xpl/mostRecentIssue.jsp?punumber=6883275

Bringsjord, S., Licato, J., Govindarajulu, N., Ghosh, R., & Sen, A. (2015). Real robots that pass tests of self-consciousness. In *Proceedings of the 24th IEEE International Symposium on Robot and Human Interactive Communication (RO-MAN 2015)* (pp. 498–504). New York: IEEE. This URL goes to a preprint of the paper. http://kryten.mm.rpi.edu/SBringsjord_etal_self-con_robots_kg4_0601151615NY.pdf

Chisholm, R. (1966). *Theory of knowledge*. Englewood Cliffs: Prentice-Hall.

Chisholm, R. (1977). *Theory of knowledge* (2nd ed.). Englewood Cliffs: Prentice-Hall.

Chisholm, R. (1987). *Theory of knowledge* (3rd ed.). Englewood Cliffs: Prentice-Hall.

Ebbinghaus, H. D., Flum, J., & Thomas, W. (1994). *Mathematical logic* (2nd ed.). New York: Springer.

Francez, N., & Dyckhoff, R. (2010). Proof-theoretic semantics for a natural language fragment. *Linguistics and Philosophy, 33*, 447–477.

Franzén, T. (2004). Transfinite progressions: A second look at completeness. *Bulletin of Symbolic Logic, 10*, 367–389.
Friedman, N., & Halpern, J. (1995). Plausibility relations: A user's guide. In *Proceedings of the Eleventh Conference on Uncertainty in AI* (pp. 175–184). https://www.cs.cornell.edu/home/halpern/papers/plausibility_manual.pdf
Galavotti, M. C. (2011). The modern epistemic interpretations of probability: Logicism and subjectivism. In D. Gabbay, S. Hartmann, & J. Woods (Eds.), *Handbook of the history of logic* (Inductive logic, Vol. 10, pp. 153–203). Amsterdam: Elsevier.
Gettier, E. (1963). Is justified true belief knowledge? *Analysis, 23*, 121–123. http://www.ditext.com/gettier/gettier.html
Govindarajulu, N. S., & Bringsjord, S. (2014). Proof verification can be hard! In *Presented at the 10th Conference of Computability in Europe (CiE)*. http://www.naveensundarg.com/papers/ProofVerificationCanBeHard.pdf
Govindarajulu, N., Licato, J., & Bringsjord, S. (2013a). Small steps toward hypercomputation via infinitary machine proof verification and proof generation. In M. Giancarlo, A. Dennuzio, L. Manzoni, & A. Porreca (Eds.), *Unconventional computation and natural computation* (Lecture notes in computer science, Vol. 7956, pp. 102–112). Berlin: Springer.
Govindarajulu, N. S., Bringsjord, S., & Licato, J. (2013b). On deep computational formalization of natural language. In A. M. H. Abdel-Fattah & K.-U. Kühnberger (Eds.), *Proceedings of the Workshop: "Formalizing Mechanisms for Artificial General Intelligence and Cognition" (Formal MAGiC) at Artificial General Intelligence 2013*. http://cogsci.uni-osnabrueck.de/~formalmagic/FormalMAGiC-Proceedings.pdf
Hájek, A. (2002/2011). Interpretations of probability. In E. Zalta (Ed.), *The stanford encyclopedia of philosophy*. https://plato.stanford.edu/entries/probability-interpret
Hamilton, E., & Cairns, H. (Eds.). (1961). *The collected dialogues of plato (including the letters)*. Princeton: Princeton University Press.
Hamkins, J. D., & Lewis, A. (2000). Infinite time turing machines. *Journal of Symbolic Logic, 65*(2), 567–604.
Hawthorne, J. (2004/2012). Inductive logic. In E. Zalta (Ed.), *The stanford encyclopedia of philosophy*. https://plato.stanford.edu/entries/logic-inductive
Hintikka, J., & Hilpinen, R. (1966). Knowledge, acceptance, and inductive logic. In P. Suppes & J. Hintikka (Eds.), *Aspects of inductive logic* (pp. 1–20). Amsterdam: North-Holland.
Ichikawa, J., & Steup, M. (2012). The analysis of knowledge. In E. Zalta (Ed.), *The Stanford encyclopedia of philosophy*. http://plato.stanford.edu/entries/knowledge-analysis
Johnson, G. (2016). *Argument & inference: An introduction to inductive logic*. Cambridge: MIT Press.
Kyburg, H. (1970). *Probability and inductive logic*. London: Macmillan.
Licato, J., Govindarajulu, N. S., Bringsjord, S., Pomeranz, M., & Gittelson, L. (2013). Analogico-deductive generation of Gödel's first incompleteness theorem from the liar paradox. In F. Rossi (Ed.), *Proceedings of the 23rd International Joint Conference on Artificial Intelligence (IJCAI–13)* (pp. 1004–1009). Beijing: Morgan Kaufmann. Proceedings are available online at http://ijcai.org/papers13/contents.php. The direct URL provided below is to a preprint. The published version is available at http://ijcai.org/papers13/Papers/IJCAI13-153.pdf, http://kryten.mm.rpi.edu/ADR_2_GI_from_LP.pdf
Pelletier, J. (1998). Automated natural deduction in thinker. *Studia Logica, 60*(1), 3–43.
Russell, S., & Norvig, P. (2009). *Artificial intelligence: A modern approach* (3rd ed.). Upper Saddle River: Prentice Hall.
Scott, D., & Krauss, P. (1966). Assigning probabilities to logical formulas. In P. Suppes & J. Hintikka (Eds.), *Aspects of inductive logic* (pp. 219–264). Amsterdam: North-Holland.
Smith, P. (2013). *An introduction to Gödel's theorems*. Cambridge: Cambridge University Press. This is the second edition of the book.

Stickel, M., Waldinger, R., Lowry, M., Pressburger, T., & Underwood, I. (1994). Deductive composition of astronomical software from subroutine libraries. In *Proceedings of the Twelfth International Conference on Automated Deduction (CADE–12)*, Nancy (pp. 341–355). SNARK can be obtained at the url provided here. http://www.ai.sri.com/~stickel/snark.html

Wiles, A. (1995). Modular elliptic curves and Fermat's last theorem. *Annals of Mathematics, 141*(3), 443–551.

Wiles, A., & Taylor, R. (1995). Ring-theoretic properties of certain Hecke algebras. *Annals of Mathematics, 141*(3), 553–572.

Wos, L. (2013). The legacy of a great researcher. In M. P. Bonacina & M. E. Stickel (Eds.), Automated reasoning and mathematics: Essays in memory of William McCune (pp. 1–14). Berlin: Springer.

Chapter 9
Mathematical and Technological Computability

Sven Ove Hansson

Abstract The study of algorithms, computations, and computability offers a major contact point between mathematics, technology, and philosophy. This chapter begins with a brief history of computations and the technical means used to support them. Summary accounts are given of two scholarly developments that provided much of the intellectual background for modern computation: attempts to express all reasoning as mathematics and attempts to reduce all of mathematics to simple, rule-bound symbol manipulation. This is followed by a discussion of the Turing machine, including a detailed explanation of why it can be said to cover all systems of rule-bound symbol manipulation. The universal Turing machine and its philosophical implications are also discussed. A two-dimensional classificatory scheme is offered for proposed constructions of computing devices with stronger computing powers than a Turing machine. This categorization serves to highlight the weaknesses of current proposals for such devices. In conclusion, it is emphasized that computation has to be understood as an intentional input-output process with high demands on reliability and lucidity. The study of computations and algorithms has much to learn from other studies of intentional human action, not least in the philosophy of technology.

9.1 Introduction

Computations, as we perform them today, provide an obvious connection between mathematics and technology. We all use technology – if nothing else an app in the phone – for our everyday calculations. Large computations, such as those underlying weather forecasts and complex scientific models, are performed on computers that do routinely what was practically impossible a generation or two ago. But is not all this rather trivial from a mathematical point of view? One might believe so, but

S. O. Hansson (✉)
Division of Philosophy, Royal Institute of Technology (KTH), Stockholm, Sweden
e-mail: soh@kth.se

S. O. Hansson (ed.), *Technology and Mathematics*, Philosophy of Engineering and Technology 30, https://doi.org/10.1007/978-3-319-93779-3_9

in fact some of the deepest problems at the very foundations of the mathematical enterprise emanated from careful investigations of seemingly trivial routines that can be executed by a machine.

In Book XI of his *Metamorphoses*, Ovid tells us of king Midas who gained the golden touch, so that everything he put his hand on turned into gold. Mathematicians do not have that ability, but they possess another, almost equally stupendous faculty: every object they touch is transformed into an abstract, purely conceptual entity. When the ancient geometers studied straight lines, the lines lost all their width but became infinitely long. When studying numbers, mathematicians constructed abstract numbers such as $\sqrt{-1}$, a figment of mathematical imagination that has the convenient property of yielding -1 when squared, but lacks the obvious connection to the real world that ordinary numbers such as 7 and $\frac{22}{7}$ have. Unsurprisingly, when mathematicians turned to machines, they transformed them as well into abstract contrivances. Mathematicians cannot put up with arbitrary limits, so just as the lines of geometry are infinitely long, the machines of mathematics have infinite capacity. On the other hand, mathematicians cannot resist an opportunity to simplify, so their machines are extremely simple as compared to the complex machines constructed by engineers.

We are going to have a close look at the Turing machine, foremost among mathematical machines, which was proposed by Alan Turing in 1937. It is astonishingly simple, but nevertheless reputed to be able to compute everything that can at all be computed. Some say that its powers are even greater than that; allegedly, it can prove every mathematical statement that it is at all possible to prove. Some have even claimed that a Turing machine can think, just like one of us. These are controversial claims, but one thing is sure: Studies of fundamental issues in both mathematics and philosophy have taken new directions through the discussions that this highly abstract machine has given rise to.

To introduce the subject we are first going to explore the nature of computation (Sect. 9.2). After that we turn to two scholarly developments that provided much of the intellectual background for Alan Turing's construction: attempts to express all reasoning as mathematical (Sect. 9.3) and to express all mathematical reasoning as computations (Sect. 9.4). We will then have a close look at the Turing machine (Sect. 9.5). We will attend to the common claim that it can compute and prove everything that can at all be computed or proved, and then consider the contrary assertion that machines can be constructed that are capable of computing what the Turing machine cannot (Sect. 9.6). Some final remarks are presented in Sect. 9.7.

9.2 The Art of Calculation

We usually see mathematics as concerned with concepts and arguments that are totally independent of material reality. Mathematics should be equally accessible to a "brain in a vat" as it is to our own embodied brains. But in practice, we rely heavily on aide-mémoires in the form of papers, blackboards, and computer screens that we fill with mathematical symbols, equations, and diagrams. This applies to all forms of mathematical activities, including elementary arithmetic. Most of us are able to make a calculation such as 35 × 75 reliably with paper and pencil, but performing it mentally is more difficult. This seems always to have been so. In ancient civilizations, calculations were performed with the help of physical objects such as pebbles, marks in sand or dust, or beads on counting boards and abacus frames (Chabert 1999). At least in some cultures, the use of marks or movable objects as aide-mémoires for numbers seems to have preceded written language.

Obviously, the reliability of calculations depends crucially on the stability and durability of the devices used to support them. Outdoors on a windy day you are well advised to use stones rather than leaves to perform your calculations. In this very elementary sense, all non-trivial calculations depend – along with the vast majority of mathematics – on our access to technology for storing information. This is a feature that mathematics shares with other expressions of human culture. For instance, literature unsuitable for learning by heart, such as novels, is equally dependent on the devices we have constructed for storing information.

Until fairly recently, the role of technology in calculations was restricted to that of recording the intermediate and final outcomes of the process. The actual operations were performed by human minds. From the eighteenth century until well into the second half of the twentieth century, large calculations were entrusted to what were then called "computers", namely people hired to perform large numbers of arithmetic operations. Logarithmic and other mathematical tables as well as astronomical tables for nautical use were obtained in this way, and so were calculations for business, administrative, and engineering purposes. A large French project in the 1790s employed three sections of workers in the calculation of mathematical tables. The first section was a small group of mathematicians who decided the exact contents of the tables and chose the mathematical formulas to be used in calculating and checking them. The second section was a handful of experts who converted these instructions into exact numerical tasks, usually series of additions and subtractions. The third section consisted of between 60 and 80 "computers" who, following these instructions, performed the large number of elementary numerical operations that were required. Many of them were unemployed female hairdressers who had left their former trade when the grandiose and ostentatious hairstyles of the Ancien Régime nobility were not longer in demand (Grattan-Guinness 1990; Grier 2005).

9.2.1 Instructions and Algorithms

A computation (or calculation[1]) performed by a human being is an activity following some *procedure*. If you guess what 35 × 75 is, and happen to give the right answer, then you did not perform a computation. Similarly, if you knew the answer because you have learned it by heart, you did not compute it. In neither case

[1]The word "calculation" is commonly used for elementary operations, and "computation" for more advanced and complex ones. There is no sharp delimitation in usage between the two terms.

was the answer obtained by means that can also be used to solve, independently, other, similar problems. The standard way to perform this calculation is to use one of the methods taught in primary school (such as long multiplication) that can be used to multiply any two numbers. (Alternatively, some method can be used that is only applicable to a smaller class of pairs of numbers.[2])

To see how essential it is that a calculation covers more than one case, consider the following instruction to "calculate" 333×999:

> Write: 332667.

Although this instruction yields the right answer, it cannot be used to multiply any other numbers. Therefore, this is not a calculation.[3]

Furthermore, for a procedure to count as a calculation, it must be expressible with an instruction that specifies exactly what to do at each stage of the operation. For instance, the following instruction for finding the square root of a positive number x does not count as a calculation since it does not tell us exactly what to do:

> Guess a number. Determine its square. If the square is larger than x, then replace your guess by a smaller number, and if is smaller than x then replace it by a larger number. Determine the square of the new guess. Continue in this way, guessing and squaring, until you find a number whose square is very close to x. This is your approximation of $\sqrt{x}$.

Surely, you can find the square root of a number by applying this rule in combination with some good, improvised mathematical thinking. However, we would not call this a computation due the element of intelligent improvisation that is required. A computation should be based on an instruction that prescribes exactly and unambiguously what to do at each stage of the process (Cleland 2002, pp. 160–161). There is an interesting parallel with the standard requirement on scientific experiments that they should be repeatable. If you have performed a well-conceived scientific experiment, and others repeat it, then they should obtain the same result. Similarly, if you have performed a calculation and someone else repeats it, then the result should be the same (and so should the whole series of operations, step by step). In both cases, repeatability ensures intersubjectivity (Hansson 1985, p. 96).

There is one more requirement on computations that needs to be mentioned: We expect a computational procedure to be sure to yield a result. An operation that can go on for ever without providing us with the answer we want would not qualify as a method for calculation.

---

[2]For instance, the following rule can be used to multiply two two-digit numbers that both end in 5 and whose first digits are either both odd or both even: Add the first digits. Divide by 2. Add their product. Write 25 afterwards. In this case, $(3 + 7)/2 + (3 \times 7) = 26$, so $35 \times 75 = 2625$. (I.e., $(10a + 5)(10b + 5) = ((a + b)/2 + ab) \times 100 + 25$.)

[3]The set of problems that can be solved with a method of calculation will have to be mathematically "natural". For instance, the above instruction can also be used to make each of the additions 332666+1, 332665+2, 332664+3, etc., but such an ad hoc collection of problems does not make it a calculation.

For example, a "perfect number" is a number that is equal to the sum of its proper divisors. The first perfect number is 6 (equal to $1+2+3$), and the second is 28 (equal to $1+2+4+7+14$). It is not difficult to construct a procedure for finding the nth perfect number: just go through the whole series of positive integers, testing each of them. However, it is currently not known whether the number of perfect numbers is finite or infinite. If it is finite, then we will reach a point when our procedure will go on for ever, without yielding an outcome and without signalling that it is useless to continue. Such a procedure would lack the property of *effectiveness*, by which is meant the ability to yield (within a finite number of steps) an output for every input.[4]

Effective procedures are also called *algorithms*. This word derives from the name of the prominent Persian mathematician Muhammad ibn Musa al-Khwarizmi (c.780–c.850), who wrote an influential treatise on calculations. His last name was latinized "Algoritmi". According to modern usage of the term, an algorithm does not have to operate on numerical expressions. It can operate on other mathematical symbols as well, and the output can also be a symbolic expression other than a number. Generally speaking, an algorithm is an effective procedure on symbols, expressed in an instruction that describes each step in precise terms that do not leave any scope for doing in more than one way.

Textbox 9.1 shows one of the most famous algorithms in the history of mathematics, namely Euclid's (fl. 300 BCE) algorithm for finding the greatest common divisor of two integers (i.e. the largest number that is a divisor of them both). This algorithm has two features that are prominent in most algorithms of later date as well. First, one of its steps comprises a *conditional rule*, i.e. an instruction that depends on the outcome of the previous step, in this case whether or not the two numbers are equal. Secondly, it contains a cyclic iteration, i.e. a part of the procedure (in this case the single step of subtraction) that has to be performed again and again until a conditional rule requires the cycle to be broken.

These two components can also be found in the algorithms described in ancient Chinese and Indian mathematical texts (Li 2015, p. 324). The presentation of a "shu" for a problem has a central role in ancient Chinese mathematics. A shu is a method for solution, usually close to what we today call an algorithm. Similarly, many Indian mathematical works focus on presenting a "pericarma", a rule that can be used to solve a problem (Li 2015, pp. 321–322 and 327). In the ancient Orient, including Babylonian, Egyptian, Indian, and Chinese mathematical traditions, the construction of algorithms was a more prominent mathematical activity than the proving of theorems (Li 2015; Ritter 2000). In contrast, mathematicians in ancient Greece focused on proving theorems, which they regarded as the most prestigious

[4]At the time of writing, it is not known if the procedure for finding perfect numbers sketched out here is effective or not. If there are infinitely many perfect numbers, then the procedure is effective, otherwise not.

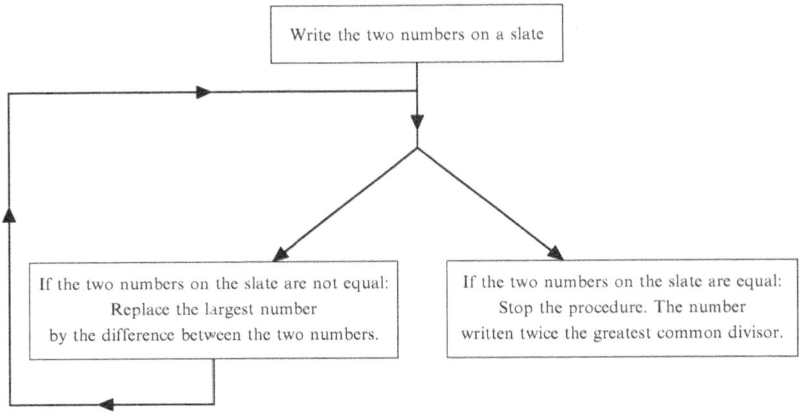

Example:

6 is the least common denominator of 114 and 48.

Textbox 9.1 Euclid's algorithm for finding the greatest common divisor of two integers

activity (but excellent algorithms were also produced, as exemplified by Euclid's algorithm and several algorithms by Archimedes, (c.287–c.212 BCE)). As we will see in Sect. 9.4.3, in the 1930s the two activities of algorithm construction and theorem proving were united in new and surprising ways.

9.2.2 The First Computing Machines

As mentioned above, the use of aide-mémoires for calculations has a long history. The use of technological means for other parts of calculatory procedures is of much later origin (depending on how we view the abacus). The Scottish mathematician John Napier (1550–1617) invented so-called calculating bones, staves based on tables for multiplication and other operations that simplified many types of calculations. He also discovered logarithms, which were implemented on slide-rules to simplify multiplication and division. In the seventeenth century, several calculating machines using rotating wheels to register numbers were introduced. Wilhelm Schickard (1592–1635) was probably the first inventor to propose such a machine and Blaise Pascal (1623–1662) and Gottfried Wilhelm Leibniz (1646–1716) the most famous ones. However, due to technical problems these machines remained rarities without much practical usage. Commercial production and widespread use of mechanical calculators only began in the second half of the nineteenth century (Swade 2011b, 2018; Lenzen 2018).

By far the most advanced computing machines to be conceived in the pre-electronic era were two constructions invented by Charles Babbage (1791–1871), the difference engine which he invented in the early 1820s and the analytical engine which he conceived in 1834. Both would have been huge mechanical constructions, and neither was completed in his life-time. The difference engine was constructed to calculate series of values for instance for logarithmic tables. The analytical machine was a general-purpose computational machine. It would be controlled with punched cards, a technology already in use for the control of automatic looms. The instructions on the punched cards – what we would now call the program – were to be based on the subdivision of complex mathematical tasks into a large number of small, simple tasks that had been developed for the organization of large-scale calculations by human computists mentioned in Sect. 9.2. (See Swade 2018 for details on Babbage's two machines.)

In the public discourse, and when applying for funding, Babbage strongly emphasized that his machines would eliminate error in the production of mathematical and astronomical tables. Error-prone humans would be replaced by what the influential science popularizer Dionysius Lardner (1793–1859) called "the untiring action and unerring certainty of mechanical agency" (Lardner [1834] 1989, p. 169). But the analytical machine had capacity for much more. The person who expressed this best was probably the mathematician and computer visionary Ada Lovelace (1815–1852) (Swade 2010).

9.2.3 Ada Lovelace's Vision

Being a woman in an environment hostile to female scholarship, Lovelace only published her thoughts as notes to a translation that she made of a text by the Italian engineer Luigi Menabrea (1809–1896). His text described the principles and operations of Babbage's analytical engine. Lovelace's notes are three times longer than the original article, but published under her initials rather than her full name in order not to draw attention to the fact that they were written by a woman. These notes are remarkable in many ways, for instance they contain the first published computer programs, written by herself (Lovelace [1834] 1989, pp. 158–170). But what is most interesting for our present purposes are her reflections on the machine and its powers.

She emphasized that although the analytical machine's operations were based on the four basic arithmetic operations, its powers were immensely extended by "the *subsequent combination of these* in every possible variety" (Lovelace [1834] 1989, p. 93n). She referred explicitly to the two mechanisms mentioned in Sect. 9.2.1, iterations and conditional instructions. She described iterations as "*cycles* of operations" (p. 150), and defined what we would today call nested cycles: "A cycle that includes *n* other cycles, successively *contained one within another*, is called a cycle of the *n*+1th order." (p. 151n). In addition, the machine was capable of following conditional instructions or, in her own words, it was able to "discover which of two or more possible contingencies has occurred, and of then shaping its future course accordingly" (p. 98n). She realized the powerfulness of such computational structures, and made the following remarkable statements:

> The Analytical Engine... is not merely adapted for *tabulating* the results of one particular function and of no other, but for *developing and tabulating* any function whatever. In fact the engine may be described as being the material expression of any indefinite function of any degree of generality and complexity (p. 115)

> [T]here is no finite line of demarcation which limits the powers of the Analytical Engine. These powers are co-extensive with our knowledge of the laws of analysis itself, and need be bounded only by our acquaintance with the latter. Indeed we may consider the engine as the *material and mechanical representative* of analysis, and that our actual working powers in this department of human study will be enabled more effectually than heretofore to keep pace with our theoretical knowledge of its principles and laws, through the complete control which the engine give us over the *executive manipulation* of algebraical and numerical symbols. (p. 121)

These passages give the impression that she was prescient enough to have a sense of the notion of a universal computer, which was precisely defined only about a century later. In fact, her assessment of the analytical engine was essentially correct; we now know that it is indeed a universal machine.[5] Perhaps even more remarkably, she also saw something that not even Babbage himself appears to have realized

[5]Gandy (1988, p. 57) showed that the functions computable with the analytical engine "are precisely those which are Turing computable."

(Swade 2010),[6] namely that the powers of his machine were not limited to numerical calculations. It could also be used to obtain "*symbolical results*" which are "not less the necessary and logical consequences of operations performed upon *symbolical data*, than are numerical results when the data are numerical." (p. 119):

> It may be desirable to explain, that by the word *operation*, we mean *any process which alters the mutual relation of two or more things*, be this relation of what kind it may. This is the most general definition, and would include all subjects in the universe. In abstract mathematics, of course operations alter those particular relations which are involved in the considerations of number and space, and the *results* of operations are those peculiar results which correspond to the nature of the subjects of operation. But the science of operations, as derived from mathematics more especially, is a science of itself, and has its own abstract truth and value, just as logic has its own peculiar truth and value, independently of the subjects to which we may apply its reasonings and processes.
> ...The operating mechanism might act upon other things besides *number*, were objects found whose mutual fundamental relations could be expressed by those of the abstract science of operations, and which should be also susceptible of adaptations to the action of the operating notation and mechanism of the engine. Supposing, for instance, that the fundamental relations of pitched sounds in the science of harmony and of musical composition were susceptible of such expression and adaptations, the engine might compose elaborate and scientific pieces of music of any degree of complexity or extent. (pp. 117–118)

> The engine can arrange and combine its numerical quantities exactly as if they were *letters* or any other *general* symbols; and in fact it might bring out its results in algebraical *notation*, were provisions made accordingly... [I]t would be a mistake to suppose that because its *results* are given in the *notation* of a more restricted science, its *processes* are therefore restricted to those of that science. (p. 144)

Interestingly, and again much ahead of her time, she ascribed this generality of the analytical engine to logic. "[T]he processes used in analysis form a logical system of much higher generality than the applications to number merely." (p. 152).

There were people at the time who believed that the analytical engine, once constructed, would be able to "think".[7] Ada Lovelace was more careful about this. In her view, the machine would be able to do "whatever we *know how to order it* to perform". It had "no power of *anticipating* any analytical relations or truth". She believed that it would be able to "*follow* analysis", but conceded that this could not be known for sure "excepting the actual existence of the engine, and actual experience of its practical results" (p. 156).

Babbage and Lovelace anticipated ideas and constructions that would not rise into prominence until well into the twentieth century. Before exploring how they were then developed, we need to have a look at two major mathematical endeavours that were instrumental in moving the art of calculation from a peripheral position in applied mathematics to a central role in the foundations of mathematics. One

---

[6]Lovelace said (p. 119) that she did not know "[w]hether the inventor of this engine had any such views in his mind whilst working out the invention."

[7]One of them was Ada Lovelace's mother, Lady Byron (1792–1869), who described it as a "thinking machine". (Quoted in Swade 2011a, p. 246.)

of these endeavours was the conversion of non-mathematical to mathematical reasoning. The other was the conversion of mathematical reasoning to symbol manipulation.

9.3 The Formalization of Reasoning

This story cannot be told without mentioning Ramon Llull (c.1232–c.1315), an excentric Majorcan theologian and philosopher who developed a method that would allegedly generate all the truths in each area of inquiry by combining the basic truths of that area. For this to be possible, all knowledge in each subject area had to be derivable from what we would now call a limited set of axioms. Today, this appears to be a strange assumption, but it seemed much more plausible at a time when Euclid's axiom-based geometry was seen as a paragon to be followed by scholars in all other disciplines. For instance, Llull assumed that all properties of God could be derived from a limited number of obvious properties, such as goodness, greatness, wisdom etc. In order to find all of God's properties, one would therefore have to systematically search for all combinations of these basic properties, and draw adequate conclusions from each such combination. The same approach could be used in all other subject areas. To obtain all the required combinations from a set of basic ideas he invented devices consisting of rotating, concentrically arranged circles that contained representations of all the basic concepts.

Llull's system and devices were immensely popular well into the eighteenth century. Jonathan Swift (1667–1745) satirized them in his *Gulliver's Travels* (1726), where he described how scholars in the academy of Lagado created new knowledge with an engine constructed to move around bits of wood with words written on them to create ever new combinations. When they found words in a row that seemed to make sense, they wrote them down. In this way, "the most ignorant person at a reasonable charge, and with a little bodily labour, might write books in philosophy, poetry, politics, law, mathematics, and theology, without the least assistance from genius or study". (ch. III:5) But others took Llull's ideas much more seriously. His ideas were among the main sources of speculations that all forms of (valid) reasoning should be reducible to some form of calculation.

Thomas Hobbes (1588–1679) repeatedly equated reasoning and computation. (It is not clear whether he was influenced by Llull, but he had access to some of his writings in the Hardwick library.[8]) In his *Leviathan* he wrote:

> When man *reasoneth*, he does nothing else but conceive a sum total, from *addition* of parcels; or conceive a remainder, from *subtraction* of one sum from another: which, if it be done by words, is conceiving of the consequence of the names of all the parts, to the name of the whole; or from the names of the whole and one part, to the name of the other part... These operations are not incident to numbers only, but to all manner of things that

[8]Forteza (1998). Cf. Hamilton (1978).

> can be added together, and taken one out of another... Out of all which we may define, that is to say determine, what that is which is meant by this word *reason*, when we reckon it amongst the faculties of the mind. For reason, in this sense, is nothing but *reckoning* that is, adding and subtracting, of the consequences of general names agreed upon for the marking and signifying of our thoughts; I say *marking* them, when we reckon by ourselves, and *signifying*, when we demonstrate or approve our reckonings to other men. (Hobbes [1651] 1839, pp. 29–30)[9]

It does not take much reflection to realize that natural languages are unsuited for such reasoning by simple addition and subtraction of concepts. Therefore, scholars striving to formalize reasoning proposed the construction of an artificial language that should reflect the structure of concepts and ideas much better than natural languages. Such a language would facilitate all scholarly pursuits, and it should therefore replace Latin as the learned language. Francis Bacon (1561–1626) was the first major proponent of such a language. He had many followers, some of whom published fairly detailed proposals for the construction of a universal language (Cram 1985; Singer 1989).

These ideas were further developed in the work of Gottfried Wilhelm Leibniz (1646–1716) (Lenzen 2018; Pombo 2010). Beginning in his youthful *Dissertio de arte combinatoria* (1666) he applied Llull's combinatorial method to characterize exhaustively the conclusions that could be drawn through traditional syllogisms from given premises. He envisaged a universal language for science and philosophy, his "characteristica universalis", that would be perfectly aligned with the structure of ideas.

> Thus I assert that all truths that can be demonstrated about things expressible in this language with the addition of new concepts not yet expressed in it – all such truths, I say, can be demonstrated *solo calculo*, or solely by manipulation of characters according to a certain form, without any labour of the imagination or effort of the mind, just as occurs in arithmetic and algebra. (Quoted in Mates 1986/1989, p. 185n.)

With the help of such a language, scholarly controversies could easily be solved:

> This being done, if controversies were to arise, there would be no more need of disputation between two philosophers than between two calculators. For it would suffice for them to take their pencils in their hands and to sit down at the abacus, and say to each other (and if they so wish also to a friend called to help): Let us calculate.[10]

In his correspondence with Damaris Masham (1659–1708), Leibniz even speculated on machines that could "imitate reason" (Jones 2014, pp. 194–195; cf. Widmaier 1986). However, not even Leibniz, one of the most prolific and inventive scholars of his times, managed to produce anything like the universal language that would be

---

[9]See also de Jong (1986) and MacDonald Ross (2007).

[10]Quo facto, quando orientur controversiae, non magis disputatione opus erit inter duos philosophos, quam inter duos computistas. Sufficiet enim calamos in manus sumere sedereque ad abacos, et sibi mutuo (accito si placet amico) dicere: *calculemus*." (Leibniz 1890, vol. 7, p. 200).

necessary for anyone – human or machine – to perform the calculations that would replace ordinary reasoning and argumentation.

In a much more modest form, these ideas were revived by the English logician George Boole (1815–1864). In his *The Laws of Thought* (1854) he developed an algebraic analysis of properties. For instance, if w denotes white things, and s denotes sheep, then ws denotes those objects that are both white and sheep, $w + s$ those objects that are either white or sheep, $w - s$ those that are white but not sheep, etc. 1 denotes "everything", and thus $1 - w$ denotes everything that is not white. Boole proposed "laws of thought" such as:

$$xy = yx$$
$$xx = x$$
$$z(x + y) = zx + zy$$
$$z(x - y) = zx - zy, \text{etc.}$$

This is the origin of the logical language that is now taught as sentential (propositional) logic. Boole's major achievement was to extend the application of algebra to objects other than numbers. As noted by Michèle Friend (2010), Boole "started to really develop the technical machinery needed to make an algebra of natural language terms, where propositions are one sort of term, amongst others." In his system, relations and functions that are expressible in natural language can be included in the formal apparatus. "When we allow symbolic representations of predicates[,] relations and functions, we can calculate out thoughts, much as we calculate out numbers." (Friend 2010, p. 174. See also Uckelman 2010). This was a major achievement, but, of course, there was still no universal language in place. Ordinary reasoning could only be replaced by calculations in the few and rather trivial cases in which simple combinations of properties were sufficient to represent the argument. It was possible to conclude that "the sheep that are not white" are the same as "the sheep that are not white sheep" (since $s(1-w) = s-ws$), but two sheep farmers arguing about the best way to manage their farms would not be much helped by Boole's algebra. Boole was himself aware of this, and in the last chapter of his great book he emphasized the need for empirical observations to acquire knowledge about the physical world; "as the cultivation of the mathematical or deductive faculty is a part of intellectual discipline, so truly is it only a part" (p. 327). In spite of these limitations, Boole's achievement was considerable. It provided a much more versatile way to represent arguments in mathematical language than what had been available previously.

In 1869 another English logician, William Stanley Jevons (1835–1882), constructed a "logical piano" that was based on Boole's logic. It had a keyboard that looked much like a piano with only white keys. By pressing the keys you could introduce premises indicating logical relations among up to four terms. Mechanical levels and pulleys made the appropriate changes on a screen, showing what these premises add up to. For instance, we can use three terms, interpreted as iron, metal, and element. If we introduce the two premises that iron is a metal and metals are elements, then the screen will show that iron is an element (Barrett and Matthew 2005). The logical piano has been hailed as "the first machine to solve Boolean logic problems faster than was possible by hand" (Gent and Walsh 2000, p. 1).

Another major achievement in the mathematization of reasoning was made by the German logician Gottlob Frege (1848–1925). In 1879, he published the *Begriffsschrift*, a book that opened up new avenues for logic. His most important innovation was predicate logic, a new formal representation of predicates, relations, and the words "all" (denoted $\forall$) and "some" (denoted $\exists$). Today, it is a standard exercise in elementary logic courses to translate natural language sentences into predicate logic. We can easily express sentences in Frege's predicate logic that had no representation in previous logical systems, with a translation process such as the following:

> Every successful team has a hardworking coach.
> For all x, if x is a successful team, then there is some y that is its coach and is hardworking.
> For all x, if Sx and Tx, then there is some y such that Cyx and Hy.
> $(\forall x)(Sx \wedge Tx \rightarrow (\exists y)(Cyx \wedge Hy))$

Frege's predicate logic is a huge advance over previous logical languages, none of which has the versatility exhibited in the above example. However, there are many everyday expressions that it cannot render, for instance adverbs ("he drove slowly"), modal sentences ("I might have come but I didn't"), and quantities intermediate between all and some ("most of his ideas go wrong"). The construction of a language in which all relations between concepts are mirrored in their logical properties is as far-fetched as ever, even with the (considerable) resources of predicate logic.

But still, predicate logic gave rise to a revolution in logic. Although it is insufficient for translating large parts of natural language, it is sufficient for expressing much – some would say all – of the natural language that is needed in mathematics. The vast majority of mathematical definitions and theorems can be expressed in predicate logic, and even more importantly: If we perform mathematical proofs very carefully in the smallest possible steps, then each step can be expressed as a statement in predicate logic, and it can be seen to follow from its predecessors according to the rules of predicate logic. Such proofs in small steps are not much liked by mathematicians – they share some of the disadvantages of looking down at your feet all the time while trying to find your way in an unknown terrain. However, predicate logic arrived at a time when mathematics was in a crisis. The possibility of appealing to such meticulous proofs rather than to intuitions expressed in natural language seemed to offer a chance to secure the foundations of mathematics.

9.4 Mathematics as Symbol Manipulation

In the nineteenth century, mathematicians attended increasingly to the foundations of their discipline. Most of the foundational work was prompted by problems in two areas of mathematics, namely analysis and geometry.

9.4.1 The Arithmetization of Analysis

Analysis is the branch of mathematics that studies differentiation, integration, and infinite series. (The near-synonym "calculus" usually refers to the less advanced parts of analysis.) Since its modern beginnings in the seventeenth century, analysis was largely based on reasoning that referred to infinitesimals, i.e. fictional numbers that were supposed to be larger than zero but smaller than all positive numbers. Pierre de Fermat (c.1607–c.1665) has been credited with their invention. Mathematicians used them in many ways. For instance, a continuous curve was described as consisting of straight line segments of infinitesimal length. Leibniz put infinitesimals to efficient use in differential calculus. We still use his notation dy/dx, which was originally thought of as the ratio between an infinitesimal change in the projection to the y-axis and an infinitesimal change in the projection to the x-axis. Some mathematicians interpreted infinitesimals as fixed quantities. Others, including Jean le Rond d'Alembert (1717–1783) interpreted them as a shorthand for a limit concept. But even with that interpretation, the concept was far from fully precise.[11]

Studies of discontinuous functions made mathematicians increasingly aware of the precarious nature of the concept of infinitesimals. In 1829, Peter Lejeune-Dirichlet (1805–1859) showed that a function could easily be constructed that is discontinuous everywhere: Let c and d be constants, and let $f(x) = c$ whenever x is rational and $f(x) = d$ whenever d is irrational. In 1861 Karl Weierstraß (1815–1897) constructed a function that is continuous everywhere but nowhere differentiable. These discoveries contributed much to the development of more rigorous foundations for calculus, using limits to define concepts such as continuity, integral, and derivative in a much more precise way. Important contributions to this development were made by Augustin-Louis Cauchy (1789–1857), Bernhard Riemann (1826–1866), and Karl Weierstraß. Perhaps the single most important step was Weierstraß's introduction of a concept of limit that replaced infinitesimals and spatial intuitions by mathematical reasoning based entirely on numbers (Edwards 1979, pp. 301–334).

In order to make calculus more rigorous, a precise account of the continuum of real numbers was crucially needed. The rational numbers can easily be "reduced"

---

[11]Much later, infinitesimals were introduced in *nonstandard analysis*, but now rigorously defined. This was largely the work of Abraham Robinson (1918–1974).

to the natural numbers, since each rational number is by definition the ratio between two natural numbers. It was much more difficult to define the real numbers in terms of the natural numbers, but in the early 1870s several mathematicians, including Richard Dedekind (1831–1916) and Georg Cantor (1845–1918) showed viable ways to do this. These constructions all made use of infinite sets of rational numbers. For instance, to define $\sqrt{2}$ we can make use of (1) the set of positive rational numbers x such that $x^2 < 2$, and (2) the set of positive rational numbers x such that $2 < x^2$. All positive rational numbers belong to one of these two sets.

Since these reconstructions of analysis redefined it in terms of numbers, they were said to "arithmetize" the subject.[12] In 1900, the French mathematician and philosopher Henri Poincaré (1854–1912) said:

> The vague idea of continuity, which we owe to intuition, resolved itself into a complicated system of inequalities referring to whole numbers.
>
> In this way, the difficulties arising from passing to the limit, or from consideration of infinitesimals, are found to be definitely clarified.
>
> Today nothing remains in analysis but integers and finite or infinite systems of integers, interrelated by a net of relations of equality or inequality.
>
> Mathematics, as it is said, has been arithmetized.[13]

In spite of the term "arithmetization" and the use of natural numbers, mathematicians of the time seem to have viewed this development not as a reduction of mathematics to the numbers 1, 2, 3..., but rather as a reduction to operations on arbitrary symbols (Jahnke and Otte 1981). We saw above that already in 1843, Ada Lovelace realized that operations on an arbitrary (finite) set of symbols could be represented as operations on natural numbers.[14] The German physicist Hermann von Helmholtz (1821–1894) expressed this insight very clearly in 1887:

> I regard arithmetic, the doctrine of the pure numbers, as a method, based on purely psychological facts, that serves to teach the consistent application of a system of signs (namely numbers) of unlimited extent and unlimited potential for refinement. To wit, arithmetic explores the question which different ways of combining these signs (calculating operations) will lead to the same final result. (Helmholtz 1887, p. 20)[15]

---

[12]This term is usually attributed to the German mathematician Felix Klein (1849–1925) who used it in a speech in 1895 (Klein 1895).

[13]"L'idée vague de continuité, que nous devions à l'intuition, s'est résolue en un système compliqué d'inégalités portant sur des nombres entiers.

Par là les difficultés provenant des passage à la limite, ou de la considération des infiniments petits, se sont trouvées définitivement éclaircies.

Il ne reste plus aujourd'hui en Analyse que des nombres entiers ou des systèmes finis ou infinis de nombres entiers, reliés entre eux par un réseau de relations d'égalité ou d'inégalité.

Les Mathématiques, comme on l'a dit, se sont arithmétisées." (Poincaré 1902, p. 120).

[14]This representability of symbols as numbers was used in masterly fashion by Kurt Gödel (1931) when he assigned a unique number to each sentence that is expressible in a logical language (Gödel numbering).

[15]"Ich betrachte die Arithmetik, oder die Lehre von den reinen Zahlen, als eine auf rein psychologischen Thatsachen aufgebaute Methode, durch die die folgerichtige Anwendung eines Zeichensystems (nämlich der Zahlen) von unbegrenzter Ausdehnung und unbegrenzter Möglichkeit der Verfeinerung gelehrt wird. Die Arithmetik untersucht namentlich,

The term "arithmetization" has not been used for long. Today, we see the reduction of analysis to sets of numbers as a reduction to sets rather than to numbers. This is probably because set theory is now a much more established discipline and it is now well-known that the natural numbers can be developed within set theory (for instance using the series $\varnothing$, $\{\varnothing\}$, $\{\{\varnothing\}\}$, $\{\{\{\varnothing\}\}\}$, ...). But what was once called "arithmetization" is still the standard method to ensure sufficient precision in mathematical analysis.

9.4.2 A New Approach to Geometry

The other area of mathematics that engendered foundational work was geometry. Non-Euclidean geometry was discovered in the 1830s, but did not catch the attention of mainstream mathematicians until the late 1860s (Freudenthal 1966). Euclid (fl.300 BCE) derived a large number of theorems for two- and three-dimensional geometry from a small set of seemingly self-evident axioms. For more than two millennia, this had been taken as the epitome of mathematical rigour. Now it had to be accepted that even these axioms were not self-evident. This led to attempts to reformulate Euclid's geometry with more rigour. In 1882 the German mathematician Moritz Pasch (1843–1930) published a new and considerably more rigorous axiomatization of Euclidean geometry. He pointed out several seemingly self-evident assumptions made by Euclid that apparently no one had noted before him, and replaced them by explicit axioms (Pasch 1882).

However, Pasch was still anxious that his axioms should be intuitively appealing. The German mathematician David Hilbert (1862–1943) took a radical new departure in an axiomatization of Euclidean geometry that he published in 1899 (Hilbert 1899). Instead of searching for axioms that expressed evident truths he wanted his axioms to be independent of any associations with intuition. The fundamental requirement was that the axioms should form a consistent system. In other words, if a statement could be derived from the axioms, the negation of that statement should not be derivable from them.

The difference between these approaches to axiomatization can be illustrated with how basic geometrical concepts such as point, line, and plane, were introduced. Euclid's *Elements* begin with these definitions:

1. A *point* is that which has no part.
2. A *line* is length without breadth.
3. The extremities of a line are points.
4. A *straight line* is a line which lies evenly with the points on itself. (Euclid 1939, pp. 436–439)

welche verschiedene Verbindungsweisen dieser Zeichen (Rechnungsoperationen) zu demselben Endergebniss führen."

Although remarkably precise, these definitions refer to spatial intuitions with words such as "part", "breadth", "extremity", and "lie evenly". These intuitions are also invoked in various ways in Euclid's proofs. Pasch found this approach to be unsatisfactory, and expressed his axiomatic ideal as follows:

> The axioms should completely include the empirical material to be dealt with mathematically, so that after the axioms have been set up there should be no need to refer back to perceptions.[16]

Hilbert went one step further. Not even when the axioms were set up should there be any appeal to empirical perceptions or spatial intuitions:

> We imagine three different systems of things: we call the things of the first system *points* and denote them $A, B, C, \ldots$; we call the things of the second system *lines* and denote them $a, b, c, \ldots$; we call the things of the third system *planes* and denote them $\alpha, \beta, \gamma, \ldots$.
>
> We think of the points, lines, and planes as being in certain relations with each other and we denote these relations with words such as "lying on", "between", "parallel", "congruent", "continuous"; the exact and complete description of these relations follows from the *axioms of geometry*.[17]

This passage is most notable for what it does not contain. There is no reference to spatial intuitions. Points, lines, and planes are presented as undefined entities, and what can be proved about them is entirely determined by the rules specifying how they relate to each other. Possibly the clearest expression of this approach was uttered by Hilbert while waiting with two colleagues for a train in a railway station in Berlin: "It should always be possible to say 'tables, chairs and beer mugs' instead of 'points, lines and planes'" (Blumenthal 1935, p. 403).

9.4.3 *Can All Mathematical Problems Be Solved in One Fell Swoop?*

At the time, Hilbert's axiomatization of Euclidean geometry must have been seen as severing geometry from empirical science. In actual fact it had the very opposite effect. Discoveries in physics in the following decades made it clear that physical space is non-Euclidean. After describing how the new view of geometry opened up

---

[16]"Die Grundsätze sollen das von der Mathematik zu verarbeitende empirische Material vollständig umfassen, so daß man nach ihrer Aufstellung auf die Sinneswahrnehmungen nicht mehr zurückzugehen braucht" (Pasch 1882, p. 17, quoted from Contro (1976) p. 286).

[17]"Wir denken drei verschiedene Systeme von Dingen: die Dinge des ersten Systems nennen wir *Punkte* und bezeichnen sie mit $A, B, C, \ldots$; die Dinge des zweiten Systems nennen wir *Gerade* und bezeichnen sie mit $a, b, c, \ldots$; die Dinge des dritten Systems nennen wir *Ebenen* und bezeichnen sie mit $\alpha, \beta, \gamma, \ldots\ldots$

Wir denken die Punkte, Geraden, Ebenen in gewissen gegenseitigen Beziehungen und bezeichnen diese Beziehungen durch Worte wie "liegen", "zwischen", "parallel", "kongruent", "stetig"; die genaue und vollständige Beschreibung dieser Beziehungen erfolgt durch die *Axiome der Geometrie*" (Hilbert 1899, p. 2).

the geometrical properties of physical space as an issue for empirical investigation, Albert Einstein wrote: "I attach special importance to the view of geometry which I have just set forth, because without it I should have been unable to formulate the theory of relativity."[18]

Hilbert's *Foundations of Geometry* was also important in another respect. He did not look for the foundations of geometry in arithmetic. Instead, he provided geometry with a fundamental axiomatization of its own. In the same way, other mathematical disciplines could be provided with independent foundations in the form of a set of axioms and precise rules for deriving theorems from these axioms. Hilbert changed his terminology and started to talk about axiomatization instead of arithmetization (Petri and Norbert 2007). A new picture of mathematics emerged, namely as the science of strictly rule-bound symbol manipulations.[19] Frege's logical language had a crucial role in this new approach, since all ordinary mathematical statements could be expressed with it. For instance, instead of writing

For all x and y, $x + y = y + x$.

we can write

$$(\forall x)(\forall y)(x + y = y + x),$$

thus eliminating natural language. Some of these fully formalized mathematical statements would be axioms, some would be definitions of new symbols in terms of those previously introduced, others steps in proofs, and yet others the outcomes of proofs. It would then, according to Hilbert, be "natural and consistent" to treat logical symbols, such as $\forall$ and the symbols for conjunction and negation, "just like the numerals and letters in algebra and to consider them, too, as signs that in themselves mean nothing, but are merely building blocks for ideal propositions." They are "just objects for the application of our rules".[20] For instance, the following proof step:

$$(\forall x)(\forall y)(x + y = y + x)$$
$$3 + 7 = 7 + 3$$

follows from a rule of substitution that is included among the rules for symbol manipulation in this system. In this way, proofs would be reduced to the manipulation of logical and mathematical symbols. Mathematics would become symbol manipulation. Although the construction of mathematical proofs would

[18]"Dieser geschilderten Auffassung der Geometrie lege ich deshalb besondere Bedeutung bei, weil es mir ohne sie unmöglich gewesen wäre, die Relativitätstheorie aufzustellen" (Einstein 1921, p. 6).

[19]Arguably, this was the realization of Ada Lovelace's above-mentioned "science of operations" that could apply to "*letters* or any other *general symbols*" as well as numbers (Lovelace [1843] 1989, pp. 117 and 144).

[20]"naturgemäß und konsequent... den Zahlzeichen und den Buchstaben in der Algebra gleichstellen und ebenfalls als Zeichen auffassen, die an sich nichts bedeuten, sondern nur Bausteine für die idealen Aussagen sind", "nur Objekte für die Anwendung unserer Regeln" (Hilbert 1928, p. 8).

continue to require mathematical acumen, checking them would be a simple routine procedure.

Hilbert raised three fundamental questions about such rigorously axiomatized mathematical systems. One of the questions was whether a mathematical system such as common arithmetic is *consistent*, by which is meant that one cannot derive both a statement and its negation from the axioms.

To introduce the other two problems we need the concept of validation. Suppose that we have a formula containing variables (such as $x, y \ldots$) and some structure containing elements that can be assigned to these variables. Let the formula be as follows:

$$(\forall x)(\forall y)(x > y \rightarrow (\exists z)(x > z > y))$$

We can apply this formula to a structure in which we interpret the variables $x, y \ldots$ as rational numbers and $>$ as "greater than". Then the formula says that if x is greater than y, then there is some number that is smaller than x but larger than y. This is obviously the case, and consequently our axiom is satisfied in this structure. But if we instead use a structure containing only natural numbers, and still interpret $>$ as "greater than", then the axiom is not satisfied, as we can see by letting $x = 3$ and $y = 2$.

Now suppose that we have an axiom system, we can call it $\mathcal{A}$, and some formula $\mathfrak{f}$. If every structure that satisfies $\mathcal{A}$ also satisfies $\mathfrak{f}$, then we can say that $\mathcal{A}$ *validates* $\mathfrak{f}$. This is conceptually quite different from saying that $\mathfrak{f}$ can be proved from $\mathcal{A}$; the former claim refers to comparisons of structures, and the latter to the constructibility of step-by-step proofs. We can therefore ask whether the formulas that can be validated from a set of axioms are the same as those that are provable from it. This question naturally divides into two. First, can all the provable formulas be validated? This is called *soundness*. And secondly, are all the formulas that can be validated provable? This is called *completeness*. Soundness is usually the easy part. It was not difficult to prove the soundness of Frege's predicate logic and mathematical systems expressible in it. The completeness part was much more difficult, and it was one of Hilbert's three questions. In 1929 Kurt Gödel (1906–1978), then a PhD student in Vienna, proved that a particular axiom system[21] is sufficient for deriving all formulas that are valid in predicate logic. Since validity is usually equated with mathematical truth, this result can be interpreted as unifying truth and provability (in predicate logic).

Hilbert's third question took another approach to validity. Suppose that we are presented with a formula (sentence) in predicate language. Is there a way to find out whether that formula is valid or not? Is there "a procedure"[22] that answers this question for all formulas that we apply it to? This is the *Entscheidungsproblem* (decision problem) which Hilbert did not hesitate to call "the main problem of mathematical logic".[23] Its importance was further enhanced by Gödel's completeness theorem.

[21] More precisely: a deductive system, i.e. the combination of a set of axioms and a set of rules for making derivations based on them.

[22] "ein Verfahren". Hilbert and Ackermann (1938), p. 91.

[23] "das Hauptproblem der mathematischen Logik", Hilbert and Ackermann (1938), p. 90.

Due to that theorem, a formula in predicate logic is valid if and only if it can be proved from the axioms. Therefore, a solution of the Entscheidungsproblem would also provide a way to determine whether a given formula is provable or not.

It was reasonable to assume that a positive solution of the Entscheidungsproblem would have an unprecedented effect on mathematics as a whole. It seemed possible to axiomatize in principle all of mathematics in predicate logic. Therefore, a method to determine the validity (truth) of everything expressible in that system would amount to a decision procedure for all mathematical statements. It would have been a philosopher's stone for mathematicians, and its discovery would have overshadowed all previous achievements in mathematics. The English mathematician Godfrey Hardy (1877–1947) denounced the idea that some system of rules could be shown to determine for any formula whether it was provable or not. "There is of course no such theorem" he said (without much argument), "and this is very fortunate, since if there were we should have a mechanical set of rules for the solution of all mathematical problems, and our activities as mathematicians would come to an end" (Hardy 1929, p. 16). He said that negative theorems were more to be expected, and that is also what happened.

In 1931 Kurt Gödel published what have come to be called his two incompleteness theorems. The first of them showed that if a consistent axiom system contains basic arithmetic, then there are statements in its language that can neither be proved nor disproved in the system itself. The second theorem showed that in such an axiom system, it is not possible to prove the consistency of the system itself. Gödel's proofs were based on what might be called the ultimate arithmetization of mathematics: he developed a method to code all formulas and sets of formulas as integers. Since the system generates proofs about numbers, it can then also generate proofs about (the number representing) itself. A sentence can be constructed that "says" about itself that it is not provable in this system, and obviously such a sentence can neither be proved nor disproved on pain of inconsistency.

Before Gödel published his incompleteness theorems, the best hope to solve the Entscheidungsproblem seemed to be that some method could be found to derive a given formula from the axioms if it was true and to derive its negation if it was not true. (Gödel's previous *complete*ness theorem had, if anything, kindled hopes that this was possible.) Gödel's new results made it clear that such a straightforward solution was not possible. It was still possible that the Entscheidungsproblem had some solution that involved a rigorous routine other than a proof. But clearly, it was now much more urgent than before to look for ways to show that the Entscheidingsproblem was insolvable.

If the Entscheidungsproblem had a solution, it would have to be a specified and well-determined routine that could be applied to all formulas in predicate logic. In other words, it would have to be an algorithm. A proof that the Entscheidingsproblem was insolvable would have to show that there exists no such algorithm. But in order to prove that, it was necessary to have a precise specification of what an algorithm is. No such specification was available, so it would have to be constructed. This is how a rather mundane problem area in applied mathematics, namely how algorithms can be constructed, became a pivotal issue in the foundations of mathematics.

9.5 Alan Turing's Machine

This brings us to the centrepiece of this chapter, namely Alan Turing's (1912–1954) paper "On Computable Numbers, with an Application to the Entscheidungsproblem". It was first presented in 1936 and then published the year after (Turing 1937a,b). It is based on an exceptional specimen of philosophical analysis. By carefully analyzing what one does when executing an algorithm, and combining this analysis with a good dose of mathematical idealization, Turing developed a simple procedure intended to cover everything that can be done with an algorithm.[24] Importantly, Turing's article was devoted to clarifying the notion of a computation that can be performed as a routine task by human beings. This resulted in tasks so well-defined that they could be performed by a certain type of machine, but his analysis was *not* an attempt to find out what types of symbolic operations can, in general, be performed by machines. The Turing machine appears "as a result, as a codification, of his analysis of calculations by humans" (Gandy 1988, pp. 83–84).[25] This is somewhat obscured to the modern reader by his frequent usage of the word "computer", which at that time referred to a human computist but is today easily misinterpreted as referring to an electronic computer.[26]

9.5.1 An Example

Before we delve into Turing's analysis in its full generality, it may be helpful to introduce some of its major components with the help of a simple example. Consider the addition of two numbers, such as 589 + 135. This is how I learned to perform that operation:

[24]In many accounts of Turing's work, this analytical work is not adequately described. Robin Gandy (1919–1995), who was Turing's graduate student, rightly called it a "paradigm of philosophical analysis" (Gandy 1988, p. 86).

[25]See for instance Turing (1937a, p. 231, [1948] 2004, p. 9), Cleland (2002, p. 166), Israel (2002, p. 196), and Sieg (1997, pp. 171–172, 2002, pp. 399–400). Misunderstandings on this are not uncommon, for instance Arkoudas (2008, p. 463) claims that "the term 'algorithm' has no connotations involving idealized human computists" and that Turing just "referred to human computers as a means of analogy when he first introduced Turing machines (e.g., comparing the state of the machine to a human's 'state of mind,' etc.)". A careful reading of Turing's 1936–7 article will show that Arkoudas's interpretation cannot be borne out by the textual evidence.

[26]Turing still used the word "computer" in this sense a decade later, see Turing ([1947] 1986, p. 116) Gandy (1988, p. 81) proposed that we use "computer" for a computing machine and "computor" for a computing human. Some authors have adopted this practice, e.g. Sieg (1994). However, the difference between the two spellings is easily overlooked. To make the difference more easily noticeable, I propose that we revive the word "computist" for a human performing calculations.

$$\begin{array}{r} \underline{1\,1} \\ 589 \\ \underline{135} \\ 724 \end{array}$$

This is a step-by-step process such that in each step, you only have to look at a part of what is recorded on the paper. In the first step, you only have to consider the column furthest to the right. You add the two numbers (9 and 5) and write down the outcome in the way indicated. Then you turn to the next column, etc. Once you know this procedure you can use it to add in principle any two numbers, even if they are very large. All you need is an instruction for what to do when you are in a column, before you proceed to the next column.

A detailed instruction for this algorithm will have to be rather long since it must cover all cases of what you can find in a column (all possible combinations such as 9+5, 1+8+3, and 1+5+1 in this example). For simplicity, we can instead consider the corresponding instruction for adding two numbers in binary notation. Mathematically, this is of course a trivial change of notation. In fact, any finite set of symbols can be encoded as series of 0 and 1, and consequently all forms of symbol manipulation can be expressed as manipulation of strings of these two (or any other two) symbols.[27]

Essentially the same algorithm can be used in binary notation. The addition 589+ 135 comes out as follows:

$$\begin{array}{r} \underline{1\,1\,1\,1} \\ 1\,0\,0\,1\,0\,0\,1\,1\,0\,1 \\ \underline{1\,0\,0\,0\,0\,1\,1\,1} \\ 1\,0\,1\,1\,0\,1\,0\,1\,0\,0 \end{array}$$

Just as in the decimal system, we begin in the right-most column, and go stepwise to the left. Each column has four spaces for symbols. When we arrive in a column, we have to look at the top three spaces, since they determine what we will have to do. We can use a simple notation for these three spaces. Then

$$\begin{bmatrix} 1 \\ 0 \\ 1 \end{bmatrix} \text{ and } \begin{bmatrix} \\ 1 \\ 1 \end{bmatrix}$$

represent the two columns furthest to the right in our example. In the rightmost column, we have to write 0 in the bottom row, move one step left and then write 1 in the top row. This be is summarized in the following short form:

[27]In this example, there are in fact three symbols, since the empty space and 0 are not treated in the same way, as can be seen from rules 1 and 18 in Textbox 9.2. It is perfectly feasible to use only two symbols; we can for instance replace each symbol space with two adjacent symbol spaces such that 00 represents an empty space, 01 represents 0 and 11 represents 1. See Sect. 9.5.2.

$\begin{bmatrix} \\ 1 \\ 1 \end{bmatrix}$: bottom 0, go left, top 1.

The 18 commands in Textbox 9.2 contain all the instructions we need to add any two binary numbers of arbitrary size. Importantly, this set of instructions can be followed by a person who does not know what the symbols 0 and 1 stand for, let alone what the binary number system is or what it means to add two numbers. You might object at this stage that someone who follows these instructions mechanically without understanding them probably runs a higher risk of making a mistake than someone who knows what she is doing. But that does not concern us here since we can allow ourselves the idealization of an error-free execution. We are not trying to find some practical way to add two numbers. Our business is to decompose the algorithm into as simple operations as possible.

From that point of view, there are still significant simplifications that can be performed in our set of instructions. Perhaps most importantly, we have required that three symbol spaces be read at the same time. This is not necessary, since we can divide the instructions into even smaller parts. Each of the rules in Textbox 9.2 can be replaced by a small series of even simpler rules, neither of which requires that we read more than one symbol space at a time. For instance, we can begin in the top row of each column (the row for carries) and read the number there. If it contains a 1, then the algorithm enters one state (with memory 1 from the top row), otherwise it goes into another state (with memory 0 from the top row). In both cases, we are instructed to go down one step. Suppose that we are performing the operation in the above example. We then have to follow the second of these instructions, i.e. we leave the top row in a state representing memory 0.

We are now in the second row of the rightmost column (and in a state with memory 0). We read the symbol in the new symbol space. What it contains determines what we will do next and what new state we will enter. In this case, since we read 1 we will be instructed to enter a state corresponding to "being on the second row and having memory 1". This state instructs us to go one step downwards, to the third row from the top in the same column. At the same time we are instructed to enter a new state that (informally speaking) carries the information that we have 1 in memory.

When we arrive in the third row, we read the symbol there, which is 1. This determines the new state (which, informally speaking, ensures that we behave as we should when we have 2 in memory). The instruction associated with this new state requires that we go one step down and write the symbol 0 in the bottom row.

After this we have to go to the top row in the column to the left and write the symbol 1 there. However, instead of doing this in one step, we can do it in five: First we take one step to the left, then three separate steps up, and finally we write the 1 in the appropriate place.

It is a fairly easy exercise to transform the instructions in Textbox 9.2 into a new set of instructions based entirely on suboperations that are so small that we only read one symbol space at a time and only move one step at a time. (It is a much more

(1) $\begin{bmatrix} \ \\ 0 \\ 0 \end{bmatrix}$: bottom 0, go left.

(2) $\begin{bmatrix} \ \\ 0 \\ 1 \end{bmatrix}$: bottom 1, go left.

(3) $\begin{bmatrix} \ \\ 1 \\ 0 \end{bmatrix}$: bottom 1, go left.

(4) $\begin{bmatrix} \ \\ 1 \\ 1 \end{bmatrix}$: bottom 0, go left, top 1.

(5) $\begin{bmatrix} 1 \\ 0 \\ 0 \end{bmatrix}$: bottom 1, go left.

(6) $\begin{bmatrix} 1 \\ 0 \\ 1 \end{bmatrix}$: bottom 0, go left, top 1.

(7) $\begin{bmatrix} 1 \\ 1 \\ 0 \end{bmatrix}$: bottom 0, go left, top 1.

(8) $\begin{bmatrix} 1 \\ 1 \\ 1 \end{bmatrix}$: bottom 1, go left, top 1.

(9) $\begin{bmatrix} \ \\ 0 \\ \ \end{bmatrix}$: bottom 0, go left.

(10) $\begin{bmatrix} \ \\ 1 \\ \ \end{bmatrix}$: bottom 1, go left.

(11) $\begin{bmatrix} \ \\ \ \\ 0 \end{bmatrix}$: bottom 0, go left.

(12) $\begin{bmatrix} \ \\ \ \\ 1 \end{bmatrix}$: bottom 1, go left.

(13) $\begin{bmatrix} 1 \\ 0 \\ \ \end{bmatrix}$: bottom 1, go left.

(14) $\begin{bmatrix} 1 \\ 1 \\ \ \end{bmatrix}$: bottom 0, go left, top 1.

(15) $\begin{bmatrix} 1 \\ \ \\ 0 \end{bmatrix}$: bottom 1, go left.

(16) $\begin{bmatrix} 1 \\ \ \\ 1 \end{bmatrix}$: bottom 0, go left, top 1.

(17) $\begin{bmatrix} 1 \\ \ \\ \ \end{bmatrix}$: bottom 1, go left.

(18) $\begin{bmatrix} \ \\ \ \\ \ \end{bmatrix}$: stop.

Textbox 9.2 An algorithm for adding two binary numbers

difficult exercise to do this in a way that requires as few states as possible.) From the viewpoint of practical application, this division into extremely small subtasks makes the addition of two numbers unnecessarily complicated. Since we are capable of perceiving three symbols simultaneously, it is obstructive to only be allowed to consider one symbol at a time. But again, what we are trying to achieve is not practical convenience but a set of suboperations that are as simple as possible. And we have now taken considerable steps in that direction. All that the computist needs to be able to do is now to (1) read (only) the symbol in her present location, (2) move one step, and (3) enter a new stage according to a simple rule that is based on the previous state and the symbol that was read in it. Could it be done simpler?

Yes, it can. One major important simplification remains. As we have presented it, this algorithm can be implemented on a checked paper that is only four squares (symbol spaces) wide. Unlimited length is required so that we can add numbers of unlimited length, but do we really need four rows?

One of the rows can easily be dispensed with. The top row, which contains the carries, is not needed since the carries are "remembered" by the states anyhow. But we can do even better than that. As shown in Fig. 9.1, if we tilt the columns then we can represent the same operation on a strip of paper that has a width of only one symbol space. It is not difficult to modify the instructions so that the operation can be performed on only one row. (In this particular case, the operation will even be simplified in this transformation, since we will only have to move in one direction, from right to left.)

We have now arrived at a mechanism with the following components:

- A tape of unlimited length with linearly ordered symbol spaces, each of which can contain a symbol from a finite alphabet (usually 0 and 1).
- A computist who always adheres to one of a finite set of specified states, and has access to ("reads") only one symbol space at each time.
- Instructions that tell the computist at each stage what to do next. Given the present state and the symbol in the accessed symbol space, (s)he is instructed (1) to write a specified symbol in the presently accessed symbol space, or to move one step to the left, or to move one step to the right, or to halt[28]; and (2) what new state to enter.

After all these simplifications, the tasks of the computist are now so "mechanical" that they can be performed by a machine. The mechanism described above is what we now call a Turing machine. We have just seen that it can be used to add any two numbers. What other algorithms can be executed by such a machine? Turing had a simple answer to that question: All algorithms. Let us now have a close look at his arguments for this bold claim.

[28]Halting is dealt with in different ways in different versions of Turing machines. A common construction is to let there be a combination of a state and an accessed symbol for which there is no instruction. When the computist arrives at that combination (s)he is assumed to halt since she has no instruction on how to proceed.

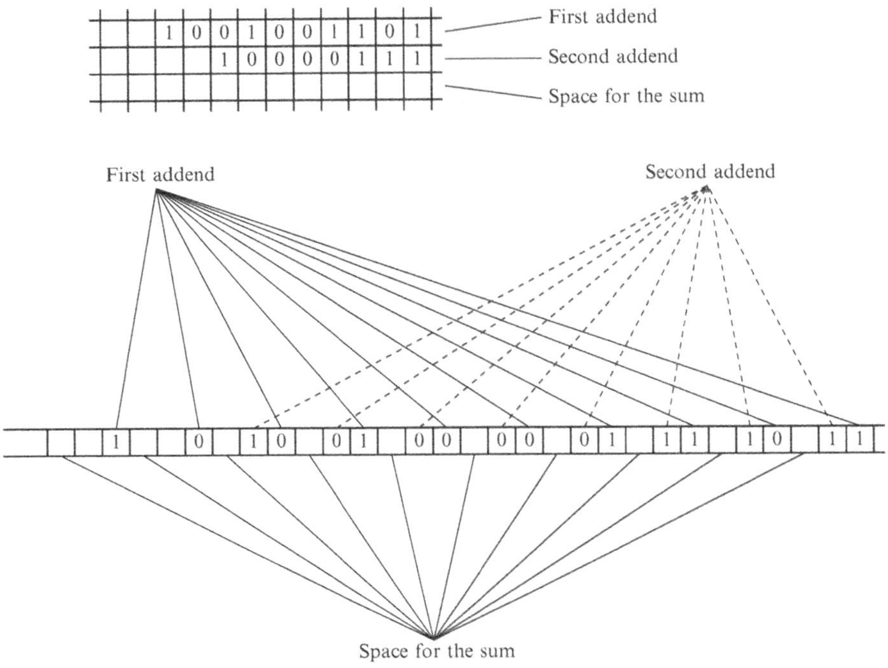

Fig. 9.1 How to represent a workspace with three rows in a workspace with only one row

9.5.2 *Turing's Analysis*

Alan Turing's analysis of computation was rather brief, and most of his article was devoted to its mathematical development. Quite a few additions are needed to make the picture complete. In the following presentation I will refer to supplements offered in particular in work by Robin Gandy (1980) and Wilfried Sieg (1994, 2002, 2009), and also add some details that do not seem to have appeared previously in the literature.

To understand Turing's analysis it is important to keep in mind that he was not concerned with *actual* computability, which depends on our resources and our mathematical knowledge. For instance, the 12 trillionth digit of π (digit number 12×10^{12}) was not actually computable in Turing's lifetime, but it seems to be so today (Yee and Kondo 2016). Mathematicians stay away from such ephemeral facts, and look beyond trivialities such as practical limitations. Like his colleagues, Turing was interested in *effective* computability.[29] A mathematical entity is effectively computable if it would be computable if we had unlimited resources, such as time

[29]Cf. Section 9.2.1. This term was apparently introduced by Alonzo Church (1936).

and paper. (The established term "effective" may be a bit confusing; the term "potential" might have been better.)

Consider a function f that takes natural numbers as arguments. If we are discussing actual computability, then we are concerned with the combination of f and a number. Suppose that we have a procedure for calculating $f(x)$ for any x. Then $f(5)$ may be actually computable whereas $f(10^{100})$ is not since its computation would require too extensive resources. (Currently this seems to be the case if $f(x)$ denotes the xth prime number.) In contrast, effective calculability is a property of the function f itself. The difference between calculating $f(5)$ and $f(10^{100})$ is a matter of practical resources, and when discussing effective computability, we should assume unlimited resources. Similarly, a mathematical statement that could only be proved with a proof that has more symbols than there are particles in the universe would still be a provable statement (Shapiro 1998, p. 276).

"Computing is normally done", said Turing, "by writing certain symbols on paper." Turing (1937a, p. 249) In saying so he implicitly acknowledged that due to limitations in human memory, we need aide-mémoires such as notes on paper to support our computations. (Cf. above, Sect. 9.2.) He did not mention that other (technological) devices than pencil and paper can be used for the same purpose, for instance an abacus or pebbles on a counting-board. However, all such devices provide us with a visual representation of symbols, and we can depict the same symbols, standing in the same relationships to each other, on paper. It therefore seems reasonable to assume that if we can perform a computation, then we can perform it with symbols on paper as the only aide-mémoire.

He continued: "We may suppose this paper is divided into squares like a child's arithmetic book." (Turing 1937a, p. 249) He did not explain why this is a reasonable assumption, but it is not difficult to justify. When we make notes to support a calculation, it is important not only which symbols we write but also in what relations they stand to each other. These relations are largely represented by their relative spatial positions. For instance, it makes a big difference if we write "$10+10=20$" or "$200+10=1$" on the paper, although the symbols are the same. The spatial relations between the symbols determine how we interpret them. Of course we could have other arrangements on the paper than the traditional checked pattern. Why not perform calculations on hexagon (honeycomb-patterned) paper? Or some other more complex tessellation? Or on two or more papers that we move over each other?

The answer is that we have no reason to believe that computing capabilities could be increased in that way. For the computational procedure to be well-defined, the impact of the relations between symbols will also have to be well-defined, and therefore these relations must be unambiguously describable. The spatial relations between symbols on a checked paper can be specified by saying that one symbol is positioned for instance "3 steps to the right and 1 step down", as compared to some other symbol. Analogous (but perhaps more complex) descriptions will have to be available for the spatial relations between symbols placed in some other pattern, if these spatial relations should be usable in a computation. Such descriptions can be written down, for instance in a long list on a checked paper. Although it may be more time-consuming to base computations on such descriptions than on a more visually appealing representation, it is difficult to see why it should not be feasible. We can therefore conclude that Turing's assumption of a checked paper does not restrict what calculations can be performed.

How large should the paper be? The answer to that question is quite simple, provided that we are concerned with effective (not actual) computability. We must be "able" (as a thought experiment) to deal with numbers of unlimited size, and therefore we must follow Turing in assuming that there is no limit to the number of squares on the tape.

Figure 9.1 showed how an operation on three rows in a squared exercise book can be squeezed into a single row. Turing took a similar step, claiming that it can always be taken:

> In elementary arithmetic the two-dimensional character of the paper is sometimes used. But such a use is always avoidable, and I think that it will be agreed that the two-dimensional character of paper is no essential of computation. I assume then that the computation is carried out on one-dimensional paper, i.e. on a tape divided into squares. (Turing 1937a, p. 249)

Turing took the step from two- to one-dimensional calculation space in a rather easy-going way, and this part of the argument is still in need of elucidation. As Robert Gandy noted, it is "not totally obvious that calculations carried out in two (or three) dimensions can be put on a one-dimensional tape" without losing any capacity (Gandy 1988, pp. 82–83). The following argument may perhaps show why this has usually not worried mathematicians: We saw above how an operation performed in three rows can easily be transferred to only one row. We have to make sure that the "states of mind" of our computist always, informally speaking, keep track of which row is currently scanned. All instructions for moving around will have to be adjusted accordingly. For instance, when we would move one step to left in the three-row system we have to move three steps to the left on the tape. Now suppose instead that we had an operation that used all the rows in a big exercise book with 60 rows. This operation could be made one-dimensional in the same way as in Fig. 9.1. The columns tilted to horizontal position would be sixty squares high, and the operation of moving one step to the left would have to be replaced by sixty moves to the left. But again, this is quite feasible, and the fact that it is impracticable does not matter in a discussion of effective computability. The same applies to an "exercise book" with, say, a thousand or a million rows.

Turing also wrote:

> I shall also suppose that the number of symbols which may be printed is finite. (Turing 1937a, p. 249)

By "symbols" he meant here types of symbols. He gave two reasons why there should only be finitely many types of symbols. First, he claimed that if there is an infinity of symbols, then there will be symbols that differ to an arbitrarily small extent and, presumably, are therefore indistinguishable. Given the limitations of human vision, this is a plausible argument, provided that there is a limit to the size of the symbols. For instance, we can require that each symbol be small enough to fit within one of the squares (symbol spaces) of the tape. We can divide each square into so many pixels that human vision cannot distinguish between two symbols if they coincide on the pixel level. And clearly, there cannot be an infinite number of pixel combinations. However, this argument refers to physical limitations, which are not preferred arguments in a mathematical context. If we presume that there are no limits to the time available to the computist, why cannot we also assume that there are no limits to her powers of perception?

Turing's second argument is much stronger. He observed that it is "always possible to use sequences of symbols in the place of single symbols" (Turing 1937a, p. 249). Thanks to the positional system we can write arbitrarily large numbers with just a few symbols (such as the ten digits $0, 1, 2, \ldots, 9$ in the decimal positional system). We can also introduce an unlimited number of variables in a mathematical language, for instance denoting them $x_0, x_1, x_2, x_3, \ldots$. Turing pointed out that strings of symbols "if they are too lengthy, cannot be observed at one glance". For instance, "[w]e cannot tell at a glance whether 9999999999999999 and 999999999999999 are the same". However, there is no need to tell this at a glance. It is sufficient that a computist can compare the two numbers digit by digit and thereby determine if they are the same.

As mentioned already in Sect. 9.5.1, when we have a finite number of symbols, then we can encode them all in binary notation (but that is a step Turing did not take in this article). Today's computers use ASCII, Unicode and other codings that assign a digital number to each symbol. It has been rigorously shown that whatever calculation can be performed by some Turing machine can also be performed by a Turing machine that only has the two symbols 0 and 1, one of which is also the symbol for a blank square (Shannon 1956, pp. 163–165).

Turing seems to have taken it for granted that only one symbol at a time can be written in a square (Turing 1937a, p. 231). That is a sensible restriction. Since there can only be a finite number of distinguishable symbols, there can only be a finite number of distinguishable combinations of symbols in a symbol space.[30] We can then treat each of these combinations as a symbol of its own. In a second step, we can encode each of these "new" symbols in a binary code with one symbol per square, as just described.

We have now, following Turing's analysis, established a minimal workspace that is sufficient for the performance of all (effective) computational procedures: An infinite tape consisting of squares in a row, each of which contains one of the two available symbols. Let us now turn our attention to the work that will be performed in that workspace. There are essentially four things that you do when computing: You *read* symbols, *write* symbols, *move* your attention (and then typically also the tip of your pen) between parts of the paper, and you *keep track* of what the rules of

[30]This argument presupposes that there is only a finite number of different positions that a symbol can have within a symbol space. This is a reasonable assumption, given the function of symbol spaces, as explained above.

this particular calculation require you to do next. Before we consider each of these activities in turn, two more general comments are in place.

In order to make his machine as simple as possible, Turing proposed that we describe the actions of the computist as "split up into 'simple operations' which are so elementary that it is not easy to imagine them further divided" (Turing 1937a, p. 250). In this style of analysis, the simplicity of the individual operations is always the top priority. A long series of small and very simple operations is always considered better than a single, somewhat more complex operation. This should be kept in mind in the discussion of all four of these activities.[31]

The other comment concerns an assumption that Turing seems to have made implicitly, namely that the operations performed by the computist take time. If each component of a computation could follow immediately upon its predecessor, so that an unlimited number of them could be performed "in no time", then we would be able to complete an infinite number of operations. This would make a big difference for what mathematical problems we could solve.[32] In fact, it would be sufficient for the operations to go successively faster in the way described by R.M. Blake in 1926: The first in an infinite series of operations takes half a second, the second operation 1/4 s, the third 1/8 etc. Then the whole infinite series will be finished in one second (Blake 1926, p. 651). Or, as Bertrand Russell said the year before Turing published his article: "Might not a man's skill increase so fast that he performed each operation in half the time required for its predecessor? In that case, the whole infinite series would take only twice as long as the first operation." (Russell 1936, p. 144). Turing's analysis tacitly excludes such unlimited acceleration of activities (although it would seem like the epitome of "effective computation"). We should assume that there is some non-zero stretch of time that each step in the calculation takes as a minimum.

Reading: Humans can perceive several symbols simultaneously. This is how we read a text; only a novice reader spells her way through a text letter by letter. But there is a limit to how much we can take in at the same time. You are just now perceiving whole words at a time in this text (and it would be very difficult to follow it if you had only one letter at a time presented to you). However, neither you nor anyone else can ingest whole pages at a time. The situation is similar for someone reading inputs or intermediate results in a computation. Turing wrote:

> We may suppose that there is a bound B to the number of symbols or squares which the computer can observe at one moment. If he wishes to observe more, he must use successive observations. (Turing 1937a, p. 250)

Suppose that the computist can simultaneously read at most ten adjacent symbols. Then we can just as well assume that she can only read one symbol at a time, but

[31] One aspect of this priority for simplicity is that each operation is assumed to affect only a minimal part of the tape. This feature can be described as a locality condition or, better, a set of locality conditions for reading, writing, and moving (Sieg 2009, pp. 584–587).

[32] Based on this omission in Turing's text, Copeland (1998) claims that Turing machines can compute Turing-incomputable functions, namely if they perform infinitely many operations in finite time.

moves across ten squares and remembers the pattern. (We will return below to how memory can be organized in a Turing machine.) The same applies to any number of adjacent symbols. Therefore, without much ado, we can assume that only one symbol space at a time can be read.

> At any moment there is just one square... which is 'in the machine'. We may call this square the 'scanned square'. The symbol on the scanned square may be called the 'scanned symbol'. The 'scanned symbol' is the only one of which the machine is, so to speak, 'directly aware'. (Turing 1937a, p. 231)

Turing was well aware that sometimes, in a calculation, you need to read something that is far away from the symbols you are currently working on. You may for instance have to pick up an intermediate result that you obtained several pages ago. But Turing pointed out that this can easily be done if we introduce a special symbol sequence adjacent to the information that may have to be retrieved later. The information can then be found by going back step by step, searching for that sequence (Turing 1937a, p. 251).

Writing: As a child I was much amused by a comic strip in which Goofy, who had for some reason temporarily become a genius, wrote poems in Sanskrit with one hand and at the same time mathematical proofs with the other. But that was a truly superhuman feat. Although we typically read more than one symbol simultaneously, it is uncommon for us humans to write more than one symbol at a time. The established procedure for writing a sequence of symbols is to write them one at a time. Unsurprisingly, Turing chose to restrict writing to one symbol at a time, just as he had done for reading. His arguments for restricting reading in this way applies to writing as well:

> We may suppose that in a simple operation not more than one symbol is altered. Any other changes can be split up into simple changes of this kind. (Turing 1937a, p. 250)

Since the spatial relations between symbols are important, it is essential to write each new symbol in the right place. Turing wrote:

> The situation in regard to the squares whose symbols may be altered in this way is the same as in regard to the observed squares. We may, therefore, without loss of generality, assume that the squares whose symbols are changed are always 'observed' squares. (Turing 1937a, p. 250)

Since he also assumed that there is at each moment only one observed square, it follows that writing will have to take place in that square. Should there be reason to write in another square than that which is observed, then that can be achieved by moving first and then writing.

In the last quotation Turing used the term "alter" for writing. This means that writing need not consist only of filling in blank squares. Overwriting is also allowed. In other words, the ideal computist performing an effective computation does not only have a pencil, she has an eraser as well.[33]

---

[33]The option of erasing a symbol to replace it by a blank square was not included in Turing's account, and it does not either seem to have had any role in later versions of the Turing machine.

Moving: Although only one symbol space at a time can be attended to, it must be possible to move around so that different symbol spaces can be read and written into.

> The machine may also change the square which is being scanned, but only by shifting it one place to right or left. (Turing 1937a, p. 231)

There are at least two ways in which longer moves can be performed in a precise manner. First, a series of instructions can, in combination, specify the number of steps in a movement. For instance, moving three steps to the right can be achieved with a series of three instructions:

(i) move one step to the right and then enter a state encoding that two steps to the right remain to be made,
(ii) move one step to the right and then enter a state encoding that one step to the right remains to be performed, and
(iii) move one step to the right.

Secondly, an iterated move can be limited by a sequence of symbols that, if read in that order, will put an end to the movement. For instance, a series of moves to the right can be discontinued as soon as three 1's in a row have been scanned.

Keeping track: We have now described the actions that a Turing machine can perform. It remains to describe how these actions are controlled. Turing wrote several years later:

> It will seem that given the initial state of the machine and the input signals it is always possible to predict all future states. This is reminiscent of Laplace's view that from the complete state of the universe at one moment of time, as described by the positions and velocities of all particles, it should be possible to predict all future states. (Turing 1950, p. 440)

This was not meant as an endorsement of a deterministic view of the universe. According to Turing, determinism is "rather nearer to practicability" for a computation than for the whole universe. (p. 440) The outcome of a computation should be exactly determined by the input and the instructions according to which the computation is performed. The application of the instructions (rules) has to be controlled by a mechanism that keeps track of "where we are" in the process. This is something that a human computist has to do throughout the process. For instance, in the addition discussed in Sect. 9.5.1, the computist must know, at each stage of the process, "I am now at a stage when I should move one step to the left", "I am now at a stage when I should write 1", etc. In addition, she will have to know what stage of the process she has to enter after completing the present one. Turing chose to call these stages *states of mind*. Furthermore:

> The behaviour of the computer at any moment is determined by the symbols which he is observing, and his 'state of mind' at that moment. (Turing 1936–7, p. 250)[34]

We therefore need a set of instructions, one for each combination of a state of mind and a scanned symbol. The instruction should tell us what to do, and what state of mind to enter after having done it. For simplicity, we can assign numbers to the states of mind so that they can easily be referred to. An instruction can then have the following form:

> If in state 12 reading 0, then write 1 and enter state 14.

In Textbox 9.3, a simple Turing machine is presented that subtracts the number 1 from any positive integer in digital notation.

How many states of mind are needed? The more complex a computation is, the more states may be required. But Turing put a limit to their numerosity:

> We will also suppose that the number of states of mind which need be taken into account is finite. (Turing 1937a, p. 250)

For this he gave two reasons. First:

> The reasons for this are of the same character as those which restrict the number of symbols. If we admitted an infinity of states of mind, some of them will be 'arbitrarily close' and will be confused. (Turing 1937a, p. 250)

This is not a very strong argument. Suppose that we have an infinite series of states of mind. We can call them $S_1, S_2, S_3, \ldots$ They can be constructed so that they all behave differently. This would make them distinguishable. And although it would take an enormous amount of time to find a state with a very high number in a table of the states, such a search task is well within the presumed capacity of a Turing machine. Turing's second argument was much stronger.

> Again, the restriction is not one which seriously affects computation, since the use of more complicated states of mind can be avoided by writing more symbols on the tape. (Turing 1937a, p. 250)

In order to keep track of how many times a specific operation has been performed, we may introduce a series of states, one for having done it once, another for having done it twice, etc. However, if we want the operation to be performable an unlimited number of times, then that solution is impossible unless we allow an infinite number of states of mind. But there is another solution. We can introduce a "counter" on the tape together with a mechanism that adds 1 to the counter each time the operation has been performed. In this way, we can keep track of an unlimited number of times that the operation has been performed, while still having a finite number of states of mind.

[34]As we saw above, Turing argued that the process could be so constructed that only one symbol at a time is observed. Consequently, "symbols" can be replaced by "symbol" in this statement. Cf. Turing (1937a), pp. 231–232, 251 and 253–254.

The machine has three states and the following instructions:
If in state 1 reading 0, then write 1 and enter state 2.
If in state 1 reading 1, then write 0 and enter state 3.
If in state 2 reading 1, then move left and enter state 1.
If in state 3 reading 0, then move left and enter state 3.
If in state 3 reading 1, then move left and enter state 3.
The machine starts in state 1, reading the rightmost digit of the number. It halts when it reaches a condition for which it has no instruction. (Alternatively,we can add instructions making it halt when reading a blank.) In this example, it subtracts 1 from 20.

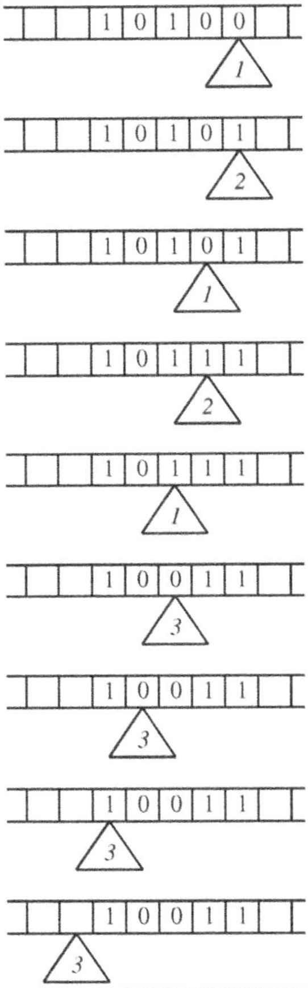

Textbox 9.3 A simple Turing machine that subtracts 1 from a number in binary notation

Another interesting argument for allowing only a finite number of states of mind was introduced by Stephen Kleene (1909–1994):

> Let us have a try at making sense out of there being a potential infinity of states of mind by a human computer with an expanding mind in applying an algorithm. So I encounter Smarty Pants, who announces to me, "I can calculate the value of a certain function $\phi(x)$, for each value of x for which it is defined, by rules already fixed which will determine my every step, so that what function $\phi(x)$ is is already determined. But *I can't tell you*, Wise Acre, *how*, because the rules have to tell how I will respond in each of an infinity of ultimately possible states of my expanding mind." I would reply, "Phooey! If you can't tell me what your method is, it isn't effective in my understanding of the term!" How can S.P. know about all those future states of his infinitely expanding – should I say exploding? – mind? (Kleene 1987, p. 493)

According to Kleene, the very notion of an algorithm or an effective computation "involves its being possible to convey a complete description of the effective procedure or algorithm by a finite communication, in advance of performing computations in accordance with it" (Kleene 1987, p. 493).[35] This is certainly an essential component of what we mean by an algorithm: it must be possible to apply unambiguously, and therefore it must also be possible to specify and communicate.

Hopefully, the arguments in this section – most of them Turing's own, but some added later on – are sufficient to show that he provided a highly convincing account of what it means for a mathematical entity to be computable, or obtainable by performing an algorithm. It should again be emphasized that Turing's analysis was aimed at determining what a human can do by following an algorithm, i.e. a fully rule-bound and deterministic procedure for symbol manipulation. The hypothetical machine that emerged from this analysis showed, as he saw it, that anything a human computist can do by just following instructions, can also be performed by a certain type of machine. Later, after several years' experience of the development of digital computers, he referred to human computing as the model on which they were based:

> The idea behind digital computers may be explained by saying that these machines are intended to carry out any operations which could be done by a human computer. The human computer is supposed to be following fixed rules; he has no authority to deviate from them in any detail. We may suppose that these rules are supplied in a book, which is altered whenever he is put on to a new job. He has also an unlimited supply of paper on which he does his calculations. He may also do his multiplications and additions on a 'desk machine', but this is not important. (Turing 1950, p. 436)

9.5.3 The Universal Machine

In addition to his path-breaking analysis of the notion of a computation, Turing's article contained another equally important achievement, namely the construction of a "universal" Turing machine that can perform all calculations performable by any

[35]Cf. Hofstadter (1979), p. 562.

other Turing machine. To see how this is possible, we can begin by noting that in order to specify a Turing machine it is sufficient to provide a list of all the rules that govern its performance. For instance, the Turing machine presented in Textbox 9.3 is specified by the five rules that are given in the box. Each of these rules can easily be converted into a list of four numbers. If we use 0 to denote "write 0", 1 for "write 1", and 2 for "go left", then the five rules in the example can be rewritten as follows:

$\langle 1, 0, 1, 2\rangle$
$\langle 1, 1, 0, 3\rangle$
$\langle 2, 1, 2, 1\rangle$
$\langle 3, 0, 2, 3\rangle$
$\langle 3, 1, 2, 3\rangle$

It is fairly easy to encode these rules in binary notation and put them on a tape in such fashion that the rule what to do in a specific situation (such as "in state 2, reading 1") can be retrieved unambiguously. The same tape can also contain the input sequence that we want to run on this machine (such as 10100 in our example). A universal Turing machine reads the first symbol of the input and then searches the tape for the instruction that is applicable in this case, executes that instruction, reads the symbol in the symbol space where it is now located, searches for the appropriate instruction, executes it, etc. Turing showed in detail how a universal machine can be constructed, providing a list of the rules determining its operations (Turing 1937a, pp. 243–246). A couple of years later, when he had experience of building digital computers, he described the universal machine as follows:

> If we take the properties of the universal machine in combination with the fact that machine processes and rule of thumb processes are synonymous we may say that the universal machine is one which, when supplied with the appropriate instructions, can be made to do any rule of thumb process. This feature is paralleled in digital computing machines such as the ACE. They are in fact practical versions of the universal machine. (Turing [1947] 1986, pp. 112–113)[36]

The construction of a universal machine makes it possible to use one and the same machine for all computations. As we saw above, this was a step essentially foreseen by Charles Babbage and Ada Lovelace, but it was nevertheless an important achievement. Turing wrote in 1948:

> The importance of the universal machine is clear. We do not need to have an infinity of different machines doing different jobs. A single one will suffice. The engineering problem of producing various machines for various jobs is replaced by the office work of 'programming' the universal machine to do these jobs. (Turing [1948] 2004, p. 414)

In his 1937 paper, Turing used the universal machine to prove that the Entscheidungsproblem is unsolvable. In order to do so he had to come up with a problem that cannot be solved by any algorithm. Note that with his encoding, every algorithm corresponds to a Turing machine that can in its turn be represented by a set of instructions on a tape (which can be run on the universal machine).

[36]Cf. Turing ([1947] 1986, p. 107, 1950, p. 436).

For any Turing machine we can ask the question: If we run this machine, starting with an empty tape, will it ever halt, or will it go on running for ever? Is there some algorithm for solving this problem for any Turing machine? If there is, then that algorithm must itself be representable as a Turing machine. Let us call that machine H. If we feed a tape representing some Turing machine into H, then (we can presume) it gives us the answer 1 if that machine halts, and 0 if it does not.

We can now construct another machine H^+ that is actually H with an extra feature at the end of the process. Whenever H prints 1, then H^+ enters a loop. Whenever H prints 0, then H^+ halts. Or, more succinctly:

- When H^+ is fed with the code for a Turing machine that halts, then H^+ does not halt.
- When H^+ is fed with the code for a Turing machine that does not halt, then H^+ halts.

Like all other Turing machines, H^+ can be represented by a code on a tape. Let us now feed the code of H^+ into itself. What will happen? It follows directly that if H^+ halts, then it does not halt, and if it does not halt, then it halts. Thus it is logically impossible to build a machine like H^+. But if H could be built, then it would be very easy to build H^+. We can therefore conclude that H cannot either be built. If there was some algorithm for solving the halting problem, then it would be possible to build H. Consequently, there is no algorithm for solving the halting problem.

9.5.4 The Reception

Alan Turing was far from the only logician in search of a precise specification of effective computability. Already in early 1934, the American logician Alonzo Church (1903–1995) speculated that a general class of number-theoretical functions, called the λ-definable functions, might coincide with the effectively computable functions (Sieg 1997). His PhD student Stephen Kleene was convinced "overnight" that this must be correct (Kleene 1981, p. 59), but others were less easily convinced. In particular Kurt Gödel, who had a very strong standing among his colleagues, considered the proposal to be quite unsatisfactory. In spite of Gödel's resistance, Church presented his proposal to the American Mathematical Society in April 1935 and published it the following year (Church 1936). But Gödel remained unconvinced.

In an appendix to his 1937 paper, Turing showed that his and Church's definitions coincided. In other words, the λ-computable functions coincided with the functions computable on a Turing machine. This meant that there were in fact three equivalent characterizations of computability, since the λ-computable functions were already known to coincide with the generally recursive functions, a class defined by Jacques Herbrand (1908–1931) and Gödel.[37]

Colleagues immediately realized that Turing's analysis was superior to the other proposals in terms of its intuitive plausibility. Kurt Gödel, who had not been convinced by Church's proposal, was persuaded by Turing's argument.[38] Alonzo Church wrote that Turing's proposal had, in comparison with his own, "the advantage of making the identification with effectiveness in the ordinary (not explicitly defined) sense evident immediately – i.e. without the necessity of proving preliminary theorems."[39] (Church 1937, p. 43). He also wrote:

> [A] human calculator, provided with pencil and paper and explicit instructions, can be regarded as a kind of Turing machine. It is thus immediately clear that computability, so defined, can be identified with (especially, is no less general than) the notion of effectiveness as it appears in certain mathematical problems (various forms of the Entscheidungsproblem, various problems to find complete sets of invariants in topology, group theory, etc., and in general any problem which concerns the discovery of an algorithm). (Church 1937, pp. 42–43)

In a textbook published in 1952, Stephen Kleene introduced the term "Church's Thesis" for the identification of effective computability with λ-computability, Turing computability and the other equivalent definitions (Soare 2007, p.708). It is now more commonly called the Church-Turing thesis. Its formal status in mathematics is not entirely clear, but a proposal by Robert Soare is worth mentioning. He compares the Church-Turing thesis to other precise mathematical explications of vague concepts that have become generally accepted, such as the definition of a continuous curve and that of an area. Soare notes that these are now "simply taken as *definitions* of the underlying intuitive concepts", thus indicating that we might think similarly of the Church-Turing thesis (Soare 1996, p. 297).

The thesis specifies what can be achieved by following exact instructions. As Kurt Gödel was eager to point out, mathematics is much more than that:

> Turing's work gives an analysis of the concept of 'mechanical procedure' (alias 'algorithm' or 'computation procedure' or 'finite combinatorial procedure'). This concept is shown to be equivalent with that of a 'Turing machine'... Note that the question of whether there exist finite *non-mechanical* procedures not equivalent with any algorithm, has nothing whatsoever to do with the adequacy of the definition of 'formal system' and of 'mechanical procedure'... Note that the results mentioned... do not establish any bounds for the powers of human reason, but rather for the potentialities of pure formalism in mathematics. (Gödel 1964, pp. 72–73)[40]

---

[37]Several other equivalent characterizations have been added to the list, first of them Emil Post's (1936) proposal, which had some elements in common with Turing's but was conceived independently. As clarified by Soare (1996, p. 300), Post's ideas were much less developed than those presented by Turing.

[38]He wrote later that he was "completely convinced only by Turing's paper". (Letter from Gödel to Georg Kreisel, May 1, 1968, quoted by Sieg (1994, p. 88)).

[39]The reference about the necessity of proving preliminary theorems refers to a technicality clarified by Sieg (1994, p. 112).

[40]See also Gödel (1958).

9.6 What Machines Can Do

When Turing wrote his famous paper, computation was still a process performed by humans. Even if the computist used a mechanical desk calculator, (s)he was always involved in every step of the process. Today, large calculations are almost invariably performed on machines. It is therefore appropriate to ask whether Turing's characterization of (effective) computations is still valid. It applies to common electronic computers as we know them, since their capacities are in principle those of a (fast) Turing machine with a finite tape. But what about other types of machines? Can we construct a machine that is capable of making computations that a Turing machine cannot perform?

9.6.1 Two Elementary Insights

Many discussions on what computations can be performed by machines have gone wrong due to the lack (or neglect) of two rather elementary insights.

First, *the notion of an effective computation by a machine is an idealization, just as the notion of an effective computation by a human* (Shapiro 1998, p. 275). In fact, no physical machine can even perform all the computations performable by a Turing machine, since the latter is an idealized machine that can operate on (finite) numbers so large that they cannot be represented in the universe.[41] Therefore, a machine with greater computing powers than a Turing machine cannot be an actual physical machine. It will have to be a hypothetical machine, although it may be describable as an idealized version of some type of physical machine. As illustrated in Fig. 9.2, the hypothetical machine will then stand in the same relationship to that physical machine as a Turing machine to an actual electronic computer. (It will, for instance, have to be absolutely error-free and provided with unlimited memory.) Importantly, even if this "idealized other machine" has greater computing powers than an "idealized electronic computer" (i.e. a Turing machine), it does not follow that the actually existing other machine has greater computing powers than the actually existing electronic computer.

The following proposal for an extension of the Church-Turing thesis to computation by machines is not untypical:

> Physical Church-Turing thesis: The class of functions that can be computed by any physical system is co-extensive with the Turing computable functions. (Timpson 2007, p. 740)

This proposed thesis is followed by a discussion of how it could be falsified by the construction of various powerful computing devices. But that is an unnecessary discussion. As it stands, the thesis is obviously false since no physical device can

[41] According to one estimate, the universe can register up to 10^{90} bits (Lloyd 2002). Obviously, the practical limitations for a computing device ever to be built are much stricter.

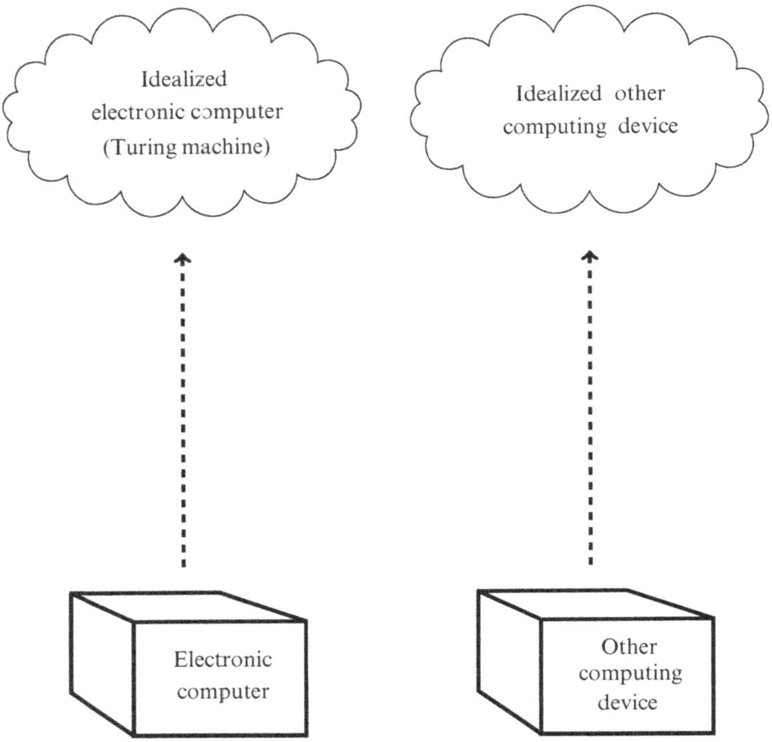

Fig. 9.2 The relationships between physical computers and idealized computers with unlimited capacities. A machine that transcends the powers of a Turing machine cannot be a physical "other computing device" but must be an "idealized other computing device" that cannot be physically realized

have the capacity of a Turing machine to produce and operate on symbol sequences of unlimited size.

The second elementary fact is that *computation is a technological operation, not just a physical event*. In technology, contrary to physics, human agency and intention are indispensable. Leaving them out can have absurd consequences, as can be seen from the so-called pancomputationalist standpoint, according to which every physical system implements every computation. (Shagrir 2012) For instance, the A4 sheet that I have in front of me represents a calculation of π, since if its long side is measured in a unit corresponding to about 9.45 cm, then it is 3,14 units long. (Pancomputationalism does not get better than this, and is not worth being taken seriously.)

Computation is a process into which an intelligent agent enters an input, and receives an output. The process has to be reliable, repeatable, and as Piccinini (2015, p. 253) pointed out, *settable* in the sense that "a user sets the system to its initial state and feeds it different arguments of the function being computed", and then receives the appropriate outputs.

9.6.2 Computativeness

We can categorize proposed methods for exceeding Turing computatibility according to two dimensions. One of these is *computativeness*, the degree to which the operation in question is or at least resembles a computation. I propose that we distinguish between three degrees of computativeness.

The highest degree requires that the process is *exactly characterized in predetermined, consecutive small steps, just like an ordinary mathematical algorithm*. This is a property that ordinary digital computers have, as noted by Turing in 1950.

> The digital computers considered in the last section may be classified amongst the 'discrete state machines'. These are the machines which move by sudden jumps or clicks from one quite definite state to another. These states are sufficiently different for the possibility of confusion between them to be ignored. Strictly speaking there are no such machines. Everything really moves continuously. But there are many kinds of machine which can profitably be *thought of* as being discrete state machines (Turing 1950, p. 439)[42]

There is an obvious problem with the idea of a machine that performs Turing-incomputable operations with this high degree of computativeness: If its operations are performed step by step in this way, then they can be checked but a human computist, and then why cannot they also be performed by a computer or by a Turing machine?

One possibility would be that the machine has capacities for parallel computing that human computists lack. This option was carefully investigated by Robin Gandy (1980). He assumed that a hypothetical physical computing device performs its operations in discrete and uniquely determined steps. Massively parallel operations are allowed, but the machine must satisfy two physical conditions: There is a lower limit on the size of its smallest parts, and there is also a limit (such as the speed of light) on the speed of signal transmission between its parts. Gandy concluded that whatever can be computed by such a machine, working on finite data according to a finite set of instructions, is Turing computable.

The second degree of computativeness is represented by an *input-output device that does not operate in describable discrete steps*. Such a device could be called a "black box", but in order to rely on it we would have to know how it works and have very good reasons to believe that it performs the computation accurately.

The major problem with such a device would be that if we rely on it, then our reliance is based on physical rather than mathematical knowledge. According to the traditional view, mathematics cannot be based on empirical observations, since mathematical knowledge must have a type of certainty that cannot be achieved in empirical science. If a mathematical result relies on a computation that we cannot follow in detail, then it may not be possible to check its validity with mathematical means. Our reliance on it would have to depend on some physical theory, and this would add a component of uncertainty that is outside of the purview of mathematics

---

[42]Charles Babbage put much effort into making his computing machines operate by switching reliably between discrete states (Swade 2011b, pp. 67–70).

– unless that theory has been "certified as being absolutely correct, unlike any existing theory, which physicists see as only an approximation to reality" (Davis 2006, p. 130).

But on the other hand, traditional belief in mathematical certainty is arguably a chimera. There is ample historical evidence that published work by highly respected mathematicians sometimes contains serious mistakes (Grcar 2013). For all that we know, the probability of a mistake in a very complex mathematical proof may be so high that its veracity is more uncertain than that of some of our physical theories. Whether we would be prepared to rely on a device with the second degree of computativeness will therefore depend on our standpoint in a highly contentious philosophical issue: Should the reliability of mathematical theorems be judged according to our best estimates of the probability of error, or should we uphold the traditional separation between mathematical and empirical knowledge?

The third and lowest degree of computativeness is represented by *physical events that cannot be harnessed in an input-output computational device*. As argued in the previous section, such a physical event is not, properly speaking, a computation or a computational event. However, much of the discussion on computations beyond the bounds of Turing computability has referred to such events. The following quotation is far from unrepresentative of the discussion:

> I can now state the physical version of the Church-Turing principle: 'Every finitely realizable physical system can be perfectly simulated by a universal model computing machine operating by finite means.' (Deutsch 1985, p. 99)

On this interpretation, any physical phenomenon which we cannot (currently) describe adequately with Turing computable functions would refute the physical version of the Church-Turing thesis. However, simulation and modelling are very different from computation. That we lack means for simulating a natural process certainly does not imply that we can use that process for making a calculation. We should therefore regard this type of events as (at most) raw material from which a computing device can be constructed.

9.6.3 Corroboration

The second dimension is *corroboration*, the degree to which the actual functioning of the potential computing device has been demonstrated. Here, again, three levels are appropriate. The highest degree of corroboration is an *actually working computer*. The next highest degree is a device for which there is a *proof of concept*, but still no working prototype. In such cases, the physical principles underlying the device are well-known and have been sufficiently demonstrated, but significant work remains to harness them in a practically useful device. Currently, quantum computation is an example of this. (It is expected to speed up some computations, but not to transcend Turing computability. See Hagar and Korolev 2007.) The third and lowest degree is *compatibility with some valid physical theory*,

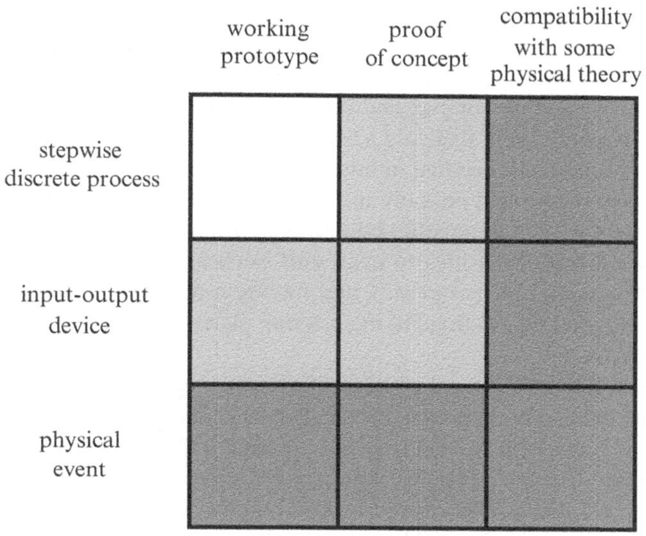

Fig. 9.3 Two dimensions for the evaluation of proposed computational devices. Computativeness is represented on the vertical and corroboration on the horizontal dimension. The white square represents actual computing devices. The light grey area represents hypothetical computing devices. The dark grey area represents vague speculations about such devices

without sufficient knowledge of the physical conditions that have to be satisfied for the device to be realizable. As was noted by Itamar Pitowsky (2007, p. 625), compatibility with a single physical theory, such as relativity theory, is "a very weak notion of physical possibility". However, since it is often referred to in these discussions we have to include it in our deliberations.

In Fig. 9.3, the two dimensions for evaluating computational devices have been combined. Let us now have a look at two of the hypothetical devices that have most frequently been discussed in the debate on whether computation transcending Turing computability is possible.

9.6.4 Two Examples

Mark Hogarth (1994) proposed what is probably the most discussed computational method intended to transcend Turing computability. Under certain conditions that are compatible with the laws of general relativity, there can be two trajectories from one point in space-time to another. One of these trajectories – we may call it the Endless Road – takes infinitely long time, whereas the other – we may call it the Shortcut – takes only finitely long time. You can then, or so it is said, start a Turing machine and send it along the Endless Road. Having done that, you take the Shortcut to the place where the two trajectories meet. There you will find out what the Turing machine achieved in infinite time. This would seem to be a nice way to solve the halting problem and a host of other problems that cannot be solved in the usual way since we cannot compute for ever and yet receive the outcome.

But does it work? Well, there are a few problems. For instance, a machine that is run for infinite time will need an infinite supply of energy. Like all other machines it will have a non-zero and possibly increasing probability of failure, which means that it is sure to malfunction within infinite time (Button 2009, pp. 778–780). There are also some additional trifles to deal with, such as locating the particular type of region in space-time (if it exists), and finding a reliable carrier that brings the machine along the Endless Road to the meeting place that was decided an infinitely long time ago.

This construction is a clear case of the lowest degree of corroboration: mere compatibility with one particular physical theory, namely, in this case, relativity theory. But if it worked, it would operate through a discrete, stepwise process, so we should place it in the rightmost square in the top row in Fig. 9.3.

Another often discussed example is based on the so-called three-body problem in classical mechanics. The problem is very simple to state: Suppose that we have three physical bodies in space. We know their masses, and we also know what positions, velocities, and directions of movement they have at a particular point in time. The three-body problem is to predict the positions of all three bodies at all future points in time. In its general form the problem has no known solution. Georg Kreisel (1923–2015) proposed that some cases of the three-body problem may lack a Turing computable solution (Kreisel 1974, p. 24).[43] But here it is essential to distinguish between the Newtonian model of mechanics, in which bodies are represented by point masses and velocity is unlimited, and the real physical world. For instance, Zhihong Xia has shown that in the corresponding five-body problem, one of the bodies could be sent off at infinite speed (Saari and Xia 1995). This will of course not happen in real life. It is an anomaly of the model. More generally speaking, mathematical representations of physical systems should not be confused with these systems themselves. Notably, "incomputability is just a property of the mathematical way of representing physical systems", not a property of the actual physical systems (Cotogno 2003, p. 186).

No practically realizable many-body constellation with uncomputable properties appears to have been presented. Furthermore, if such a constellation were to be brought about, it would not be an input-output device but just a physical phenomenon which could not be simulated by a computable function. We must therefore put this type of example in the right-most square at the bottom line of Fig. 9.3.

These are just two examples, but they are among the most promoted ones. Most of the proposals for computations beyond the limit of Turing computability fail in a very elementary respect: No other proof is given of their realizability than that

[43]Cf. Smith (2006).

they are compatible with a particular physical theory. In addition, most of them lack the input-output relationship and the settability that are characteristic of anything that anyone, outside of this debate, would call a computer. Whether more promising proposals will come up in the future is, of course, an open issue.

9.7 Conclusion

Let us summarize some of the main themes discussed in this chapter. Mathematicians in ancient civilizations were engaged in two major pursuits. One was to prove theorems, i.e. general statements about mathematical subject matter. The other was to construct algorithms, procedures for solving various classes of problems. An algorithm is a rule-bound and completely determinate procedure on symbols that can be performed “mechanically”. Algorithms were invented for simple tasks such as the basic arithmetic operations, but also for a wide variety of more advanced tasks. In a sense, algorithms are the technology of mathematics.

Beginning in ancient Greece, theorem-proving became the dominant activity in European mathematics. The construction of algorithms was a subsidiary and less esteemed activity. But at least since the thirteenth century, scholars have worked hard to find ways to reduce all form of human reasoning to simple procedures in the style of an algorithm. Major intellectuals such as Francis Bacon and Gottfried Wilhelm Leibniz were deeply engaged in these activities, and considerable efforts – including the construction of logic-friendly artificial languages – were spent on the project. But not much success was registered until scholars turned to the more limited task to encode mathematical reasoning, rather than reasoning in general, in a strictly formalized system.

In the second half of the nineteenth century, logicians developed new and more powerful logical languages. Although still insufficient for most forms of human reasoning, the new languages were sufficient to encode mathematical reasoning. Mathematical axioms and theorems could be expressed as logical formulas, without any need for natural language. Proofs could take the form of lists of such logical statements, beginning with the axioms and ending with the theorem. Each item on the list would have to follow from its predecessors according to a set of derivation rules. These rules carried instructions for simple, rule-bound symbol manipulations, just like classical algorithms.

These achievements came at a most timely occasion since two of the foremost mathematical disciplines, analysis and geometry, had severe foundational problems. The new logic offered a way to put these and other mathematical disciplines on firm foundations. The problem how to construct algorithms moved from the periphery of mathematical research to a central role in the foundations of the discipline.

In 1937 Alan Turing provided a characterization of routine symbol manipulations. Every such operation that a human can perform can be reduced to a set of very simple, truly “mechanical” operations. These operations were in fact also mechanical in another sense: They can be performed by a machine. A digital computer can do everything that a human can do routinely (and do it much faster).

Mathematical operations such as computations and proofs have important features in common with technological processes. They are *intentional*, contrary to most other physical events. If a storm brings together a pile of six pebbles with another pile that has eight pebbles, it has not performed a computation – and neither have I if I just raked together two piles of pebbles without reflecting on their numbers. A physical process that takes place independently of anyone's intentional action is neither a technological nor a mathematical process. Unfortunately, physical events involving no one's intentions have often been confounded with computations.

Furthermore, both mathematical and technological processes are required to be *reliable*. This requirement is usually stricter in mathematics than in technology. A mathematical process such as a computation has to yield the right result on each and every occasion when it is implemented according to the instructions.

A third property of interest is *lucidity*. It is an advantage if we know how a technological process works, not only *that* it works. However, this is not an absolute criterion in technology, and reliable technologies have been used without much understanding of why and how they work (Norström 2013). In mathematics, to the contrary, lucidity is considered to be an absolute criterion. We expect to have full access to computations and proofs so that we can check them. This creates problems for proposals to perform computations in physical systems that we cannot follow stepwise as we can with ordinary digital computers.

Criteria such as intentionality, reliability, and lucidity have to be taken into account in the analysis of devices that may potentially be used for computational and other mathematical purposes. Although mathematics and technology are distinctly different activities, the study of algorithms and computations has much to learn from studies of intentional human action that have been performed not least in the philosophy of technology.

References

Arkoudas, K. (2008). Computation, hypercomputation, and physical science. *Journal of Applied Logic, 6*, 461–475.

Barrett, L., & Connell, M. (2005). Jevons and the logic 'piano'. *The Rutherford Journal, 1*. http://www.rutherfordjournal.org

Blake, R. M. (1926). The paradox of temporal process. *Journal of Philosophy, 23*(24), 645–654.

Blumenthal, O. (1935). Lebensgeschichte. In D. Hilbert (Ed.), *Gesammelte Abhandlungen* (Vol. 3, pp. 388–429). Berlin: Springer.

Boole, G. (1854). *The laws of thought*. London: Walton and Maberly.

Button, T. (2009). SAD computers and two versions of the Church-Turing thesis. *British Journal for the Philosophy of Science, 60*(4), 765–792.

Chabert, J.-L. (Ed.). (1999). *A history of algorithms: From the pebble to the microchip*. New York: Springer Science & Business Media.

Church, A. (1936). An unsolvable problem of elementary number theory. *American Journal of Mathematics, 58*(2), 345–363.

Church, A. (1937). [Review of] A.M. Turing. On computable numbers, with an application to the Entscheidungsproblem. *Journal of Symbolic Logic, 2*(1), 42–43.
Cleland, C. E. (2002). On effective procedures. *Minds and Machines, 12*(2), 159–179.
Contro, W. S. (1976). Von Pasch zu Hilbert. *Archive for History of exact Sciences, 15*(3), 283–295.
Copeland, B. J. (1998). Even Turing machines can compute uncomputable functions. In C. S. Calude, J. Casti, & M. J. Dinneen (Eds.), *Unconventional models of computation* (pp. 150–164). Singapore: Springer.
Cotogno, P. (2003). Hypercomputation and the physical Church-Turing thesis. *British Journal for the Philosophy of Science, 54*(2), 181–223.
Cram, D. (1985). Universal language schemes in 17th-century Britain. *Histoire Epistémologie Langage, 7*(2), 35–44.
Davis, M. (2006). The Church-Turing thesis. Consensus and opposition. In A. Beckmann, U. Berger, B. Löwe, & J. V. Tucker (Eds.), *Logical Approaches to Computational Barriers. Second Conference on Computability in Europe, CiE 2006, Proceedings*, Swansea, June 30–July 5, 2006 (Lecture notes in computer science, Vol. 3988, pp. 125–132). Springer.
de Jong, W. R. (1986). Hobbes's logic: Language and scientific method. *History and Philosophy of Logic, 7*(2), 123–142.
Deutsch, D. (1985). Quantum theory, the Church-Turing principle and the universal quantum computer. *Proceedings of the Royal Society A, 400*, 97–117.
Edwards, C. H. (1979). *The historical development of the calculus*. New York: Springer.
Einstein, A. (1921). *Geometrie und Erfahrung*. Berlin: Springer.
Euclid. (1939). Elements. In I. Thomas (Ed. and Transl.) *Greek mathematical works, volume I. Thales to Euclid* (Loeb classical library, Vol. 335). Cambridge, MA: Harvard University Press.
Forteza, B. (1998). La influencia de Francisco Suárez sobre Thomas Hobbes. *Convivum, 11*, 40–79.
Frege, G. (1879). *Begriffsschrift, eine der aritmetischen nachgebildete Formelsprache des reinen Denkens*. Halle: Louis Nebert.
Freudenthal, H. (1966). The main trends in the foundations of geometry in the 19th century. *Studies in Logic and the Foundations of Mathematics, 44*, 613–621.
Friend, M. (2010). Boole: From calculating numbers to calculating thoughts. In F. Ferreira, B. Löwe, E. Mayordomo, & L. M. Gomes (Eds.) *Programs, proofs, processes* (Lecture notes in computer science, Vol. 6158, pp. 172–179). Berlin/Heidelberg: Springer.
Gandy, R. (1980). Church's thesis and principles for mechanisms. In J. Barwise, H. J. Keisler, & K. Kunen (Eds.), *The Kleene symposium* (pp. 123–148). Amsterdam: North-Holland Publishing Company.
Gandy, R. (1988). The confluence of ideas in 1936. In R. Herken (Ed.), *The universal Turing machine: A half-century survey* (pp. 55–111). Oxford: Oxford University Press.
Gent, I. P., & Walsh, T. (2000). Satisfiability in the year 2000. *Journal of Automated Reasoning, 24*(1–2), 1–3.
Gödel, K. (1931). Über formal unentscheidbare Sätze der Principia Mathematica und verwandter Systeme I. *Monatshefte für Mathematik und Physik, 38*, 173–198.
Gödel, K. (1958). Über eine bisher noch nicht benützte Erweiterung des finiten Standpunktes. *Dialectica, 12*, 280–287.
Gödel, K. (1964). Postscriptum. In M. Davis (Ed.), *The undecidable: Basic papers on undecidable propositions, unsolvable problems and computable functions* (pp. 71–73). Hewlett: Raven Press.
Grattan-Guinness, I. (1990). Work for the hairdressers: The production of de Prony's logarithmic and trigonometric tables. *Annals of the History of Computing, 12*(3), 177–185.
Grcar, J. F. (2013). Errors and corrections in mathematics literature. *Notices of the AMS, 60*(4), 418–425.
Grier, D. A. (2005) *When computers were human*. Princeton: Princeton University Press.
Hagar, A., & Korolev, A. (2007). Quantum hypercomputation–hype or computation? *Philosophy of Science, 74*, 347–363.
Hamilton, J. J. (1978). Hobbes's study and the Hardwick library. *Journal of the History of Philosophy, 16* (4), 445–453.

Hansson, S. O. (1985). Church's thesis as an empirical hypothesis. *International Logic Review, 16* (2), 96–101.
Hardy, G. H. (1929). Mathematical proof. *Mind, 38*(149), 1–25.
von Helmholtz, H. (1887). Zählen und Messen, erkenntnisstheoretisch betrachtet. In *Philosophische Aufsätze, Eduard Zeller zu seinem fünfzigjährigen Doctorjubiläum gewidmet* (pp. 17–52). Leipzig: Fues.
Hilbert, D. (1899). *Grundlagen der Geometrie*. Leipzig: Teubner.
Hilbert, D. (1928). *Die Grundlagen der Mathematik. Hamburger Mathematische Einzelschriften* (Heft 5). Wiesbaden: Springer.
Hilbert, D., & Ackermann, W. (1938). *Grundzüge der theoretischen Logik* (2nd ed.). Berlin: Springer.
Hobbes, T. ([1651] 1839). *The English works of Thomas Hobbes of Malmesbury* (Vol. 3, Leivathan, W. Molesworth, Ed.). London: Bohn.
Hofstadter, D. R. (1979). *Gödel, Escher, Bach: An eternal golden braid*. New York: Basic Books.
Hogarth, M. (1994). Non-Turing computers and non-Turing computability. *PSA, 1994*(1), 126–138.
Israel, D. (2002). Reflections on Gödel's and Gandy's reflections on Turing's thesis. *Minds and Machines, 12*(2), 181–201.
Jahnke, H. N., & Otte, M. (1981). Origins of the program of 'Arithmetization of Mathematics'. In H. Mehrtens, H. Bos, & I. Schneider (Eds.), *Social history of nineteenth century mathematics* (pp. 21–49). Boston: Birkhäuser.
Jones, M. L. (2014). Reflection: Reason, calculating machines, and efficient causation. In T. M. Schmalz (Ed.), *Efficient causation. A history* (pp. 192–197). New York: Oxford University Press.
Kleene, S. C. (1981). Origins of recursive function theory. *Annals of the History of Computing, 3*(1), 52–67.
Kleene, S. C. (1987). Reflections on Church's thesis. *Notre Dame Journal of Formal Logic, 28*(4), 490–498.
Klein, F. (1895). Über Arithmetisirung der Mathematik. In *Nachrichten von der Königl. Gesellschaft der Wissenschaften zu Göttingen. Geschäftliche Mitteilungen* (Heft 2, pp. 82–91).
Kreisel, G. (1974). A notion of mechanistic theory. *Synthese, 29*, 9–24.
Lardner, D. ([1834] 1989). Babbage's calculating engine. *Edinburgh Review, 59*, 263–327. Reprinted In: M. Campbell-Kelly (Ed.), *The works of Charles Babbage* (Vol. 2, pp. 118–186). London: William Pickering.
Leibniz, G. W. (1890). Undated manuscript. In C. I. Gerhardt (Ed.), *Die philosophischen Schriften von Gottfried Wilhelm Leibniz* (Vol. 7, p. 200). Berlin: Weidmannsche Buchhandlung.
Lenzen, W. (2018). Leibniz and the calculus ratiocinator. In S. O. Hansson (Ed.), *Technology and mathematics*. Cham: Springer.
Li, W. (2015). On the algorithmic tradition in the history of mathematics. In D. E. Rowe & W.-S. Horng (Eds.), *A delicate balance: Global perspectives on innovation and tradition in the history of mathematics* (pp. 321–341). Cham: Springer.
Lloyd, S. (2002). Computational capacity of the universe. *Physical Review Letters, 88*(23), 237901.
Lovelace, A. A. ([1843] 1989). "Notes by the translator", appendix in "Sketch of the analytical engine. *Scientific Memoirs, 3*, 666–731. Reprinted on In M. Campbell-Kelly (Ed.), *The works of Charles Babbage* (Vol. 3, pp. 89–179). London: William Pickering.
MacDonald Ross, G. (2007). Leibniz's debt to Hobbes. In P. Phemister & S. Brown (Eds.), *Leibniz and the English-speaking world. Activity, passivity and corporeal substances in Leibniz's philosophy* (pp. 19–33). Dordrecht: Springer.
Mates, B. (1986/1989). *The philosophy of Leibniz: Metaphysics and language*. New York: Oxford University Press.
Norström, P. (2013). Engineers' non-scientific models in technology education. *International Journal of Technology and Design Education, 23*, 377–390.
Pasch, M. (1882). *Vorlesungen über neuere Geometrie*. Leipzig: Teubner.

Petri, B., & Schappacher, N. (2007). On arithmetization. In C. Goldstein, N. Schappacher, & J. Schwermer (Eds.), *The shaping of arithmetic after C.F. Gauss's Disquisitiones Arithmeticae* (pp. 343–374). Berlin: Springer.

Piccinini, G. (2015). *Physical computation: A mechanistic account*. Oxford: Oxford University Press.

Pitowsky, I. (2007). From logic to physics: How the meaning of computation changed over time. In S. B. Cooper, B. Löwe, & A. Sorbi (Eds.), *Computation and Logic in the Real World. Third Conference on Computability in Europe, CiE 2007, Proceedings*, Siena, June 18–23, 2007 (Lecture notes in computer science, Vol. 4497, pp. 621–631). Berlin: Springer.

Poincaré, H. (1902). Du rôle de l'intuition et de la logique en mathématiques. In E. Duporcq (Ed.), *Compte-rendu du deuxième Congrès international des mathématiciens tenu à Paris du 6 au 12 août 1900* (pp. 115–130). Paris: Gauthier-Villars.

Pombo, O. (2010). Three roots for Leibniz's contribution to the computational conception of reason. In F. Ferreira, B. Löwe, E. Mayordomo, & L. M. Gomes (Eds.), *Programs, proofs, processes* (Lecture notes in computer science, Vol. 6158, pp. 352–361). Berlin/Heidelberg: Springer.

Post, E. L. (1936). Finite combinatory processes – Formulation 1. *Journal of Symbolic Logic, 1*(3), 103–105.

Ritter, J. (2000). Egyptian mathmatics. In H. Selin (Ed.), *Mathematics across cultures: The history of non-Western mathematics* (pp. 115–136). Dordrecht: Kluwer Academic.

Russell, B. (1936). The limits of Empiricism. *Proceedings of the Aristotelian Society, 36*, 131–150.

Saari, D. G., & Xia, Z. (1995). Off to infinity in finite time. *Notices of the American Mathematical Society, 42*, 538–546.

Shagrir, O. (2012). Computation, implementation, cognition. *Minds and Machines, 22*, 137–148.

Shannon, C. E. (1956). A universal Turing machine with two internal states. In C. E. Shannon & J. McCarthy (Eds.), *Automata studies* (pp. 157–165). Princeton: Princeton University Press.

Shapiro, S. (1998). Incompleteness, mechanism, and optimism. *Bulletin of Symbolic Logic, 4*(3), 273–302.

Sieg, W. (1994). Mechanical procedures and mathematical experience. In A. George (Ed.), *Mathematics and mind* (pp. 71–117). New York: Oxford University Press.

Sieg, W. (1997). Step by recursive step: Church's analysis of effective calculability. *Bulletin of Symbolic Logic, 3*(2), 154–180.

Sieg, W. (2002). Calculations by man and machine: Conceptual analysis. In W. Sieg, R. Sommer, & C. Talcott (Eds.), *Reflections on the foundations of mathematics: Essays in honor of Solomon Feferman* (Lecture notes in logic, Vol. 15, pp. 390–409). Natick: A. K. Peters.

Sieg, W. (2009). On computability. In A. D. Irvine (Ed.), *Philosophy of mathematics* (pp. 536–630). Amsterdam: Elsevier.

Singer, T. C. (1989). Hieroglyphs, real characters, and the idea of natural language in English seventeenth-century thought. *Journal of the History of Ideas, 50*(1), 49–70.

Smith, W. D. (2006). Church's thesis meets the N-body problem. *Applied Mathematics and Computation, 178*, 154–183.

Soare, R. I. (1996). Computability and recursion. *Bulletin of Symbolic Logic, 2*(3), 284–321.

Soare, R. I. (2007). Computability and incomputability. In S. B. Cooper, B. Löwe, & A. Sorbi (Eds.), *Computation and Logic in the Real World. Third Conference on Computability in Europe, CiE 2007, Proceedings*, Siena, June 18–23, 2007 (Lecture notes in computer science, Vol. 4497, pp. 705–715). Berlin: Springer.

Swade, D. (2010). Pioneer profile: Ada Lovelace. *Resurrection. The Bulletin of the Computer Conservation Society, 53* 2010/11.

Swade, D. (2011a). Calculating engines: Machines, mathematics, and misconceptions. In R. Flood, A. Rice, & R. Wilson (Eds.), *Mathematics in Victorian Britain* (pp. 239–259). Oxford: Oxford University Press.

Swade, D. (2011b). Pre-electronic computing. In C. B. Jones & J. L. Lloyd *Dependable and historic computing. Essays dedicated to Brian Randell on the occasion of his 75th birthday* (Lecture notes in computer science, Vol. 6875, pp. 58–83). Berlin: Springer.

Swade, D. (2018). Mathematics and mechanical computation. In S. O. Hansson (Ed.), *Technology and mathematics*. Cham: Springer.

Timpson, C. G. (2007). What is the lesson of quantum computing? In S. B. Cooper, B. Löwe, & A. Sorbi (Eds.) *Computation and Logic in the Real World. Third Conference on Computability in Europe, CiE 2007, Proceedings*, Siena, June 18–23, 2007 (Lecture notes in computer science, Vol. 4497, pp. 739–741). Berlin: Springer.

Turing, A. (1937a). On computable numbers, with an application to the Entscheidungsproblem. *Proceedings of the London Mathematical Society, 42*, 230–265.

Turing, A. (1937b). On computable numbers, with an application to the Entscheidungsproblem. A Correction. *Proceedings of the London Mathematical Society, 43*, 544–546.

Turing, A. ([1947] 1986). Lecture to the London mathematical society on 20 February 1947. In B. E. Carpenter & R. W. Doran (Eds.), *A.M. Turing's ACE report of 1946 and other papers* (pp. 106–124). Cambridge: MIT Press.

Turing, A. ([1948] 2004). Intelligent machinery. In B. Jack Copeland (Ed.), *The essential Turing: Seminal writings in computing, logic, philosophy, artificial intelligence, and artificial life plus the secrets of Enigma* (pp. 410–432). Oxford: Oxford University Press.

Turing, A. (1950). Computing machinery and intelligence. *Mind, 59*, 433–460.

Uckelman, S. L. (2010). Computing with concepts, computing with numbers: Llull, Leibniz, and Boole. In F. Ferreira, B. Löwe, E. Mayordomo, & L. M. Gomes (Eds.), *Programs, proofs, processes* (Lecture notes in computer science, Vol. 6158, pp. 427–437). Berlin/Heidelberg: Springer.

Widmaier, R. (1986). Korrespondenten von GW Leibniz: 8. Damaris Masham, geb. Cudworth geb. 18. Januar 1658 in Cambridge-gest. 20. April 1708 in Oates. *Studia Leibnitiana, 18*(2), 211–227.

Yee, A. J., & Kondo, S. (2016). 12.1 Trillion Digits of Pi. And we're out of disk space.... http://www.numberworld.org/misc__runs/pi-12t/. Downloaded June 26, 2017.

Chapter 10
On Explaining Non-dynamically the Quantum Correlations Via Quantum Information Theory: What It Takes

Mauro Dorato and Laura Felline

Abstract In this paper, we argue that quantum information theory can provide a kind of non-causal explanation ("causal account" here stands quite generally both for dynamical and for mechanistic account of causal explanation) of quantum entanglement. However, such an explanation *per se* does not rule out the possibility of a dynamical explanation of the quantum correlations, to be given in terms of some interpretations (or alternative formulations) of quantum theory. In order to strengthen the claim that it can provide an explanation of the quantum correlations, quantum information theory should inquire into the possibility that the quantum correlations could be treated as "natural", that is, as phenomena that are physically fundamental. As such, they would admit *only* a structural explanation, similarly to what happened in crucial revolutionary episodes in the history of physics.

10.1 Introduction

On the wake of the remarkable success enjoyed by quantum theory in its application to computation theory and cryptography, many philosophers and physicists have recently explored the idea that information can also play a privileged foundational role. Such thesis has been articulated in a variety of ways, starting from the claim that an analysis of the information-processing capabilities of quantum systems can provide a deeper understanding of some of the most puzzling quantum phenomena, to the claim that Quantum Information Theory is the right framework for the formulation of quantum theory, to the claim that such a theory is about quantum information. With respect to the wave-particle duality, for instance, Bub has argued that: "quantum mechanics [ought to be regarded] as a theory about the representation and manipulation of information constrained by the possibilities and

M. Dorato (✉) · L. Felline
Department of Philosophy, Communication and Media studies, Rome, Italy
e-mail: mauro.dorato@uniroma3.it; lfelline@uniroma3.it

S. O. Hansson (ed.), *Technology and Mathematics*, Philosophy of Engineering and Technology 30, https://doi.org/10.1007/978-3-319-93779-3_10

impossibilities of information-transfer in our world (a fundamental change in the aim of physics), rather than a theory about the behavior of non-classical waves and particles." (Bub 2005, 542).

Bub, together with many other philosophers and physicists, relied in particular on axiomatic reconstructions of quantum theory in terms of information-theoretic principles as the right framework for the reformulation of quantum theory in terms of information. In view of the many progresses axiomatic reconstructions of quantum theory achieved in the study of nonlocality, in fact, it is natural to raise the question whether such theories can somehow explain the kind of nonlocality displayed by the quantum correlations. In this paper, we investigate this question by discussing and evaluating a remarkable theorem by Clifton, Bub and Halvorson (2003) (CBH henceforth).

In addition, we argue that, on the one hand, axiomatic reconstructions of quantum theory *can* provide a genuine explanation of one aspect of nonlocality, in virtue of its counterfactual dependence on the core principles of quantum theory. On the other hand, however, this explanation *per se* does not account for the occurrence of quantum correlations. As we will show, explaining quantum correlations in terms of quantum information theory would require a structural explanation (Dorato and Felline 2011), which rules out the possibility of other causal or dynamical accounts of the quantum correlations. A fully structural explanation of nonlocality could therefore only be achieved if the quantum correlations turned out to be fundamental or "natural", in the sense of being non-caused or non-dynamically explainable. In this sense of natural, vertical motion was natural in Aristotelian physics, inertia became natural in Newtonian physics, length contractions and free fall became natural in special and general relativity respectively, *in virtue of a replacement of previous dynamical explanations by explanations given in terms of a new spatiotemporal structure providing a structural explanation* (Dorato 2014). Our conclusion will be that (at least so far) axiomatic reconstructions of quantum theory cannot show that quantum correlations are fundamental or natural in this sense.

In the first section of the paper we analyze axiomatic reconstructions of quantum theory's account of nonlocality in terms of CBH's characterization theorem. In the second section, we illustrate the sense in which CBH's approach can provide an explanation of nonlocal entangled states by showing how they are counterfactually dependent on the core principles of quantum theory. In Sect. 10.3 we introduce the notion of structural explanations as explanations that render a causal/dynamical account superfluous. We finally evaluate CBH's claim that axiomatic reconstructions of quantum theory makes a dynamical interpretation of quantum theory explanatorily irrelevant by translating it into the question whether their explanations can be equated to structural explanations in the sense discussed in the first two sections.

10.2 Quantum Entanglement and Axiomatic Reconstructions of Quantum Theory

Axiomatic reconstructions aim at finding few physical principles from which it is possible to derive the Hilbert structure of quantum theory:

> [t]heorems and major results of physical theory are formally derived from simpler mathematical assumptions. These assumptions or axioms, in turn, appear as a representation in the formal language of a set of physical principles. (Grinbaum 2007)

By default, the principles at the basis of axiomatic reconstructions of a physical theory do not have a foundational role within the theory, nor are they required to be ontologically prior to its theorems. Their only role is to provide an axiomatic basis for the deduction of the theory: "nothing can be generally said about their ontological content or the ontic commitments that arise from these principles." (Grinbaum 2007, 391). The same, of course, is valid in the specific case of axiomatic reconstructions of quantum theory and its information-theoretic principles about the possibilities and impossibilities of information transfer. Within quantum information theory in general, and in axiomatic reconstructions of quantum theory specifically, information is meant in the physical sense, a notion of quantity of information cashed out in terms of the resources required to transmit messages – measured classically by Shannon entropy or, in quantum theory, by Von Neumann entropy.

Some of the advocates of axiomatic reconstructions of quantum theory interpret its success as evidence for the fact that information technologies and therefore information-theoretic principles possess a central role also in the *ontology* of quantum theory. In different forms, this position is held by some of the philosophers/physicists working in this field, defending the claim that quantum theory is about the epistemological state of the observer or about the claim that the world is, at its bottom, just information, while offering different ontological interpretations of what is meant by information. Timpson (2013) calls such contributions to the foundations of quantum theory the 'direct' approaches to quantum information theory and contrasts it to the 'indirect' approach, which, more humbly, aims at learning something useful about the structure or axiomatics of quantum theory by reflecting on quantum information-theoretic phenomena. In this paper, we shall mainly discuss the indirect approach.

10.2.1 CBH Axiomatic Reconstruction of Quantum Theory

CBH prove a theorem that characterizes quantum theory by proposing three "fundamental information theoretic laws of nature" (CBH 2003, 1562) or, less ambitiously, three principles concerning the impossibility of information transfer:

1. *No superluminal information transfer via measurement.* It states that merely performing a local (non-selective) operation[1] on a system *A* cannot convey any information to a physically distinct system.
2. *No broadcasting.* This constraint states the impossibility of perfectly broadcasting the information contained in an unknown physical state. Broadcasting is a generalization of the process of cloning which, in turn, is a process that starts with a system in any arbitrary state $|\alpha\rangle$ and ends up with two systems, each in the state $|\alpha\rangle$. While cloning applies only to pure states, broadcasting generalizes also to mixed states. In quantum mechanics, broadcasting is possible for a set of states ρ_i iff they are commuting.
3. *No bit-commitment.* The bit commitment is a cryptographic protocol in which one party, Alice, supplies an encoded bit to a second party, Bob, as a warrant for her commitment to the value 0 or 1. The information available in the encoding should be insufficient for Bob to ascertain the value of the bit at the initial commitment stage. However, such information should be sufficient, together with further information supplied by Alice at a later stage – the 'revelation stage', when she is supposed to "open" the commitment by revealing the value of the bit – for Bob to be convinced that the protocol does not allow Alice to cheat by encoding the bit in a way that leaves her free to reveal either 0 or 1 at will. As an illustration of how this cheating strategy should work, consider this example from (Timpson 2013):

> Consider a spin-1/2 system: a 50/50 mixture of spin-up and spin-down in the z-direction is indistinguishable from a 50/50 mixture of spin-up and spin-down in the x-direction—both give rise to the maximally mixed density operator ½ **1**. Alice might associate the first type of preparation with a 0 commitment and the second with a 1 commitment. Bob, when presented with a system thus prepared will not be able to determine which procedure was used. Alice also needs to keep a record of which preparation procedure she employed, though, to form part of the evidence with which she will convince Bob of her probity at the revelation stage. Thus, for a 0 commitment, Alice could prepare a classically correlated state of the form:
>
> 0 commitment: (1)
>
> whilst for a 1 commitment, she could prepare a state
>
> 1 commitment: (2)
>
> System 2 is then sent to Bob.
>
> At the revelation stage, Alice declares which bit value she committed to, and hence which preparation procedure she used. The protocol then proceeds in the following way: If she committed to 0, Alice and Bob both perform σ_z measurements and Alice declares the result she obtains, which should be perfectly correlated with Bob's result, if she really did prepare state ρ_0. Similarly, if she committed to 1, Alice and Bob both perform σ_x measurements and Alice declares her result, which again should be perfectly correlated with Bob's result, if she really prepared state ρ_1. If the results reported by Alice and obtained by Bob don't correlate, then Bob knows that Alice is trying to mislead him. The trouble with this otherwise attractive protocol is that Alice is able to cheat freely by making use of what is known as an EPR cheating strategy. Thus, rather than preparing one of the states ρ_0 or ρ_1 at the commitment stage, Alice can instead

[1] Selective measurements operations are here obviously not considered, given that in such operations the statistics in general changes due to a change of the ensemble under study.

> prepare an entangled state, such as the Bell state $|\varphi^+\rangle_{12}$. The reduced density operator for Bob's system will still be ½ **1**, but Alice can now simply wait until the revelation stage to perform a suitable measurement on her half of the entangled pair and prepare Bob's system at a distance in whichever of the two different mixtures she chooses (pp. 212–213).[2]

By asserting the impossibility of such a secure cryptographic protocol, the no bit-commitment principle assures the stability of entangled states also in macroscopic or nonlocal processes and forbids that entangled states decay in macroscopic or nonlocal states. Schrödinger contemplated the possibility of such a theory in (1936). Within this kind of theory (which, following Timpson, we shall call *Schrödinger-type theory*), the EPR cheating strategy would not be applicable (given that entangled states would not be stable enough) and the secure bit-commitment would be in general possible.

The CBH Characterization Theorem, therefore, demonstrates that the basic kinematic features of a quantum-theoretic description of physical systems (i.e. noncommutativity and entanglement) can be derived from the three information-theoretic constraints.

The formal model utilized by quantum information theory in order to derive such a result is the C*-algebra. This is an abstract representation of the algebra of observables which can represent both classical (particle and field) and quantum mechanical theories.

As far as quantum mechanics is concerned, the algebra $\mathcal{B}(\mathcal{H})$ of all bounded operators on a Hilbert space $\mathcal{H}$ is a C*-algebra. A quantum system A is therefore represented by a C*-algebra $\mathcal{A}$ and a composite system A+B is represented by the C*-algebra $\mathcal{A} \vee \mathcal{B}$. Observables are represented by self-adjoint elements of the algebra and a quantum state is an expectation-valued functional over these observables, with the constraint that two systems A and B are physically distinct when "any state of $\mathcal{A}$ is compatible with any state of $\mathcal{B}$, i.e., for any state ρ_A of $\mathcal{A}$ and for any state ρ_B of $\mathcal{B}$, there is a state ρ of $\mathcal{A} \vee \mathcal{B}$ such that $\rho|_{\mathbf{A}} = \rho_A$ and $\rho|_{\mathbf{B}} = \rho_B$" (Bub 2004 p. 5).

The CBH theorem proves that quantum theory – which they take to be a theory formulated in C_*-algebraic terms in which the algebras of observables pertaining to distinct systems commute, the algebra of observables on an individual system is noncommutative, and which allows space-like separated systems to be in entangled states – can be derived from the assumption of the three information-theoretic constraints. More exactly, it is demonstrated that (see e.g. Bub 2004, pp. 246–247): (1) the commutativity of distinct algebras follows from the first constraint (no superluminal information transfer via measurement) it follows the commutativity of distinct algebras: if the observables of distinct algebras commute, then the no superluminal information transfer via measurement constraint holds (the converse result is proved in (Halvorson 2003)). Commutativity of distinct algebras is meant to represent no-signalling; (2) cloning is always allowed by classical (i.e.

[2]In the following we follow closely CBH's and Timpson's treatments.

commutative) theories and, if any two states can be (perfectly) broadcast, then the algebra is commutative. Therefore, from the second constraint, no broadcasting, follows the noncommutativity of individual algebras. (3) if $\mathcal{A}$ and $\mathcal{B}$ represent two quantum systems (i.e., if they are individually noncommutative and mutually commuting), there are nonlocal entangled states on the C*-algebra $\mathcal{A} \vee \mathcal{B}$ they generate.

However, Bub argues, we still cannot identify quantum theories with the class of noncommutative C*-algebras. It is at this point that the third information-theoretic constraint, the no unconditionally secure bit-commitment, is introduced, 'to guarantee entanglement maintenance over distance'.

It has been argued that the role of no bit-commitment is in this sense somewhat ambiguous (see e.g. Timpson 2013). The first suggested motivation for the need of the no bit-commitment is in fact the following: in the account so far provided, the existence of nonlocal entangled states follows directly from the choice of the C*-algebra and from its formal properties. On the other hand, "in an information-theoretic characterization of quantum theory, the fact that entangled states can be instantiated nonlocally, should be shown to follow from some information-theoretic principle." (Bub 2004, p. 6). It seems, in other words, that the role of the no bit-commitment is to provide an information-theoretic ground, in the context of C*-algebra, to the origin of entanglement, which, otherwise, would be a mere consequence of the choice of the mathematical machinery used by the theory. This suggestion is made clearer in (Clifton et al. 2003):

> So, at least mathematically, the presence of nonlocal entangled states in the formalism is guaranteed, once we know that the algebras of observables are nonabelian. What does not follow is that these states actually occur in nature. For example, even though Hilbert space quantum mechanics allows for paraparticle states, such states are not observed in nature. In terms of our program, in order to show that entangled states are actually instantiated, and—contra Schrödinger—instantiated nonlocally, we need to derive this from some information-theoretic principle. This is the role of the 'no bit-commitment' constraint. (p. 10)

But if the mathematical structure of reference is a C*-algebra, it would seem that the function of the third principle would be to reassess the occurrence of entangled states. But, as Timpson argues, the idea of positing a principle in order to "rule in" something which is already part of the theory is quite peculiar: "ruling states *in* rather than *out* by axiom seems a funny game. Indeed, once we start thinking that some states may need to be ruled *in* by axiom then where would it all end? Perhaps we would ultimately need a separate axiom to rule in every state, and that can't be right." (Timpson 2013, p. 206) On the other hand, given that the problem seems to rise from the existence of other weaker algebras where entanglement could not follow from the first two principles, the no bit-commitment could be seen as a constraint on this more general context. But in this case, it is still to be proved that no bit-commitment would succeed, given that so far there is no proof that it would guarantee in this more general context the stability of nonlocal entanglement.

In other occasions Bub suggests a slightly different role for the no bit-commitment. We have already seen that the no bit-commitment is incompatible with Schrödinger-type theories that, even if not in violation of the no information via measurement and no broadcasting principles, eliminate nonlocal entanglement by assuming, for instance, its decay with distance. About this, Timpson argues that also this argument is anyway dubious, since "a Schrödinger-type theory is only an option in the sense that we could arrive at such a theory by imposing further requirements to eliminate the entangled states that would otherwise occur naturally in the theory's state space." (Timpson 2013, p. 207).

In (Hagar and Hemmo 2006, n.12 and 19) the no bit-commitment is interpreted as a dynamical constraint, meant to rule out dynamical theories (such as GRW) which, while coherent with the first two principles, implies a decay of entanglement at the macroscopic level. Timpson also considers this option (2004, Ch. 9) and rejects it as in evident contrast with CBH's explicit ambitions of being concerned only with the "kinematic features of a quantum-theoretic description of physical systems" (Bub 2004, p. 1). Anyway, as noticed by Hagar and Hemmo, also in this interpretation the no bit-commitment has a controversial status. The problem lies in the fact that if the no bit-commitment applies merely to cryptographic procedures where the entangled states utilized are states of microsystems, then it is redundant (since also GRW complies with it); otherwise (i.e., if it also applies to situations where the entanglement concern also massive systems) it would be unwarranted, and quantum information theory would end up being a no-collapse theory. To see why, recall the previous illustration of the bit-commitment procedure. In standard quantum mechanics, the no bit-commitment holds since entanglement is stable also at a distance, so that Alice can always cheat by sending to Bob a particle in entangled state. Given the well-known result of the Aspect experiment, we know that in such a situation a Schrödinger type theory (postulating a decay of the entangled state) is not empirically adequate. This is the reason why, in such a case, the no bit-commitment is justified. On the other hand, in this kind of situation the no bit-commitment is respected also by the GRW theory, since the entangled state is stable at microscopic scale (also when the particles are far). GRW violates the no bit-commitment just in case the entangled state concerns a massive system, since in this case the entanglement decays very quickly. In other words, a secure bit-commitment shall always be possible in principle via a set up that requires Alice to encode her commitment in the position state of a massive enough system (Hagar and Hemmo 2006, §3.2). But in this case, also standard quantum mechanics implies an effective decay of the system (and therefore an effective violation of the no bit-commitment), due to decoherence. And at the moment there is no empirical ground for deciding which of the two approaches (GRW's collapsed state or standard quantum theory with decoherence) is the correct one. But then it follows that if the no bit-commitment is meant to ensure the stability of entanglement also at a distance, then it is uninformative; if it is meant to ensure the stability of entanglement also for massive bodies, then it is not supported by empirical grounds. (Hagar and Hemmo 2006, §3.2).

Finally, there is another feature that seems to testify against the interpretation of the no bit-commitment as a dynamical principle. So far, we have utilized Hagar and Hemmo's treatment of the no bit-commitment in order to show how, even if taken as a dynamical principle, it is not able to provide an information-theoretic ground to the occurrence of entangled states. But another obvious consequence of taking the no bit-commitment seriously in virtue of its active role in ruling out dynamical reduction models *à la* GRW would mean to forbid collapse not only for nonlocal entangled states, but also in massive bodies. In other words, what we would end up with would be a genuine no-collapse theory, which is clearly not what CBH had in mind with their third information-theoretic principle.

In a word, the conclusions that we have reached with respect to the effectiveness of the no bit principle in providing an information-theoretic ground to entanglement are as follows: as a kinematic principle, the no bit-commitment has a dubious role: either it is redundant (in the context of the C*-algebra); or it is unconvincing. As a dynamical principle, either it applies merely to cryptographic procedures where the entangled states utilized are states of microsystems, in which case it is, again, redundant, or it also applies to situations where massive systems are concerned, in which case it would be unfounded, and it would make quantum information theory correspond to a no-collapse theory.

Given the dubious role of the no bit-commitment principle, and for reasons of illustrative simplicity and clarity, in the rest of this paper we follow Timpson's analysis and consider non-locality as following from the first two principles only.

In sum, the fundamental three theses defended by Bub on the significance of the CBH theorem are as follows:

- A quantum theory is best understood as a theory about the possibilities and impossibilities of information transfer, as opposed to a theory about the mechanics of non-classical waves or particles.
- Given the information-theoretic constraints, any "mechanical" theory of quantum phenomena that tries to offer a dynamical account of the measuring instruments that are responsible for the observed phenomena must be empirically equivalent to a quantum theory.
- Assuming that the information-theoretic constraints are in fact satisfied in our world, no mechanical theory of quantum phenomena that includes an account of measurement interactions can be acceptable, and the appropriate aim of physics at the fundamental level then becomes the representation and manipulation of information. (Bub 2004)

10.3 How Do Axiomatic Reconstructions of Quantum Theory Explain?

We are now in the position of presenting CBH's explanation of non-locality and to show that this derivation provides the basis for an explanation of (an aspect of) quantum nonlocality, i.e. of the existence of nonlocal entangled states.[3] Following the results illustrated in the previous section, the explanation of non-locality follows three steps:

[3]Part of the results of this section are exposed in more details in (Felline 2016).

1. The 'no superluminal information transfer' principle entails the commutativity of distinct algebras: if the observables of distinct algebras commute, then the 'no-superluminal information transfer' constraint holds. Commutativity of distinct algebras is meant to represent 'no signalling'. A theory violating this principle would display strong non-locality and superluminal signalling;
2. The 'no broadcasting' principle entails the non-commutativity of individual algebras. Cloning is always allowed by classical theories and if any two states can be (perfectly) broadcast, then the algebra is commutative. A theory violating this principle is therefore a classical theory with commutative individual algebras;
3. If **A** and **B** are two individually non-commutative sub-algebras but mutually commuting algebras, there are nonlocal entangled states on the C*-algebra **A V B** that they generate.

It has been sometimes argued that axiomatic reconstructions of quantum theorys explain by providing Deductive-Nomological explanations or explanations by unifications, as they unify the laws of quantum theory under the few principles that play the role of axioms. For instance, Flores (1999) characterizes explanations in theories of principle[4] (and therefore in axiomatic reconstructions of quantum theory) as providing explanations by unification.

As a first reaction to these claims, we must stress that a logical derivation of P from laws of nature is not always explanatory in science. This is exactly why the conjecture that principle theories provide Deductive-Nomological explanations allows Brown and Pooley (2006) to conclude that Special Relativity, as a principle theory, lacks explanatory power (Felline 2011). In the same way, the claim that axiomatic reconstructions of quantum theorys provide Deductive-Nomological explanations hides the real explanatory contribution of these theories. On the other hand, it is clear that the explanatory power of axiomatic reconstructions of quantum theory is deeply entangled with the highly unifying power of the theories that they aim to achieve; however, according to the view we propose, unification is one virtue of explanations rather than its essence. As we will argue, the most distinctive contribution of axiomatic reconstructions of quantum theory in understanding the quantum world can be captured neither by the Deductive-Nomological, nor by the unificationist approaches.

Felline (2016) proposes an alternative account of explanation in axiomatic reconstructions of quantum theory and in particular of quantum entanglement. In order to account for the explanation of quantum entanglement, she borrows from Mark Steiner's account of explanation in mathematics. Steiner's central idea

[4]"We can distinguish various kinds of theories in physics. Most of them are constructive. They attempt to build up a picture of the more complex phenomena out of the materials of a relativity simple formal scheme from which they start out. Along with this most important class of theories there exists a second, which I will call 'principle-theories.' These employ the analytic, not synthetic, method. The elements which form their basis and starting-point are not hypothetically constructed but empirically discovered ones, general characteristics of natural processes, principles that give rise to mathematically formulated criteria which the separate processes or the theoretical representations of them have to satisfy" (Einstein 1919, p. 228).

is that "to explain the behavior of an entity, one deduces its behavior from its *characterizing property*,[5] i.e. a "property unique to a given entity or structure within a family or domain of such entities or structures." (Steiner 1978, p. 143) According to Steiner's account, an explanatory proof

> makes reference to a characterizing property of an entity or structure mentioned in the theorem, such that from the proof it is evident that the result depends on the property. It must be evident, that is, that if I substitute in the proof a different object of the same domain, the theorem collapses; more, I should be able to see as I vary the object how the theorem changes in response. In effect, then, explanation is not simply a relation between a proof and a theorem; rather, a relation between an array of proofs and an array of theorems, where the proofs are obtained from one another by the 'deformation' prescribed above. (Steiner 1978, 144)

According to Felline's account, the definition of a "characterizing property" applies also to the principles of the axiomatic reconstructions of quantum theory. The principles' function, in fact, is to "isolate" quantum theory from a family of other physical theories representable by *C**algebra. More precisely, the CBH's principles isolate quantum theory from the family of all theories that can be represented with a *C**-algebra.

Moreover, CBH's explanation of nonlocality consists, as in Steiner's account, in the derivation of the explanandum from the principles 'no superluminal signals' and 'no broadcasting'.

Third, a crucial part of CBH explanation consists in showing that, and how, the theorem/explanandum changes when the characterizing property is changed. CBH show that if the no broadcasting condition is dropped, then one has a classical phase space theory while, if the no-superluminal signals principle is dropped, one has a theory where distinct and distant physical systems are not kinematically independent, i.e. a strongly nonlocal theory.

Let us scrutinize more in depth the epistemic content of this kind of explanation. A central concern of axiomatic reconstructions of quantum theory is the question "How does the quantum world differ from the classical one?".

Many physicists have faced this question and provided their answer (the discretization of the energy levels of oscillators for Planck, the discretization of angular momentum and the Principle of Complementarity according to Bohr, while for de Broglie the characterizing feature of quantum theory was the wave nature of matter, and for Schrödinger it was entanglement, for Dirac it is superposition and so on). Axiomatic reconstructions of quantum theory addresses the question 'how does the quantum differ from the classical?' with a new perspective, i.e. with the axiomatization approach. Thanks to this formal approach, the explanations just seen show how the mathematical structure of the theory constrains the kind of properties that are admissible within the theory and that the explanandum is a consequence of such constraints. In order to understand why this kind of explanation is especially powerful in axiomatic reconstructions of quantum theory, it is first of all useful to

[5]The objection that this is really a form of Hempelian derivation will be dealt with below.

resort in more details to Einstein's (1919) well-known dichotomy between theories of principle and constructive theories already mentioned in note 1.

As illustrated by the well-known case of the special theory of relativity, theories of principle often provide a more general picture of the structural features of the world. This is due in general to their analytic method, which starts from general phenomenological laws, leading to conclusions that are both independent of the details of the constituents of the physical systems under study, and of wider, more general application.

In the same way, according to CBH, the three information-theoretic principles "constrain the law-like behavior of physical systems" (Clifton et al. 2003, p. 24) and quantum theory "can now be seen as reflecting the constraints imposed on the theoretical representations of physical processes by these principles" (pp. 24–25).

With respect to this constraining function, the notion of Shannon information (or von Neumann entropy) is especially useful, as it allows to abstract away from assumptions about the constitution of bodies and the dynamical details underlying the occurrence of the correlations. By singling out the axiomatic structure of the theory from the details that a constructive theory would require, axiomatic reconstructions of quantum theorys make explicit the connections and relations of dependence between the elements of the theory, and between quantum theory and the rest of our scientific theories. For instance, according to CBH reconstruction, nonlocal entanglement depends on non-commutativity and kinematic independence.

To sum up, axiomatic reconstructions of quantum theory searches for an answer to the above question: "how does the quantum world differ from the classical one?" in the different constraints on quantum and classical information processing. The information-theoretic approach invites us to look at physical systems as tools for the transfer and manipulation of information, and the difference between quantum and classical systems, more specifically, lies in the different resources that quantum systems provide for information processing tasks.

Before we conclude this section, let us notice that the explanations provided by reconstructions of quantum theory in terms of information-theoretic principles might be seen as a particular case of 'what-if-things-had-been-different' explanations, with a counterfactual dependence structure that is made explicit by the deformation of the principles and the derivation of its consequences. This fact suggests that the model of explanation presented here naturally fits those kinds of general accounts of explanation that attribute a central role to the counterfactual dependence between explanans and explanandum (See Morrison 1999, but also Reutlinger 2012; Pincock 2014) and that include as special cases also causal or mechanistic theories of explanations that attribute a central role to counterfactual dependence in their definition of "mechanism" (Craver 2007; Glennan 2010).

Finally, notice that such a counterfactual approach is distinct from a Deductive-Nomological model of explanation. Of course, logical/mathematical derivations are involved also in counterfactual explanations and are therefore necessary, but they are not sufficient to grasp the essence of this approach. There are at least two distinguishing elements: the Deductive-Nomological model neglects the specific role of laws as characterizing properties, and the fact that the explanation is not constituted by one, but by an array of derivations, which provide the counterfactual information referred to above.

10.4 Are Explanations in Axiomatic Reconstructions of Quantum Theory Structural?

In this section, we are going to see what it takes for an axiomatic reconstructions of quantum theory to provide a complete explanation of quantum non-locality – i.e. not only an explanation of the existence of entangled quantum states, but also of the occurrence of non-local quantum correlations. In order to investigate this issue, we first introduce the notion of structural explanation and show that an information-theoretic explanation of quantum phenomena must belong to this variety; then we argue that the explanations provided by axiomatic reconstructions of quantum theory per se do not yet provide structural explanations of quantum correlations and discuss what additional assumptions will be required to provide such a kind of explanation.

In the literature, we find other kinds of non-causal explanations of physical phenomena, namely structural explanation (Hughes 1989; Bokulich 2009; Clifton 1998; Dorato and Felline 2011). As a first approximation, a structural explanation is an explanation of a physical phenomenon in terms of its "representative" in the mathematical model. This representative is linked by a set of relations to other members of the model, and the phenomenon is an exemplification of the network. The often discussed, paradigmatic example of a structural explanation is the geometrical explanation of length-contraction in special relativity. Not only is such an explanation independent of metaphysical assumptions about the nature of Minkowski's spacetime – and therefore of the substantivalism/relationism dispute – but also of any assumption about the mechanical details and physical composition of the systems underlying the phenomena to be explained (Lange 2013a, b; Janssen 2002a, b, 2009). A structural explanation, if successful, renders a dynamical[6] account of length contraction not just superfluous, but also wrongheaded.

Dorato and Felline (2011) have argued that the formal structure of quantum theory provides a structural explanation of quantum nonlocality in terms of the Hilbert structure that is used to represent quantum states.

What distinguishes a structural explanation of quantum nonlocality from other non-causal (mathematical) explanations of the kind given by axiomatic reconstructions of quantum theory is that the former consists in showing that the explanandum (the quantum correlations in our case) could not be possibly explained by any causal explanation because it is part of the "causally fundamental" structure of the world. By "causally fundamental" structure of the world we intend to refer to phenomena

[6]See note 1.

or relations thereof that are to be regarded as "natural", i.e., as such that cannot be in turn accounted for, or inferred by, the behavior or laws of "underlying" entities.

Some historical considerations may help us to formulate an alternative account of a non-causal explanation, in which the quantum correlations could be regarded as natural in this sense. In fact, there exist often neglected but deep analogies between the discovery of quantum correlations and previous major transitions that characterized the history of physics.

As is well-known, in Kuhn's view, scientific revolutions are accompanied by radical shifts in the kind of phenomena that are regarded as in need of an explanation (Kuhn 1970, p. 104). In our case, the ongoing debates in the interpretation of quantum theory could be usefully described in terms of the different fundamental commitments about what one should take as explanatory primitive and what instead should be explained. This shift may apply both to the measurement problem and to nonlocal quantum correlations, the two major conceptual innovations with respect to classical physics. If these correlations were to be regarded as natural in virtue of their fundamentality *vis à vis* the quantum world – rather than an explanandum to be accounted for by dynamical laws – they would become an explanans, i.e. the fundamental ground for explaining why the macroscopic world does not appear to be entangled, something that classical physics had been taking for granted!

The same radical switch of explanatory perspective took place when inertial motion replaced previous dynamical explanations of Aristotle's "violent motions", when Einstein's kinematical treatment of the relativistic effects replaced previous attempts to derive them from the Lorentz covariance of dynamical laws governing the inner behavior of rods and clocks, and when Einstein's postulation of a curved spacetime superseded previous explanations of gravity involving a force. In fact, while one of the main problems of Aristotelian physics was to give some sort of dynamical account that could explain why bodies continue in their motion despite the absence of a "motor", in Newtonian physics the continuation of motion became the natural, primitive state of bodies and forces have been introduced to explain deviation from rectilinear, inertial motion. Later, the introduction of affine spaces codified in a geometrically precise way the new role given by the principle of inertia to rectilinear motion. Likewise, dynamical attempts to explain contractions and dilations were superseded by geometrical explanations in terms of Minkowski fourdimensional geometry and in the case of general relativity the gravitational force was geometrized away thanks to the introduction of Riemannian manifold with a variable curvature. Explanations like these count as structural because they show how the explanandum is the manifestation of a fundamental structure of the world that is accounted for only by the different geometrical structures defining the appropriate spacetimes.[7]

If this pattern of scientific change could also be extended to the nonlocal quantum correlations, what kind of structural explanation could we advance in order to replace actual or possible causal models of nonlocal correlations and treat them as we now treat inertia, the speed of light and free fall? Can we regard the axiomatic

[7]For more details, see Dorato (2014).

reconstructions of quantum theory explanation of entanglement depicted above as a form of structural explanation that can also explain the occurrence of quantum correlations?

In order to answer this question, one should keep in mind that the most puzzling issues related to quantum phenomena emerge when the attempt is made to account for *how* such phenomena occur. The same applies to quantum nonlocality, when the latter is understood as the occurrence of nonlocal correlations: how do such correlations occur? Or, in other words, what are the entities and processes that "underlie" or produce their occurrence? Notice that the traditional, minimal interpretation of Shannon (and von Neumann), in which information is a measure of the amount of correlation between systems, does not rule out such a (possibly causal) account, which is therefore also compatible with the axiomatic reconstructions of quantum theory.

To defend the stronger claim that quantum information theory's explanation of non-locality is the only game in town, a further argument is required. For instance, information immaterialism (Zeilinger 1999, 2005) adds to the claim that the quantum state is about quantum information the radical metaphysical view that information (the "immaterial") is the fundamental subject matter of physics.

Under this assumption, other mechanical explanations of quantum phenomena are ruled out. In fact, in this case, the information-theoretic structure is not an epistemic tool for measuring the amount of correlation between unknown systems regarded as black boxes, but *is all there is*. In virtue of its fundamentality, it is the complete description of the world, since in principle there is no "reality" underlying these correlations. Such a description would therefore also automatically provide the basis for a structural explanation of quantum phenomena.

Admittedly, the ontological picture behind information immaterialism is controversial to say the least (e.g. Timpson 2010, 2013). While here we cannot discuss it in details, this immaterial ontology may not be so lethal to the explanatory power of quantum information theory. Structural explanations are independent of the ontology "underlying" the explanandum, and this independence includes as a special case the "software-without-hardware" ontology of information immaterialism: the explanatorily relevant facts here are part of the mathematical properties of the structure of which the explanandum is a manifestation. This, of course does not rescue Zeilinger's information immaterialism from its independent problems but, under a structural account of explanation, the explanatory power of the theory might remain intact.

By avoiding the complications of Zeilinger's bold immaterialism, also CBH argue that the conceptual problems of quantum theory dissolve as soon as one interprets the quantum state as quantum information. However, within Zeiliger's immaterialism, the rejection of a deeper explanation of quantum phenomena follows from the fact that immaterialism itself provides an information-based explanation of such phenomena. CBH, instead, ground their epistemological analysis on the claim that, exactly as special relativity regarded as a theory of principle made Lorentz's theory (a constructive theory) explanatorily superfluous, in the same way their theorem render any alternative interpretation of quantum theory explanatorily superfluous. As a consequence, they argue that quantum theory is best understood as a theory of principle in Einstein's sense (1919), involving just the possibilities and impossibilities of information processing. In this sense, although the ontology at the basis of quantum information theory and quantum theory is still uncertain, we can still endorse a structural explanation of quantum phenomena, since – for epistemological reasons – information is to be considered a fundamental physical quantity.

In any case, for the success of their project it is crucial to show that, as a consequence of the CBH theorem, information must be taken as a physical primitive. The way CBH argue for this conclusion is to conjecture that the CBH theorem makes any constructive mechanical interpretation of quantum theory in principle empirically underdetermined.

> You can, if you like, tell a mechanical story about quantum phenomena (via Bohm's theory, for example) but such a story, if constrained by the information-theoretic principles, can have no excess empirical content over quantum mechanics, and the additional non-quantum structural elements will be explanatorily superfluous. (Bub 2005, p. 14)

As a first comment, note that a structural explanation of the quantum correlations is stronger than the account provided by axiomatic reconstructions of quantum theory. In such theories, and in Bub (2005) in particular, possible mechanical or dynamical accounts of the quantum correlations are not excluded but only deemed in principle empirically equivalent to whatever is derived in terms of the quantum informational principles. According to a structural explanation of the quantum correlations instead, no explanation deriving from theories that are empirically equivalent to standard quantum theory is possible, for the simple reason that the quantum correlations do not need in principle any dynamical or mechanical explanation.

An even more serious problem, though, derives from the premise of CBH argument stating that all constructive interpretations of quantum theory are empirically equivalent. Many criticisms to the quantum information theory's reconstruction program hinge exactly on this point. Hagar and Hemmo (2006), for instance, argue that quantum information theory is not sufficient and a further account in terms of a constructive and mechanical quantum theory is instead necessary.

For instance, in principle collapse and no-collapse theories have incompatible empirical predictions. In the case of GRW-type theories, such an incompatibility is at the moment practically untestable but it could become testable sooner than expected. The problem is that such predictions concern the detection of superpositions in macrosystems – and in these cases even collapse theories predict an effective collapse due to environmental decoherence. However, the fact that so far we have not been able to properly isolate a macrosystem in such a way as to control decoherence, does not make the two kinds of theories in principle empirically equivalent. On these assumptions, a mechanical story about the dynamics of quantum systems is therefore still possible and, according to Hagar's and Hemmo's, is still needed to explain the *un*observed.

10.5 Conclusions

In conclusion, we want to suggest that Bub's approach can be reconciled with Hagar and Hemmo's more constructive account. On the one hand, considering the nonlocal correlations as wholly natural in the stronger sense suggested in Sect. 10.3 sounds rather plausible to us (and it is plausible even to a Bohmian rejecting the formulation of the theory in terms of a quantum potential). But a conceptual move consisting in considering quantum correlations as fundamental as inertia, the speed of light and free fall, renders an account as to why the macroscopic world is not an entangled mess even more indispensable. In Newtonian mechanics, forces are introduced to explain a "deviation" from natural inertial motion and in the general theory of relativity the geodesic deviation equation is introduced to explain "deviation" from the naturally local inertial trajectories. What explains a "deviation" from the naturally entangled states of the microworld, in such a way that the macroworld appears to be non-entangled?

References

Bokulich, A. (2009). How scientific models can explain. *Synthese, 180*(1), 33–45.

Brown, H., & Pooley, O. (2006). Minkowski space-time: A glorious non-entity. In D. Dieks (Ed.), *The ontology of Spacetime* (pp. 67–89). Amsterdam: Elsevier.

Bub, J. (2000). Quantum mechanics as a principle theory. *Studies in the History and Philosophy of Modern Physics, 31B*, 75–94.

Bub, J. (2004). Why the quantum? *Studies in the History and Philosophy of Modern Physics, 35B*, 241–266 arXiv:quant-ph/0402149 v1.

Bub, J. (2005). Quantum theory is about quantum information. *Foundations of Physics, 35*(4), 541–560 arXiv:quant-ph/0408020 v2.

Bub, J. (2015). Quantum entanglement and information, Edward N. Zalta (ed.). *The stanford encyclopedia of philosophy* (Summer 2015 Edition). http://plato.stanford.edu/archives/sum2015/entries/QM-entangle/

Clifton, R. (1998). Structural explanation in quantum theory. Eprint: http://philsci-archive.pitt.edu/archive/00000091/00/explanation-in-quantumtheory.pdf, unpublished.

Clifton, R., Bub, J., & Halvorson, H. (2003). Characterizing quantum theory in terms of information theoretic constraints. *Foundations of Physics, 33*(11), 1561 Page refs. to arXiv:quant-ph/0211089.

Craver, C. F. (2007). *Explaining the brain*. Oxford: Oxford University Press.

Dorato, M. (2014). Causal *versus* structural explanations in scientific revolutions. *Synthese, 2014*, 1–21, https://doi.org/10.1007/s11229-014-0546-7

Dorato, M., & Felline, L. (2011). Scientific explanation and scientific structuralism. In A. Bokulich & P. Bokulich (Eds.), *Scientific structuralism* (pp. 161–176). Dordrecht: Springer.

Einstein, A. (1919). Time, space, and gravitation. *The London Times*, November 28. Page references to reprint (under the title: What is the theory of relativity?) In A. Einstein (Ed.), Ideas and ppinions (pp. 227–232). New York: Crown Publishers, 1954.

Flores, F. 1999, "Einstein's Theory of Theories and Types of Theoretical Explanation", International Studies in the Philosophy of Science, 13, 2, (pp. 123–134).

Felline, L. (2011). Scientific explanation between principle and constructive theories. *Philosophy of Science, 78*(5), 989–1000.

Felline, L. (2015). Mechanisms meet structural explanation. *Synthese*, 1–16.
Felline, L. (2016). It's a matter of principle: Scientific explanation in information-theoretic reconstructions of quantum theory. *Dialectica, 70*(4), 549–575.
Felline, L. (2018). Mechanisms meet structural explanation. *Synthese, 195*(1), 99–114.
Flores, F. (1999). Einstein's theory of theories and types of theoretical explanation. *International Studies in the Philosophy of Science, 13*(2), 123–134.
Glennan, S. (2010). Mechanisms, causes, and the layered model of the world. *Philosophy and Phenomenological Research, 81*(2), 362–381.
Grinbaum, A. (2007). Reconstruction of quantum theory. *British Journal for the Philosophy of Science, 58*(3), 387–408.
Hagar, A., & Hemmo, M. (2006). Explaining the unobserved—Why quantum mechanics ain't only about information. *Foundations of Physics, 36*(9), 1295–1324 Page refs. to http://arxiv.org/abs/quant-ph/0512095.
Hughes, R. I. G. (1989). Bell's theorem, ideology, and structural explanation. In J. Cushing & E. McMullin (Eds.), *Philosophical consequences of quantum theory* (pp. 195–207). Indiana: Notre Dame.
Janssen, M. (2002a). Reconsidering a scientific revolution, the case of Einstein versus Lorentz. *Physics in Perspective, 4*, 421–446.
Janssen, M. (2002b). COI stories: Explanation and evidence in the history of science. *Physics in Perspective, 10*, 457–520.
Janssen, M. (2009). Drawing the line between kinematics and dynamics in special relativity. In *Studies in history and philosophy of modern physics, 40*(1) 26–52.
Körner, S. (1969). *Fundamental questions of philosophy*. Harmondsworth: Penguin Books.
Kuhn, T. (1970). *The structure of scientific revolutions* (Second Enlarged ed.). Chicago: Chicago University Press.
Lange, M. (2013a). How to explain the Lorentz transformations. In S. Mumford & M. Tugby (Eds.), *Metaphysics and science*. Oxford: Oxford University Press.
Lange, M. (2013b). What makes a scientific explanation distinctively mathematical? *British Journal for the Philosophy of Science, 64*, 485–511.
Morrison, M. (1999). Models as autonomous agents. In M. Morgan & M. Morrison (Eds.), *Models as mediators: Perspectives on natural and social science* (pp. 38–65). Cambridge: Cambridge University Press.
Pincock, C. (2014). Abstract explanations in science. *The British Journal for the Philosophy of Science* 66(4), 857–882. axu016.
Reutlinger, A. (2012). Getting rid of interventions. *Studies in History and Philosophy of Science Part C: Studies in History and Philosophy of Biological and Biomedical Sciences, 43*(4), 787–795.
Schrödinger, E., & Born, M. (1935). Discussion of probability relations between separated systems. *Mathematical Proceedings of the Cambridge Philosophical Society, 31*(4), 555–563.
Steiner, M. (1978). Mathematical explanation. *Philosophical Studies, 34*(2), 135–151.
Timpson, C. G. (2010). Information, immaterialism, instrumentalism: Old and new in quantum information. In A. Bokulich, G. Jaeger, (Eds.) *Philosophy of quantum information and entanglement*, 208–227. Cambridge: Cambridge University Press.
Timpson, C. G. (2013). *Quantum information theory and the foundations of quantum mechanics*. Oxford: Oxford University Press.
Zeilinger, A. (1999). A foundational principle for quantum mechanics. *Foundations of Physics, 29*(4), 631–643.
Zeilinger, A. (2005). The message of the quantum. *Nature, 438*, 743.

Chapter 11
Universality, Invariance, and the Foundations of Computational Complexity in the Light of the Quantum Computer

Michael E. Cuffaro

Abstract *Computational complexity theory* is a branch of computer science dedicated to classifying computational problems in terms of their difficulty. While computability theory tells us what we can compute in principle, complexity theory informs us regarding what is feasible. In this chapter I argue that the science of *quantum computing* illuminates complexity theory by emphasising that its fundamental concepts are not model-independent, but that this does not, as some suggest, force us to radically revise the foundations of the theory. For model-independence never has been essential to those foundations. The fundamental aim of complexity theory is to describe what is achievable in practice under various models of computation for our various practical purposes. Reflecting on quantum computing illuminates complexity theory by reminding us of this, too often under-emphasised, fact.

11.1 Introduction

Computational complexity theory is a branch of computer science that is dedicated to classifying computational problems in terms of their difficulty. Unlike computability theory, whose object is to determine what we can compute in principle, the object of complexity theory[1] is to inform us with regards to which computational problems are actually feasible. It thus serves as a natural conceptual bridge

[1]There are a number of sciences (for example: complex systems theory, the study of Kolmogorov complexity, and so on) which are referred to as complexity theories. Unless otherwise noted, any occurrence of 'complexity theory' in what follows should be understood as referring in particular to *computational* complexity theory, and any conclusions made should be taken as pertaining only

M. E. Cuffaro (✉)
Rotman Institute of Philosophy, University of Western Ontario, London, ON, Canada
e-mail: mcuffar@uwo.ca

S. O. Hansson (ed.), *Technology and Mathematics*, Philosophy of Engineering and Technology 30, https://doi.org/10.1007/978-3-319-93779-3_11

between the study of mathematics and the study of technology, in the sense that computational complexity theory informs us with respect to which computational procedures may reasonably be expected to be technologically realisable.

Quantum computer science is the study of algorithms and other aspects of computer systems whose construction involves an explicit appeal to various features of quantum physical theory. Strikingly, there are quantum algorithms that appear to significantly outperform algorithms which do not take advantage of quantum resources. What distinguishes, quantitatively, quantum from classical computation is not the number of problems that can be solved using one or the other model. Rather, what distinguishes the quantum from the classical model of computation is that the number of problems solvable *efficiently*—i.e. the number of problems whose solution is feasibly realisable—in the former model appears to be larger than the number of problems solvable efficiently in the latter. The study of quantum computer science therefore advances the goal of complexity theory in the sense that it adds to our knowledge of the class of feasibly realisable computational procedures.

More generally, as I will argue below, the study of quantum computation illuminates the very nature and subject matter of complexity theory. Yet it does not do so in a way that is often claimed. In particular it is not uncommon to come across statements in the philosophical and scientific literature to the effect that advances in quantum computing force a fundamental revision of the foundations of complexity theory (Hagar 2007; Nielsen and Chuang 2000; Bernstein and Vazirani 1997). According to this view it is the traditional aim of complexity theory to understand the nature of concepts such as that of a 'tractable problem' *in themselves*; i.e., apart from the manner in which they are implemented under particular models of computation. Model-independence, in turn, is taken to rest upon an 'extended' or 'strong' version of the Church-Turing thesis, or alternately, upon an 'invariance' thesis. And because quantum computers seemingly violate these theses, it is concluded that complexity theory's foundations must be somehow rebuilt.

As I will argue, however, model-independence is not and never has been at the core of computational complexity theory. Its foundations are therefore not shaken by the advent of quantum computing. Complexity theory is fundamentally a practical science, whose aim is to guide us in making distinctions in practice among tractable and intractable problem sets. The model-independence of complexity-theoretic concepts is not a necessary condition for realising this aim. Quantum computation indeed illuminates the subject matter of complexity theory. But it does not do so by overturning its foundations. Rather, quantum computing illuminates complexity theory by reminding us of its practical nature.

This is both a virtue of the theory as well as a reason for increased philosophical attention to it. Science does not always or only, or perhaps ever, progress through the absolute identification of fundamental entities, be they abstract or concrete.

to it. See Müller (2010) for a discussion of the difficulties associated with formulating machine-independent concepts in the context of Kolmogorov complexity.

Complexity theory furnishes us with a particularly striking illustration that scientific progress—even in the mathematical sciences—is, in fact, often built upon pragmatically justified foundations and conceptual structures.[2] There is a general philosophical lesson in this, which in different contexts has been profitably analysed by some (for example, Carnap 1980 [1950], 1962, ch. 1), though in my view too few, philosophers.

In the next section we will briefly review, from a historical perspective, the foundations of computability theory. Section 11.3 will then connect the foregoing discussion to the foundations of computational complexity theory, and will introduce the theory's basic concepts. In Sect. 11.4 we will discuss the 'universality of Turing efficiency' thesis, as well as the closely related 'invariance thesis'. Section 11.5 will introduce the basic concepts of quantum computing. In Sect. 11.6 we will discuss quantum computing's significance for the conceptual foundations of complexity theory. We will then conclude.

11.2 The *Entscheidungsproblem* and the Origins of the Church-Turing Thesis

With his second incompleteness theorem, Gödel demonstrated that any ω-consistent formalisation of number theory, whose formulas are primitively recursively definable, and which is rich enough to permit arithmetisation of syntax, cannot prove its own consistency.[3] For such a capability would be incompatible with Gödel's first incompleteness theorem, by which he demonstrated that within any such formalisation there are sentences neither provable nor refutable from the axioms. Finding a general and effective procedure for determining whether a given formula in such a system is one of these sentences, however, remained an open question.

This was the Decision Problem—in German: the *Entscheidungsproblem*—for validity, originally posed for first-order logic by Hilbert and Ackermann (1928, Pt. III); that is, to describe an 'effective procedure' by which one can decide whether an arbitrarily given expression of first-order logic is provable from the axioms.[4]

[2]See also Dean (2016a), who reviews the arguably insurmountable problems that face any attempt to regard an algorithm as a mathematical object in the light of computer science practice.

[3]An ω-consistent theory is such that it is both consistent and satisfies a syntactic analogue of soundness (see Dawson 2007, p. 504). A primitively recursively definable formula is such that it can be built up from a finite number of successive basic operations. Arithmetisation of syntax refers to a procedure by which every sentence in a formal system is encoded uniquely into a natural number (called its 'Gödel number').

[4]The reason for Hilbert and Ackermann's focus on the special case of first order logic is that it is the most restricted example of a general logic adequate for representing higher arithmetic, i.e. number theory. Its study was to constitute the first step in the development of a more encompassing logical framework for mathematics (Dawson 2007, p. 500). Note that Hilbert and Ackermann additionally posed a parallel Decision Problem for satisfiability. In the sequel, unless otherwise indicated, I

Informally, an effective computational procedure consists of a finite number of precise finite-length instructions guaranteed to produce some desired result in a finite number of steps if followed exactly by a human being using nothing other than paper and pencil. An example of an effective procedure is the truth-table method as applied to sentential logic. Famously, Church and Turing were independently able to show that the *Entscheidungsproblem* for first-order logic could not be solved; i.e., no effective calculational procedure for determining the validity of an arbitrarily given expression in first-order logic exists.

Turing, to whom we will restrict our attention, showed this partly by means of a penetrating philosophical analysis of the notion of effective computation.[5] Turing (1936–7, pp. 249–51) argued that it is essential to the idea of carrying out a computation that the computer uses a notebook from which she reads, and onto which she writes, various symbols related to her work. These symbols, as they must be distinguishable from one another, are chosen from a finite alphabet. At any given moment during a computation, the computer will find herself in one of a finite number of relevant states of mind which summarise her memory of the actions she has performed up until that point along with her awareness of what she must now do (pp. 253–4). The actions that are available to her are characterised by a finite set of elementary operations, such as 'read the next symbol' from the notebook, 'write symbol *a*' to the notebook, and so on. Turing then argued that one could design an automatic machine, which he called an α-machine, to instantiate each of these essential features of the practice of human computation (see Fig. 11.1). In doing so he identified the extension of the concept 'effectively calculable' with that of 'computable by α-machine'. This identification is known as Turing's thesis, which he proved (p. 263ff) to be equivalent with Church's independently arrived at thesis that the class of effectively calculable functions is identical with the class of λ-definable functions (Church 1936). For this reason it is also called the Church-Turing thesis.

Turing then addressed the *Entscheidungsproblem* in an indirect way (Turing 1936–7, pp. 259–63, 1938). He first showed that it is impossible to determine, for a given α-machine, whether it is 'circle-free'; i.e. whether it is not the case that it never outputs more than a finite number of symbols. He then showed that if the *Entscheidungsproblem* were solvable, one could determine, for any given α-machine, whether it is circle-free. Since this contradicts the first result, the *Entscheidungsproblem* is unsolvable.

---

will take the Decision Problem or *Entscheidungsproblem* to refer exclusively to the problem for validity.

[5] In what follows it must be kept in mind that computation, at the time of the publication of "On Computable Numbers," generally referred to an activity performed by human beings; a computer was a person employed to carry out computations.

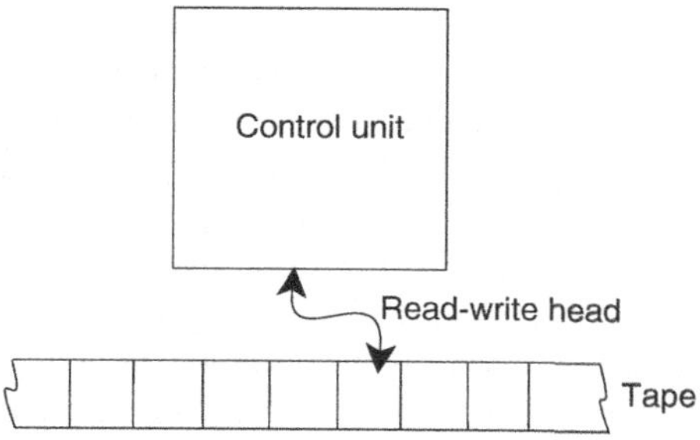

Fig. 11.1 A version of what is now called a 'Turing machine'. The control unit houses the machine's 'state of mind', which in general changes after every operation of the read-write head. The read-write head reads, writes, and moves back and forth along portions of a one-dimensional tape (the machine's 'notebook'). Such a machine is an idealised representation of the components involved in human computation

11.3 Efficient Computation

The period just discussed, during which the seminal papers by Church, Gödel, Turing, and others were published, is the period of the birth of computer science in the modern sense. It was to be nearly three more decades before the particular branch of modern computer science that furnishes the subject matter for this chapter, computational complexity theory, took shape with the work of Cobham (1965), Edmonds (1965), Hartmanis and Stearns (1965), and others. Yet one of its key questions was anticipated significantly earlier by none other than Gödel. Revisiting the *Entscheidungsproblem* in a letter he wrote to von Neumann in 1956, Gödel asked for von Neumann's opinion concerning the number, $\varphi(l)$, of steps needed, in the worst case, to decide whether some arbitrarily given formula of first-order logic has a proof of length l. In his letter Gödel asks[6]: "how fast does $\varphi(l)$ grow for an optimal [Turing] machine?" He notes that "One can show that $\varphi(l) \geq Kl$" (for some constant K), and then asserts:

> If there actually were a machine with $\varphi(l) \sim Kl$ (or even only with $\sim Kl^2$), this would have consequences of the greatest magnitude. That is to say, it would clearly indicate that, despite the unsolvability of the Entscheidungsproblem, the mental effort of the mathematician in the case of yes-or-no questions could be completely [Gödel's Footnote: Apart from the postulation of axioms] replaced by machines. One would indeed have to simply select an l so large that, if the machine yields no result, there would then also be no reason to think further about the problem (Gödel 1956, p. 10).

To illustrate: take some proposition F of first-order logic and consider testing to see whether F has a proof, Ψ, of length l. Let l be a number of steps far too large for any unaided human being to survey in a lifetime, but small enough that a machine could survey them all relatively quickly. Gödel's point is that, from the machine's

[6]In the following quotations I have replaced the variable n with l.

perspective, the Kl (or perhaps Kl^2) steps needed to discover whether Ψ exists is not very much greater than the l steps that would be needed to survey it. We would expect, therefore, that the machine will give us an answer to the question of whether F has a proof of length l in a reasonable amount of time. By assumption, however, surveying a proof of length $\geq l$ is beyond the practical capabilities of any human being. So if the machine yields a negative result, then we can conclusively say that, for the practical purposes of unaided human computation, F is unprovable. Indeed there would be no reason to bother with the practical computational purposes of unaided human mathematicians at all; if such a machine existed we could henceforth consider such questions exclusively with respect to it.

There is an additional, deeper, point that is implicit here as well. Gödel's question to von Neumann is stated in the context of the *Entscheidungsproblem*, where it is assumed that the procedure to be used by a human mathematician to answer the question of whether F can be proved is an effective one, in the sense described in the previous section. Recall that following an effective procedure requires no ingenuity on the part of the person doing the following; it is a purely mechanical procedure which, if followed exactly, is guaranteed to give one a result in a finite number of steps. It is precisely for this reason that we can model it with a machine. In general, however, theorem proving is an activity which we do take to require insight and ingenuity. We take there to be more to the process of discovering a proof of a particular theorem than blindly following a set of rules; we need insight into the 'essential nature' of the problem at hand in order to guide us to the most likely route to a solution, and we need ingenuity to proceed along this route in a skillful, efficient, way. Or so one could object. Be that as it may, if we could in fact build a machine to discover, in only Kl (or Kl^2) steps, whether any given proposition of first-order logic has a proof of length l, it would make, not just human beings themselves, but the ingenuity and insight associated with their activities in this context, dispensable.

Implicit in the above considerations is the idea that neither $\varphi(l) \sim Kl$ nor $\varphi(l) \sim Kl^2$ yields a significantly greater number than l from the point of view of a machine. This is consistent with the ideas of modern complexity theory, where in fact any decision problem (i.e., yes-or-no question) for which a solution exists whose worst-case running time is bounded by as much as a polynomial function of its input size, n, is considered to be a 'tractable' (a.k.a. 'feasible', 'efficiently solvable', 'easy', etc.[7]) problem. Indeed, these ideas are not just consistent; one way to motivate the modern complexity-theoretic identification is to begin with essentially Gödel's assertion that problems which require only Kn or Kn^2 steps to solve are tractable.[8] Combine this with the computer programmer's intuition that an efficient program, to which one adds a call to an efficient subroutine, should

[7] I will be using these terms interchangeably below.

[8] Note that although Gödel's letter to von Neumann anticipates this and other ideas of modern complexity theory, I am not claiming that it actually influenced the theory's development. As far as I am aware, Gödel's letter was unknown prior to its translation and publication in Sipser (1992).

continue to be thought of as efficient (Arora and Barak 2009, p. 27), and we naturally arrive at the conclusion that the set of efficiently solvable problems just is the set of problems solvable in a polynomial number of time steps. This 'polynomial principle' is generally considered to be at the heart of the theory of computational complexity. We will discuss it in more detail (and critically) in Sect. 11.6.

In the context of the Turing machine (TM) model, the set of decision problems solvable in polynomial time is referred to as the class P.[9] More formally, we can conceive of a decision problem as one whose goal is to yield a yes-or-no answer to the question of whether a given string x of length n is a member of the 'language' L. For example, the decision problem for determining whether a given number is prime can be represented as the problem to determine, for an arbitrarily given binary string, whether it is a member of the language {10, 11, 101, 111, 1011, 1101, 10001, 10011, ... } (the set of binary representations of prime numbers). Now call a given language L a member of the class DTIME($T(n)$) if and only if there is a Turing machine[10] for deciding membership in L whose running time, $t(n)$, is 'on the order of $T(n)$', or in symbols: $O(T(n))$. Here, $T(n)$ represents an upper bound for the growth rate of $t(n)$ in the sense that, by definition, $t(n)$ is $O(T(n))$ if for every sufficiently large n, $t(n) \leq k \cdot T(n)$ for some constant k.[11] So for any language L in, for example, DTIME(n^2), there is a TM that will take no more than kn^2 steps to decide membership in L. We can now formally characterise P as (Arora and Barak 2009, p. 25):

$$\mathrm{P} = \bigcup_{k \geq 1} \mathrm{DTIME}(n^k). \tag{11.1}$$

Note that the class DTIME($T(n)$) is defined, strictly speaking, to be a set of languages. Below I will sometimes use statements of the form: '(decision) problem R is in DTIME($T(n)$)', which is shorthand for the assertion that the language L_R, associated with R, is decidable in $O(T(n))$ steps.

We have just seen that L is in P if and only if one can construct a polynomial-time TM that will decide, for any given x, whether $x \in L$. Now suppose that one is presented with a proof that $x \in L$. If one can *verify* this proof using a polynomial-time TM M, then we say that L is a member of the complexity class NP.[12] More formally (Arora and Barak 2009, p. 39),

$$L \in NP \text{ whenever: } \quad x \in L \Leftrightarrow \exists u \text{ s.t. } M(x, u) \stackrel{poly}{=} \text{'yes'}, \tag{11.2}$$

---

[9]It is also sometimes referred to as PTIME, in order to emphasise the distinction between it and PSPACE, the class of problems solvable using space resources bounded by a polynomial function of n.

[10]The 'D' in DTIME stands for 'deterministic'. It contrasts with 'nondeterministic time', which I will introduce later.

[11]The qualification 'for every sufficiently large n' can be rephrased as the assertion that there exists some finite $n_0 \geq 1$ such that $t(n) \leq k \cdot T(n)$ whenever $n \geq n_0$.

[12]NP stands for nondeterministic polynomial time. The reason for this name will become clear shortly.

where u is string (usually called a 'certificate') whose length is given by a polynomial function of the length, n, of x, and $M(x, u) \stackrel{poly}{=}$ 'yes' asserts that the machine M accepts x, given u, in polynomial time.[13]

The restricted form of the *Entscheidungsproblem* described above by Gödel is certainly in NP; given a proposition x, and a proof u of x whose length is $\leq l$, one can obviously verify this in polynomial time. Indeed, the problem also happens to be 'NP-complete' (Hartmanis 1993).[14] NP-complete problems are the hardest problems in NP, in the sense that if we have in hand a solution to an NP-complete problem, we can easily convert it into a solution to any other problem in NP. That is, a language $L \in$ NP is in the class NP-complete if and only if a procedure for deciding L can be converted, in polynomial time, into a procedure for deciding L', for any $L' \in$ NP. More concisely, $L \in$ NP is NP-complete if and only if $\forall L' \in NP$, L' is polynomial-time reducible, in the above sense,[15] to L (Arora and Barak 2009, p. 42).

The proposition that there exists a general solution to the restricted *Entscheidungsproblem* which requires no more than Kl^2 steps to carry out—call this the 'Gödelian conjecture'[16]—does not amount merely to the proposition that this decision problem is in NP. Recall that the restricted *Entscheidungsproblem* is the problem to decide whether an arbitrarily given formula x has a proof of length l; it is not merely the problem of verifying this fact about x given a certificate u. The Gödelian conjecture, therefore, amounts to the claim that the restricted *Entscheidungsproblem* is in P. But since this problem is known to be NP-complete, the Gödelian conjecture, if correct, amounts to the claim that P = NP.[17]

Interestingly, there has been no proof or disproof to date of the statement that P = NP. Partly due to the intuitive implausibility of its consequences—that "the mental effort of the mathematician in the case of yes-or-no questions could be completely replaced by machines" (Gödel 1956)—the statement is generally believed to be false. Besides this there are further, mathematical, reasons to believe that P $\neq$ NP (Aaronson 2013a, p. 67). I will not mention these here as the P = NP question is not our focus. I will only say that the project to prove or disprove P = NP is a worthwhile one, not so much because the outcome is in doubt, but because a formal proof would likely enlighten us with regards to just what it is that insight and ingenuity contribute to the practice of mathematics.

---

[13] u must be of polynomial length in n to ensure that M can read u in polynomial time.

[14] Gödel himself gives no indication that he realises this in his letter.

[15] What I have described above is actually called a Karp reduction. It is a weaker concept than the related one of Cook reduction. We will not discuss the distinction here. For more on this, see Aaronson (2013a, p. 58).

[16] Gödel does not himself actually conjecture this, although he comes close to doing so: "it seems to me ... to be totally within the realm of possibility that $\varphi(l)$ grows slowly." (Gödel 1956, p. 10).

[17] Strictly speaking it only entails that NP $\subseteq$ P. But since obviously P $\subseteq$ NP, it would follow that P = NP.

Our discussion of the Turing machine model of computation has thus far focused on the standard, i.e., deterministic, case. A standard TM is such that its behaviour at any given moment in time is wholly determined by the state that it finds itself in plus whatever input it receives. The machine can be fully characterised, that is, by a unique transition function over the domain of states and input symbols. One can, however, generalise the TM model by allowing the machine to instantiate more than one transition function simultaneously.[18] Upon being presented with a given input in a given state, a *nondeterministic Turing machine* (NTM) is allowed to 'choose' which of its transition functions to follow (see Fig. 11.2). Exactly how this choice is made is left undefined, and for the purposes of the model can be thought of as arbitrary. We say that an NTM accepts a string x if and only if there exists a path through its state space that, given x, leads to an accepting state. It rejects x otherwise. We define the class NTIME($T(n)$), analogously to DTIME($T(n)$), as the set of languages for which there exists an NTM that will decide, in $O(T(n))$ steps, whether a given string x of length n is in the language L.

Recall that above I characterised NP as the set of languages for which one can construct a polynomial-time TM to verify, for any x, that $x \in L$, given a polynomial-length certificate u for x. One can alternatively characterise NP as the set of languages for which there exists a polynomial-time NTM for determining membership in L:

$$\text{NP} =_{df} \bigcup_{k \geq 1} \text{NTIME}(n^k). \tag{11.3}$$

This definition is the source of the name NP, in fact, which stands for 'nondeterministic polynomial time'.

Definitions (11.2) and (11.3) are equivalent. Given a language L and a polynomial-time NTM that decides it, then for any $x \in L$, there is by definition a polynomial-length sequence of transitions of the NTM which will accept x. One can use this sequence as a certificate for x, and verify it in polynomial-time using a (deterministic) TM. Conversely, suppose there is a TM M_D that, given a polynomial-length certificate u for x, can verify in polynomial time that $x \in L$. Then one can construct a polynomial-time NTM M_N that will 'choose' certificates from among the set of possible polynomial-length strings (e.g., by randomly writing one down). Upon choosing a certificate u, M_N then calls M_D to verify x given u, and transitions to 'yes' only if M_D outputs 'yes' (Arora and Barak 2009, p. 42).

For an NTM, no attempt is made to define how such a computer chooses, at any given moment, whether to follow one transition function rather than another. In particular, it is not assumed that any probabilities are attached to the machine's choices. Indeed, under Turing's original conception (1936–7, p. 232), these are

[18]The idea of a machine with an ambiguous transition function can be found in Turing (1936–7). Turing calls this a 'choice machine' (p. 232), and notes its extensional equivalence with the automatic (i.e. deterministic) machine (p. 252, footnote ‡).

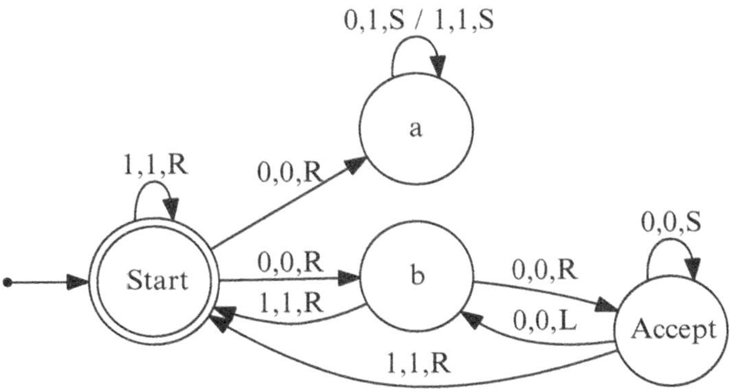

$\delta_1(Start, 0)=(a, 0, R)$ $\delta_2(Start, 0)=(b, 0, R)$

$\delta_1(Start, 1)=(Start, 1, R)$ $\delta_2(Start, 1)=(Start, 1, R)$

etc. etc.

Fig. 11.2 A nondeterministic Turing machine (NTM) is such that, for a given state and a given input, the state transitioned to is not predetermined; at any given step the machine is able select from more than one transition function (in this case, δ_1 and δ_2). The machine depicted accepts binary strings ending in '00', since there exists a series of transitions for which, given such a string, the machine will end in the 'Accept' state. But it is not guaranteed to do so. The machine additionally is guaranteed to reject any string not ending in '00'. In the diagram, an edge from s_1 to s_2 with the label α, β, P is read as: In state s_1, the machine reads α from its tape, writes β to the tape in the same position, moves its read/write head along the tape to the position P with respect to the current tape position (L = to the left, R = to the right, S = same), and finally transitions to state s_2

thought of as the choices of an external operator. They are thus arbitrary from the machine's point of view. In a *probabilistic Turing machine* (PTM), on the other hand, we characterise the computer's choices by associating a particular probability with each of its transitions (see Fig. 11.3).

Like TMs and NTMs, PTMs have associated with them a number of complexity classes. The most important of these is the class BPP (bounded-error probabilistic polynomial time). This is the class of languages such that there exists a polynomial-time PTM that, on any given run, will correctly determine whether or not a string x is in the language L with probability $\geq 2/3$. The particular threshold value of 2/3 is inessential to this definition. It is chosen in order to express the idea of a 'high probability'.[19] But any threshold probability $p_{min} \geq 1/2 + n^{-k}$, where k is a constant, will suffice for the definition of BPP. For given a polynomial-time PTM that correctly determines whether or not $x \in L$ with probability p_{min}, re-

[19]Note that the high probability requirement constitutes a key conceptual difference between BPP and NP. The latter demands only that it is *possible* for an NTM to arrive at a correct solution.

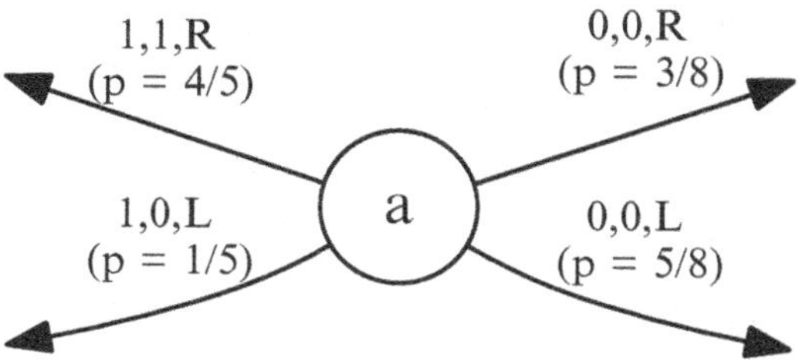

Fig. 11.3 One node in a PTM. Given an input of 1 in the state **a**, the machine will write 1 to the tape and move right with probability 4/5, or write 0 and move left with probability 1/5. On an input of 0 it will write 0 and move right with probability 3/8, or write 0 and move left with probability 5/8. For a given state and a given input, edge probabilities must add up to 1. We can imagine that the machine's choices are made in accordance with these probabilities by repeatedly 'flipping a coin'

running it a number of additional times that is no more than polynomial in n and taking the majority answer will yield a correct result with probability close to 1 (Arora and Barak 2009, p. 132). Since, as I mentioned above, the time it takes to run a polynomial-time algorithm a polynomial number of times is still polynomial, varying p_{min} in this way will do nothing to alter the set of languages contained in BPP.

11.4 The Universality and Invariance Theses

The Church-Turing (C-T) thesis claims nothing about the efficiency of any particular model of computation. Nor does it carry with it any implications concerning physically possible computing machines in general (see Turing 1950, §§3, 5, 6.7). Both Church's and Turing's theses are, as we saw earlier, theses concerning the limits of effective procedures. Despite this, the C-T thesis is often misrepresented in this regard in the philosophical and even in the scientific literature (for further discussion of the reasons for the confusion, see Copeland 2015; Timpson 2013; Pitowsky 1990). In more informed literature, however, these re-interpretations of the C-T thesis are explicitly distinguished from it. The thesis (I) that any reasonable model of computation can be simulated with at most a polynomial number of extra time steps by a PTM is often called the 'strong' C-T thesis (see, e.g., Nielsen and

Chuang 2000, p. 140).[20,21] The thesis (II) that a physical instantiation of a TM can simulate any physically possible machine that realises a finite instruction set and that works on finite data is often called the 'physical' C-T thesis (Andréka et al. 2018; Piccinini 2011). But confusingly, (II) is also sometimes called the 'strong' thesis (Goldin and Wegner 2008), and (I) is sometimes called the 'physical' thesis (Hagar 2007).

So as not to contribute to the confusion arising from this ambiguous labelling, and more importantly, to discourage any erroneous inferences to the intended scope of Church's and Turing's original theses themselves, I will, following Copeland (2015), refer to (II) as 'Thesis M'. I will refer to (I), the subject of this section, as the 'universality of Turing efficiency thesis'. For it follows from the truth of (I) that the set of problems efficiently solvable in general, i.e., on any reasonable digital machine model $\mathfrak{M}$, is identical with the set of problems efficiently solvable on a PTM. Formally this can be expressed as:

$$\bigcup \text{Poly}_{\mathfrak{M}} = \text{BPP}. \tag{11.4}$$

In other words, the thesis implies that the set of problems solvable in polynomial time does not grow beyond BPP if we allow ourselves to vary the underlying model.[22]

A further closely related notion is what van Emde Boas (1990, p. 5) has called the 'invariance thesis'. This states that any reasonable machine model can simulate any other reasonable machine model with no more than a polynomial slowdown.

---

[20] Some textbooks state (I) as a thesis about the TM rather than the PTM model (see, e.g., Arora and Barak 2009, p. 26). I will follow Nielsen and Chuang (2000), in order to leave open the possibility that P $\subsetneq$ BPP, and also because BPP constitutes a more natural contrast (see Footnote 22 below) with its quantum analogue, BQP, which we will introduce in the next section. Until recently, P $\subsetneq$ BPP was thought to be very likely true, however evidence (e.g., Agrawal et al. 2004) has been mounting in favour of the conjecture that in fact P = BPP. Whether (I) is formulated with respect to TMs or PTMs makes little difference to what follows. A TM can be thought of as a special case of a PTM for which transition probabilities are always either 0 or 1.

[21] The qualification 'reasonable' will be explained shortly.

[22] There is a slight complication that I am glossing over here, namely that what it means for a machine to constitute a solution to a problem varies across computational models. In particular a TM solution to a problem is required to yield a correct answer with certainty, whereas (as I mentioned previously) a PTM solution in general need only yield a correct answer with high probability. Implicit in (11.4), therefore, is an appeal to the more general criterion for solvability corresponding to that appropriate to a PTM rather than to a TM. This subtle distinction regarding what it means to solve a problem under various models of computation is one reason, that I alluded to in Footnote 20 above, for expressing the universality thesis in terms of BPP rather than P. For as we will see in the next section, a quantum computer, like a PTM, is a probabilistic machine and is subject to the same criterion for success. Expressing the universality thesis in terms of BPP thus allows for a more straightforward analysis of the quantum model's significance for the thesis. A similar remark applies to the invariance thesis, which I now introduce.

(See also: Goldreich 2008, p. 33, who names it differently.[23]) The invariance thesis implies the universality thesis, but not vice versa. Note that in the context of both the universality and invariance theses, 'reasonable' is typically understood as *physically realisable*. Reasonable models include variants of the TM model, for example, but do not include models which employ unbounded parallelism.[24] This will be discussed further in Sect. 11.6.

There are reasons for believing in the truth of both the universality and invariance theses. Neither the standard variations on the Turing model, such as adding more tapes, increasing the number of squares readable or writable at a given moment, and so on (Arora and Barak 2009), nor the alternative reasonable universal (classical) models of computation that have been developed since Turing's work, are faster than PTMs by more than a polynomial factor, and all appear to be able to simulate one another efficiently in this sense (van Emde Boas 1990).

Over the last three decades, however, evidence has nevertheless been mounting against universality and invariance, primarily as a result of the advent of quantum computing (Aaronson 2013a, chs. 10, 15). We will discuss quantum computing in more detail in the next section.

11.5 Quantum Computation

Consider the (non-quantum) machine depicted in Fig. 11.4. This simple automaton has two possible states: $\{0, 1\}$. It has one possible input (omitted in the state-transition diagram), which essentially instructs the machine to 'run'. This can be implemented, for example, by a button connected to the machine's inner mechanism. At the end of any given run, the machine will either remain in the state it was previously in or else transition to the opposite state, with equal probability. One can imagine that the machine also includes a small door which, when opened, reveals a display indicating what state the machine is in. A typical session with the machine consists in: (a) opening the door to record the machine's initial state; (b) pushing the 'run' button one or more times; (c) opening the door to reveal the machine's final state.

Let us suppose that between the initial and final opening of the door, our experimenter pushes the button twice. Given that the initial reading was 0, what is the probability that the final reading is 0 as well? This is given by:

[23]For the purposes of our discussion of the invariance thesis we will not be distinguishing between TMs and PTMs but will be taking the former to be a special case of the latter (this is motivated in Footnote 20 above). We will understand the thesis, then, as asserting that any reasonable probabilistic machine model is efficiently simulable by any other reasonable probabilistic machine model. PTMs and quantum computers are both examples of probabilistic models.

[24]The parallel random access machine (PRAM) model, for example, is excluded.

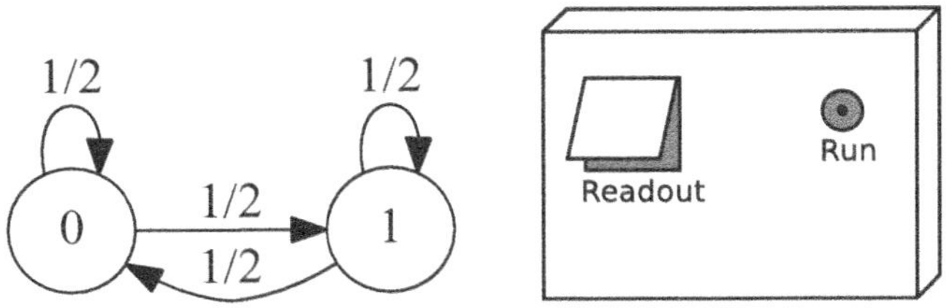

Fig. 11.4 A simple automaton which, when run, randomly transitions to one of two possible states. In the state-transition diagram at the left, edge labels represent probabilities for the indicated transitions

$$\begin{aligned}\Pr(C^2 0 \to 0) &= \Pr(C0 \to 0) \times \Pr(C0 \to 0) \\ &+ \Pr(C0 \to 1) \times \Pr(C1 \to 0) \\ &= 1/2, \end{aligned} \tag{11.5}$$

where $C^n \psi \to \phi$ signifies that the computer is run n times after beginning in the state ψ, and ends in ϕ. Equation (11.5) illustrates that there are two possible ways for the computer to begin and end in the state 0 after two runs. Either it remains in 0 after each individual run, or else it first flips to 1 and then flips back. From Fig. 11.4, one can easily see that:

$$\Pr(C^2 0 \to 0) = \Pr(C^2 0 \to 1) = \Pr(C^2 1 \to 0) = \Pr(C^2 1 \to 1) = 1/2, \tag{11.6}$$

and indeed we have that $\Pr(C^n \psi \to \phi) = 1/2$ for any n.

The internal state (what is revealed by opening the door) of the simple machine pictured in Fig. 11.4 is describable by a single binary digit, or 'bit'. In general the internal state of any classical digital computer is describable by a sequence of n bits, and likewise for its inputs and outputs. A bit can be directly instantiated by any two-level classical physical system, for example by a circuit that can be either open or closed. In a *quantum* computer, on the other hand, the basic unit of representation is not the bit but the qubit. To directly instantiate it, we can use a two-level quantum system such an electron (specifically: its spin). The qubit generalises the bit. Like a bit, it can be 'on', i.e. in the state $|0\rangle$, or 'off', i.e. in the state $|1\rangle$. In general, however, the state of a qubit can be expressed as a normalised linear superposition:

$$|\psi\rangle = \alpha|0\rangle + \beta|1\rangle, \tag{11.7}$$

where the 'amplitudes' α and β are complex numbers such that $|\alpha|^2 + |\beta|^2 = 1$. We refer to $|\psi\rangle$ as the 'state vector' for the qubit.[25]

[25]The modulus squared (or 'absolute square'), $|c|^2$, of a complex number c is given by $c\bar{c}$, where $\bar{c}$ is the complex conjugate of c. $|\psi\rangle$ is normalised when $|\alpha|^2 + |\beta|^2 = 1$.

An important difference between qubits and bits is that not all states of a qubit can be observed directly; in particular, one never observes a qubit in a linear superposition (aside from the trivial case in which one of α, β is 0).[26] According to the Born rule, a qubit in the state (11.7), when measured, will be found to be in the state $|0\rangle$ with probability $|\alpha|^2$, and in the state $|1\rangle$ with probability $|\beta|^2$. For example, consider a simple one-qubit quantum machine that implements the following transitions:

$$Q|0\rangle \rightarrow \frac{i}{\sqrt{2}}|0\rangle + \frac{1}{\sqrt{2}}|1\rangle \equiv |\chi\rangle, \tag{11.8}$$

$$Q|1\rangle \rightarrow \frac{1}{\sqrt{2}}|0\rangle + \frac{i}{\sqrt{2}}|1\rangle \equiv |\xi\rangle. \tag{11.9}$$

If the machine begins in the state $|0\rangle$, and the button is pushed once, it will transition to $|\chi\rangle$. Then with probability $|\frac{i}{\sqrt{2}}|^2$, opening the door will reveal $|0\rangle$, and with probability $|\frac{1}{\sqrt{2}}|^2$ it will reveal $|1\rangle$.

Since $|\frac{i}{\sqrt{2}}|^2 = |\frac{1}{\sqrt{2}}|^2 = 1/2$, a series of 'one-push' experiments with this quantum machine will produce identical statistics as will a series of one-push experiments with the classical machine depicted in Fig. 11.4. Things become more interesting when we consider two-push experiments. If the machine is in the initial state $|0\rangle$, then after the first push the machine will effect the transition (11.8). If, before opening the door, we push the button again, the machine will make the following transition:

$$\begin{aligned} Q\Big(\frac{i}{\sqrt{2}}|0\rangle + \frac{1}{\sqrt{2}}|1\rangle\Big) &= \frac{i}{\sqrt{2}}Q|0\rangle + \frac{1}{\sqrt{2}}Q|1\rangle \\ &= -\frac{1}{2}|0\rangle + \frac{i}{2}|1\rangle + \frac{1}{2}|0\rangle + \frac{i}{2}|1\rangle = i|1\rangle. \end{aligned} \tag{11.10}$$

Since $|i|^2 = 1$, opening the door will find the machine in the state $|1\rangle$ with certainty. Likewise, if the machine begins in $|1\rangle$, a two-push experiment will find it in the state $|0\rangle$ with certainty. A state transition diagram for the quantum machine is given in Fig. 11.5.[27]

[26] To be more precise: one never observes a qubit in a linear superposition with respect to a particular measurement basis. Generally, in quantum computing, measurements are carried out in the computational, i.e. $\{|0\rangle, |1\rangle\}$, basis. In this basis the superposition $(|0\rangle + |1\rangle)/\sqrt{2}$, for example, can never be the result of a measurement. If one measures in the $\{|+\rangle, |-\rangle\}$ basis, however, then such a result is possible, since $|+\rangle = (|0\rangle + |1\rangle)/\sqrt{2}$. On the other hand, a measurement in the $\{|+\rangle, |-\rangle\}$ basis will never yield the result $|0\rangle = (|+\rangle + |-\rangle)/\sqrt{2}$ even though a result of $|0\rangle$ is possible in the computational basis.

[27] Note that overall phase factors have been abstracted away from in Fig. 11.5. Two normalised state vectors which differ only in their overall phase factor yield all of the same probabilities for outcomes of experiments and are considered as equivalent according to quantum theory. For

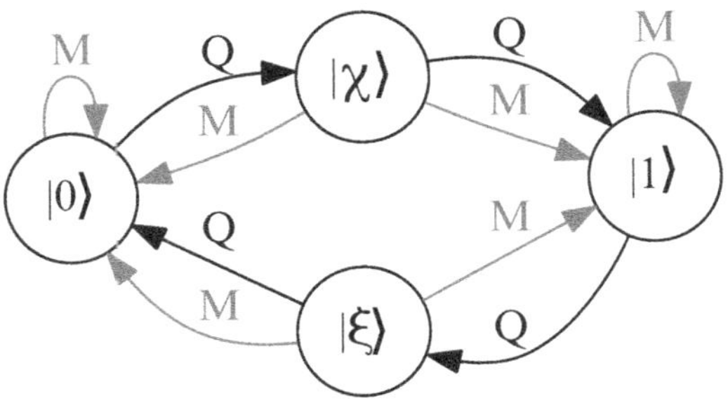

Fig. 11.5 A simple quantum computer. With each button push, the machine deterministically oscillates, via the transition Q, between the states $|0\rangle, |\chi\rangle, |1\rangle, |\xi\rangle$ in the manner depicted. When the door is opened, the machine undergoes the 'measurement' transition M. This results, when the computer is in one of the states $|\chi\rangle$ and $|\xi\rangle$, in a reading of $|0\rangle$ or $|1\rangle$ with equal probability. Opening the door when the machine is in either $|0\rangle$ or $|1\rangle$ has no effect on the computer's state

The probabilities for outcomes of two-push experiments with the quantum computer Q are significantly different from those associated with two-push experiments on C. This is despite the fact that if one performs two (or in general n) repetitions of a *one*-push experiment (i.e. in which one opens the door after every button push), the resulting statistics will be identical for both C and Q. One can think of a one-push experiment with C or Q as instantiating a 'maximally noisy' (i.e. completely useless) NOT-gate. With a two-push experiment on Q, however, we have instantiated a perfect NOT-gate. We cannot do anything analogous with C.

The foregoing was a simple—almost trivial—illustration of some of the basic differences between classical and quantum computation. But by taking advantage of these and other differences, researchers have been able to develop quantum algorithms to achieve results that seem impossible for a classical computer. Quantum computers cannot solve non-Turing-computable problems (see Hagar and Korolev 2007). However, as we will discuss shortly, quantum computers are able to efficiently solve problems that have no known efficient classical solution. This apparent ability of quantum computers to outperform classical computers is known as 'quantum speedup'.

A fascinating question, assuming that they indeed have this ability,[28] regards exactly which physical features of quantum systems are responsible for it. We will not be discussing this question further here.[29] Rather, let us return to Gödel's

example, all of these express the same physical state: $|\psi\rangle, -|\psi\rangle, i|\psi\rangle, -i|\psi\rangle$. *Local* phase factors, however, are important; $\frac{|0\rangle+|1\rangle}{\sqrt{2}}$ and $\frac{|0\rangle-|1\rangle}{\sqrt{2}}$, for example, are different states.

[28] There is strong evidence (some of which we will discuss shortly), however there is still no proof, that quantum computers can efficiently solve more problems than classical computers can.

[29] In (11.10), the partial amplitudes contributing to the $|0\rangle$ component of the state vector cancel each other out. Many quantum algorithms include similar transitions, leading some to view quantum interference as the source of quantum speedup (Fortnow 2003), although others have questioned

question regarding the resources required to solve the restricted form of the *Entscheidungsproblem*. Could a quantum computer be used to solve NP-complete problems such as this one efficiently? It turns out that a quantum computer can yield a performance improvement over a standard TM with respect to such problems. Recall (see Definition 11.2) that a language L is in NP if there is a TM, C, such that $x \in L$ if and only if there is a certificate u whose length is polynomial with respect to the length of x, that if fed to C can be used by C to verify x's membership in L in polynomial time. If in addition, in polynomial time, for any given x, C can itself either find such a suitable certificate, or determine that one does not exist, then L is also in P.

Let the question be, 'Does the string x of length n have a "proof", i.e. a certificate, of length $\leq n^k$?', for some constant k. The number of possible certificates u_i is $N = 2^{n^k}$. Assuming the space of certificates is unstructured, C will need $O(N)$ steps to decide whether x is in L; the computer will run through the possible certificates u_i one by one, testing each in turn to see if it is a valid certificate for x, moving on to the next certificate if it is not. Using a quantum computer and Grover's quantum search algorithm (Grover 1996), however, only $O(\sqrt{N})$ steps are required. It turns out that this is the best we can do (Bennett et al. 1997). But while this quadratic speedup is impressive, the overall running time of the quantum computer remains exponential in the length, n, of x. Quantum computers, therefore, do not appear to allow us to affirm the Gödelian conjecture.[30]

However there is evidence that the class of languages efficiently decidable by a quantum computer is larger than that corresponding to either a deterministic or probabilistic classical computer. To be more precise, define the class BQP, analogously to the class BPP, as the class of languages such that there exists a polynomial-time quantum computer that will correctly determine, with probability $\geq 2/3$, whether or not a string x is in the language L. The question of whether a quantum computer can outperform a classical computer amounts to the question of whether BQP is larger than BPP. It is clear that BPP $\subseteq$ BQP; one invocation of the transition (11.8), for example, followed by a measurement, can serve to simulate a classical 'coin flip', and in polynomial time this procedure can be used

whether interference is a truly quantum phenomenon (Spekkens 2007). The fact that some quantum algorithms appear to spawn parallel computational processes has led to the idea of 'quantum parallelism' as the primary contributing mechanism (Duwell 2007, 2018), and to the related but distinct idea that this processing occurs in parallel physical universes (Hewitt-Horsman 2009; for a criticism see Aaronson 2013b; Cuffaro 2012). Others view quantum entanglement (Cuffaro 2017, 2018; Steane 2003), or quantum contextuality (Howard et al. 2014), as providing the answer. Still others view the structure of quantum logic as the key (Bub 2010).

[30]Note that above it was assumed that the space of certificates is unstructured. However it is possible that a given NP-complete language L possesses non-trivial structure that can be exploited to yield further performance improvements (Cerf et al. 2000). Therefore we cannot rule out that L is efficiently decidable by a classical computer, let alone by a quantum one.

to simulate any of a given PTM's transition probabilities.[31] As for the evidence for strict containment—i.e. for BPP $\subsetneq$ BQP—this comes mainly from the various quantum algorithms that have been developed.

Shor's quantum algorithm (Shor 1997) for integer factorisation is a spectacular example. The best known classical factoring algorithm is the number field sieve (Lenstra et al. 1990), which requires $O(2^{(\log N)^{1/3}})$ steps to factor a given integer N. Popular encryption mechanisms such as RSA (Rivest et al. 1978) rely on the assumption that factoring is hard. Yet Shor's algorithm requires only a number of steps that is polynomial in $\log N$—an exponential speedup over the number field sieve. There are also provable 'oracle' separations between the classical and quantum computational models. An oracle is a kind of imaginary magic black box, to which one puts a question chosen from a specified set, and from which one receives an answer in a single time step. For example, Simon's problem (Simon 1994) is that of determining the period of a given function f that is periodic under bitwise modulo-2 addition. One can define an oracle $\mathcal{O}$ for evaluating arbitrary calls to f. Relative to $\mathcal{O}$, Simon's quantum algorithm requires $O(n)$ steps, while a classical algorithm requires $O(2^n)$. This is an exponential speedup.

None of these results are absolutely conclusive. On the one hand, not every complexity-theoretic proposition relativises to an oracle. The result that IP = PSPACE, for example, does not hold under certain oracle relativisations (PSPACE is the class of problems solvable using polynomially many space resources; IP is the class of problems for which an affirmative answer can be verified using an interactive proof). Further, there are oracles relative to which P = NP, as well as oracles relative to which P $\neq$ NP. Oracles are important tools that, among other things, help to simplify and clarify the conceptual content of complexity-theoretic propositions, however they cannot be used to resolve these and other questions (for a discussion, see Fortnow 1994). Nor can they definitively show that BPP $\subsetneq$ BQP. Simon's problem, for instance, might contain some hitherto unknown structure, obscured by the relativisation of the problem to an oracle, that could be exploited by a classical algorithm. Regarding Shor's algorithm, on the other hand, unlike Simon's, it does not make essential use of an oracle. Yet this nevertheless does not conclusively demonstrate that BPP $\subsetneq$ BQP, for it is still an open question whether factoring is in BPP. While most complexity theorists believe this to be false, their conviction is not as strong as their conviction, for example, that P $\neq$ NP—for the factoring problem does have *some* internal structure, which is in fact exploited by the classical number field sieve algorithm (Aaronson 2013a, 64–66).

While none of the individual pieces of evidence are conclusive, taken together they nevertheless do lend a great deal of plausibility to the thesis that quantum computers can solve more problems efficiently than can classical computers. That said, it is not the place here to evaluate this evidence. For the purposes of our discussion we will assume that this thesis is true. In the next section we will consider its consequences.

[31]Rather than Q, one typically uses a 'Hadamard gate' (H) for this purpose, which acts as follows:

$$H|0\rangle \rightarrow \frac{1}{\sqrt{2}}|0\rangle + \frac{1}{\sqrt{2}}|1\rangle, \qquad H|1\rangle \rightarrow \frac{1}{\sqrt{2}}|0\rangle - \frac{1}{\sqrt{2}}|1\rangle.$$

11.6 Quantum Computing and the Foundations of Computational Complexity Theory

If BPP $\subsetneq$ BQP, then it follows that the universality of Turing efficiency thesis is false. Some authors view the consequences of this for complexity theory to be profound. Bernstein and Vazirani (1997), for example, take it that the theory "rests upon" this thesis (p. 1411), and that the advent of quantum computers forces us to "re-examine the foundations" (p. 1412) of the theory. The sense in which complexity theory rests upon universality is expressed by Nielsen and Chuang (2000), who write that the failure of the universality thesis implies that complexity theory cannot achieve an "elegant, model independent form" (p. 140). Of this, Hagar (2007) writes:

> To my mind, the strongest implication [of the violation of universality] is on the autonomous character of some of the theoretical entities used in computer science, ... given that quantum computers may be able to efficiently solve classically intractable problems, hence re-describe the abstract space of computational complexity, computational concepts and even computational kinds such as 'an efficient algorithm' or 'the class NP', will become machine dependent, and recourse to 'hardware' will become inevitable in any analysis of the notion of computational complexity. (pp. 244–5).

Given that the universality of Turing efficiency thesis states that any reasonable model of computation can be simulated with at most a polynomial number of extra time steps by a (probabilistic) *Turing machine*, however, the reader may be understandably confused by the claim that this thesis grounds the model-independence of complexity-theoretic concepts to begin with. At most, it seems that only a very weak sense of model-independence follows from universality. The truth of (11.4), that is, implies that any assertion, of the form 'language L is decidable efficiently by an instance of the reasonable machine model $\mathfrak{M}$', is replaceable by the assertion that 'language L is decidable efficiently by a PTM'. And since 'by a PTM' is thus made universally applicable, it can be omitted from all such statements in the knowledge that it is implicit (cf. Nakhnikian and Salmon 1957). But while this yields a kind of model-independence in the sense that one need not explicitly mention the PTM model when speaking of the complexity-theoretic characteristics of L, it remains the case, nevertheless, that a reference to the PTM model is implicit in one's assertions about L.

To illustrate just how weak such a notion of model-independence is, note that, based on it, one could argue that, although quantum computing refutes the model-independence consequent upon the universality of *Turing* efficiency, it at the same time provides a replacement for it (cf. Deutsch 1985, Bernstein and Vazirani 1997, p. 1413). Nielsen and Chuang (2000) write that if the universality of Turing efficiency thesis were true, that it would be

> great news for the theory of computational complexity, for it implies that attention may be restricted to the probabilistic Turing machine model of computation. After all, if a problem has no polynomial resource solution on a probabilistic Turing machine, then the [universality of Turing efficiency] implies that it has no efficient solution on any computing device. Thus, the [universality of Turing efficiency] implies that the entire theory of computational complexity will take on an elegant, model-independent form if the notion of efficiency is identified with polynomial resource algorithms (p. 140).

Putting aside for the moment the somewhat strange comment that an expansion of our knowledge of the extent of the space of efficiently decidable languages is 'bad news' for complexity theory, note that quite the same argument could be made if one replaces 'probabilistic Turing machine' with 'quantum computer' and 'universality of Turing efficiency' with 'universality of quantum efficiency'. For although BPP in (11.4) is now replaced with BQP, we have analogously given a characterisation, in terms of a single machine model, of $\bigcup \text{Poly}_{\mathfrak{M}}$. And yet if a computer based on the principles of *quantum physics* can be taken to ground an absolute model-independent characterisation of complexity-theoretic concepts, then the right conclusion to draw is that this is not a satisfactory notion of model-independence.[32]

One could, perhaps, counter that the model-independence consequent on the universality thesis actually stems from the nature of the Turing model itself. Assuming that one is convinced by Turing's philosophical analysis, the Turing model does, after all, represent the conceptual essence of effective computation (cf. Hartmanis and Stearns 1965, p. 285). Be that as it may, there is no reason to think that such a model must also be the most efficient one.[33] The model-independence of complexity theory thus would turn out to be a contingent fact. This in itself is not a criticism. Nevertheless in that case it is not clear just what model-independence would contribute to the ground of complexity theory in the theoretical sense. A 'universality of quantum efficiency thesis' would be, perhaps, less metaphysically satisfying from the point of view of a computer scientist, but in itself would do just as much theoretical work as the universality of Turing efficiency thesis.

BPP $\subsetneq$ BQP also implies the failure of the invariance thesis. Because of its supposed relation to the Church-Turing thesis, it is universality and not invariance that has received the lion's share of attention from philosophers (an exception is Dean). But unlike the universality thesis, there is a sense of model-independence built right into the very statement of invariance. After all, it amounts to a direct claim that the particular machine model under consideration, since it can be efficiently simulated by any other reasonable model, is irrelevant for the purposes of providing a characterisation of the complexity of a problem. Note also that the statement of invariance itself makes no reference to the Turing model, so it is not susceptible to the same sort of criticism I directed at the universality thesis above. It is true that the domain of 'reasonable' or physically realisable models does not, for example, include the 'unreasonable' parallel computational models, thus the invariance thesis cannot provide model-independence in a truly absolute sense. Still, the study of efficient algorithms in particular is mainly concerned with reasonable models. So

[32]One could, however, ground a *relative* notion of model-independence based on quantum principles. I will discuss this further below.

[33]For a discussion of some possible justifications, and the problems that go along with them, for the choice of the multi-tape TM as the benchmark for complexity-theoretic analyses, see Dean (2016c).

invariance, if true, arguably provides absolute model-independence in the only sense that matters.[34]

Van Emde Boas takes the invariance thesis (he does not mention the universality thesis) to, as a consequence, constitute a foundational thesis for complexity theory:

> The fundamental complexity classes P and NP became part of a *fundamental hierarchy*: LOGSPACE, NLOGSPACE, P, NP, PSPACE, EXPTIME, ... And again theory faced the problem that each of these classes has a *machine-dependent* definition, and that efficient simulations are needed before one can claim that these classes are in fact *machine-independent* and represent fundamental concepts of computational complexity. It seems therefore that complexity theory, as we know it today, is based on the [assumption that the invariance thesis holds] (van Emde Boas 1990, p. 5, ellipsis in original).

I will be criticising this statement. Before I do, however, it is important to note that it is clear that the simplifying assumption of invariance can be profoundly useful; its truth would imply that one can restrict one's attention to the (reasonable) model of one's choice when inquiring into the complexity-theoretic characteristics of particular problems. Further, and independently of this, studies such as van Emde Boas's, of the extent to which one model can simulate another, illuminate the structure of complexity theory by allowing one to characterise the space of machine models in terms of various 'equivalence classes'. Van Emde Boas, for instance, defines the models comprising the 'first machine class' as those for which the invariance thesis holds. The 'second machine class' is defined with respect to a different 'parallel computation thesis' (1990, p. 5), which I will not further describe here. Note that van Emde Boas is careful to point out the partly conventional and partly empirical (he does not use these words) nature of such theses:

> The escape in defending the Invariance Thesis ... is clearly present in the word *reasonable* ... For example, when in 1974 it was found that a RAM model with unit-time multiplication and addition instructions (together with bitwise Boolean operations) is as powerful as a parallel machine, this model (the MBRAM) was thrown out of the realm of reasonable (sequential) machines and was considered to be a "parallel machine" instead. The standard strategy seems to be to adjust the definition of "reasonable" when needed. The theses become a guiding rule for specifying the right classes of models rather than absolute truths and, once accepted, the theses will never be invalidated. This strategy is made explicit if we replace the word *reasonable* by some more neutral phrase. (van Emde Boas 1990, pp. 5–6).

There is much to commend in this statement. But note first that it is not clear that the 'standard strategy' will work in the face of quantum computing. On the one hand, one would be hard-pressed to argue that quantum computers are not physically realisable machines. On the other hand, the 'quantum parallelism thesis' (Duwell 2007, 2018; Pitowsky 1990, 2002) is quite controversial (Cuffaro 2012; Steane 2003), so it is not obvious that quantum computers should be classed as parallel rather than sequential computers. This said, even if one takes the invariance thesis to be violated by quantum computing, the idea of it, not as an absolute truth but as a 'guiding rule'—a sort of intensional principle for characterising the extensions

[34] See Dean (2016c) for a discussion of ways to circumscribe the space of operations that should be allowable in a reasonable model.

corresponding to different equivalence classes of models—remains, and it remains a highly illuminating methodological principle for studying the relations between computational models.

To illustrate what I mean by this,[35] note that 'quantum computer' is actually an umbrella term for a number of (universal) computational models: the quantum Turing model (Deutsch 1985), the quantum circuit model (Deutsch 1989), the cluster-state model (Briegel et al. 2009), the adiabatic model (Farhi et al. 2000), and many others. To date, all of these models have been found to be computationally equivalent in the sense that they all yield the same class of problems, BQP (see, for example, Aharonov et al. 2007; Raussendorf and Briegel 2002; Nishimura and Ozawa 2009). Thus it seems as though a third *quantum* machine class, in addition to van Emde Boas's first and second machine classes, exists. Fascinatingly, the differences between the first and this third machine class turn out to be related to the differences in the physics used to describe the machines which comprise them. The physics, in turn, informs our judgements regarding which of these equivalence classes should be deemed as 'reasonable'. On the basis of these judgements we are then enabled make conclusions with regards to what is feasible for us to accomplish in the real world (cf. Dean 2016a, pp. 30, 56). If it were not for the existence of quantum computers, one would be warranted in the belief that only a single reasonable machine class exists. Quantum computing teaches us that there are at least two.

Invariance, thought of as a guiding rule or *methodological principle*, rather than as a thesis, is what is driving these investigations; through the search for equivalence classes we carve out and illuminate the structure of the space of computational models. This yields a notion of *relative* model-independence among the machine models comprising a particular class. To be clear, the existence of relative model-independence within the conceptual structure of complexity theory is itself not strictly speaking necessary for the theory. It is true that the theory would arguably be far less interesting if every abstract model of computation were different from every other; for one thing there would be no unified notion of 'classical computation' to compare quantum computation with—a quantum computer would be just another model among many. Yet one can still imagine what such a complexity theory would look like: a theory describing the computational power of various abstract models of computation and their interrelations. This is not so alien that it would be unrecognisable from our modern point of view.[36] In fact the early period

[35]I am indebted to Scott Aaronson and to Sona Ghosh for independently prompting the discussion in this and the next two paragraphs.

[36]The very notion of an abstract model of computation presupposes *some* notion of complexity-theoretic invariance, of course, without which it would be impossible to unify various physical instantiations of a model under a single concept. I am perfectly ready to concede that if complexity-theoretic invariance failed to hold in this minimal sense then this would be disastrous for modern complexity theory. But then I cannot see how it would be possible to revise complexity theory in light of this; it would seem impossible to have a theory of complexity, or indeed any theory of any subject, if even basic abstraction were not possible.

of complexity theory, before the introduction of the polynomial principle (to be discussed below) looked much like this. Representative of this period are results such as that by Hartmanis and Stearns (1965), for example, who prove that the multi-tape TM model is capable of a quadratic speedup with respect to the single-tape TM model. Such analyses are indeed still present in the modern theory.

The fact that relative model-independence does exist, on the other hand, arguably tells us something deep, or anyway something, about how computer science connects up with the world. The invariance principle (rather than thesis) is a vitally important part of computational complexity theory partly for this reason. And as a methodological principle, it fulfils this role whether it is successful in its search for equivalence classes of computational models or not. For the lack of any relative model-independence within the theory would arguably also tell us something about computer science's relation to the physical world. A further, still methodological, role of invariance is as a simplifying principle. For from a practical perspective the theory would be exceedingly unwieldy, even if it would not strictly speaking be impossible to develop, if no equivalence classes of abstract computational models existed.

And yet none of this seems to imply or depend upon model-independence in the sense of the first of my quotes of van Emde Boas above. Indeed it is not clear what one gains from model-independence in that sense. LOGSPACE, NLOGSPACE, P, NP, PSPACE, EXPTIME, and other complexity classes are each of them classes of *languages*, after all. To compare any two of them is to compare one class of languages with another; they are thus already machine-independent in that sense. For this reason it is a meaningful question to ask whether the class P is large enough to include all of the languages in NP, irrespective of how one defines either of them in terms of an underlying machine model. On the other hand, one *can* define P as the class of languages which are efficiently decidable by a TM. And one *can* define NP as the class of languages which are efficiently decidable by an NTM. And to be sure, deeper insight is gained by reflecting on how one translates the definition of NP given in terms of the NTM model (11.3), into the alternative definition of NP given in terms of the standard TM model (11.2). For then one sees clearly that the statement that P = NP amounts to the assertion that the restricted form of the *Entscheidungsproblem* is efficiently solvable. But in this case it is by reflecting on the characteristics of a *particular model* that one gains this insight, namely, the Turing machine model insofar as it represents the conceptual essence of *human* digital computation.[37]

I am not alone in my skepticism. The idea that it is the fundamental goal of complexity theory to get at some metaphysical notion of an independently existing *thing* called 'efficient computation' is certainly not shared by all complexity theorists. For example, Fortnow (2006) writes:

> By no means does computational complexity "rest upon" a [universality of Turing efficiency] thesis. The goals [sic.] of computational complexity is to consider different notions

[37]This is one reason why the question whether P = NP remains interesting even if $P \subsetneq BQP$.

> of efficient computation and compare the relative strengths of these models. Quantum computing does not break the computational complexity paradigm but rather fits nicely within it.

Fortnow's statement refers to the universality thesis; however it can clearly equally well be asserted as a counter to the claimed foundational status of the invariance thesis. A quick glance at the practice of complexity theorists seems to confirm that Fortnow is right, for since the advent of quantum computing in the mid-1990s, complexity theory appears to have continued on in much the same way as before. Classic textbooks such as Papadimitriou's (1994) excellent reference, written before Shor's (1994) breakthrough in quantum factoring, continue to be cited frequently in modern scholarly work; more modern textbooks such as Arora and Barak's (2009) book include a chapter on quantum computation but otherwise present the subject in much the same way as similar textbooks always have done; BQP is just one of many classes in Aaronson's (2016) 'complexity zoo'. Despite the fact that the prospects for a practicable and scalable quantum computer are improving significantly every year (Veldhorst et al. 2015), and despite the fact that most computer scientists believe that BPP $\subsetneq$ BQP and thus that the universality and invariance theses are false, complexity theory—as a discipline—does not appear to be in crisis. Complexity theory as a whole has grown—many new and important questions have arisen regarding exactly how BQP relates to other complexity classes—but the basic conceptual framework within which we ask these questions remains much the same as before.

But if model-independence is not constitutive of complexity theory, what is? Built into the definition of both the universality and invariance theses is the more basic idea that an algorithm is efficient if and only if it requires no more than a polynomial number of steps to complete. As we have seen, the roots of this idea go back at least as far as Gödel's letter to von Neumann, although from the modern perspective, its main sources are the seminal articles of Cobham (1965) and Edmonds (1965). I will call it the 'polynomial efficiency principle' or 'polynomial principle' for short.[38] Unlike either the universality or invariance theses, there is no question that the polynomial principle is de facto foundational with respect to the modern framework of complexity theory, in the sense that the conceptual structure of the theory—the definitions of and relations between its most important complexity classes such as P, NP, BPP, BQP, and so on—depend crucially upon the principle.

And yet there is a different sense in which it can be said to be controversial. The goal of the polynomial principle is to capture our pre-theoretic notion of what it means for an algorithm to be efficient. Expressing this, Edmonds writes:

[38] Forms of this principle are often referred to as the Cobham-Edmonds thesis (e.g., see Dean 2016b). Unfortunately, this terminology is not always consistent. In Goldreich (2008, p. 33), for example, the Cobham-Edmonds thesis is the name given to what we have here called the invariance thesis.

> ... my purpose is ... to show as attractively as I can that there is an efficient algorithm [for maximum matching]. According to the dictionary, "efficient" means "adequate in operation or performance." This is roughly the meaning I want—in the sense that it is conceivable for maximum matching to have no efficient algorithm. Perhaps a better word is "good."
> I am claiming, as a mathematical result, the existence of a *good* algorithm for finding a maximum cardinality matching in a graph. (Edmonds 1965, p. 420, emphasis in original).

One could argue, however, that the polynomial principle fails to achieve this goal. For example, a problem for which the best algorithm requires $O(n^{1000})$ steps to complete is considered to be tractable according to the principle, while a problem for which the best algorithm requires $O(2^{n/1000})$ steps is considered to be intractable. Yet despite these labels, the 'intractable' problem will take fewer steps to solve for all but very large values of n. Strictly speaking, of course, since it is defined asymptotically, the principle does not yield an incorrect answer even in such cases. However problems we are faced with in practice are invariably of bounded size, and an asymptotic measure—the preceding example illustrates this—seems to at least sometimes be ill-suited for their analysis. A further reason for doubting the polynomial principle is that it is a measure of worst-case complexity. Yet it does not seem implausible that an average-case measure might give us better insight into just how 'good' a given algorithm is.

All of this may be granted. And yet growth rates such as the above are extremely rare in practice. Generally speaking, polynomial-time algorithms have growth rates with small exponents, and the simplification made possible by the use of an asymptotic measure generally does more good than it does harm; "For practical purposes the difference between algebraic and exponential order is often more crucial than the difference between finite and non-finite." (Edmonds 1965, p. 451). We also generally do not know in advance how the inputs to a particular problem will be distributed, and in such circumstances average case complexity analyses are impracticable (see Papadimitriou 1994, pp. 6–7).

What the arguments for and against the polynomial principle illustrate is that its goal is not so much to provide an absolute or metaphysical distinction between good and bad algorithms. What these arguments show us is that the purpose of the principle is to guide us in making such distinctions in practice. In particular, what the arguments *for* the principle amount to is the—empirical—claim that the polynomial principle has been highly successful, in the sense that it has tended to provide us with extraordinarily good guidance for the problems with which we are generally faced. Aaronson sums this up as follows:

> Of the big problems solvable in polynomial time—matching, linear programming, primality testing, etc.—most of them really *do* have practical algorithms. And of the big problems that we think take exponential time—theorem-proving, circuit minimization, etc.—most of them really *don't* have practical algorithms. So, that's the empirical skeleton holding up our fat and muscle (2013a, p. 54).

The precise way in which the polynomial principle aids us in the search for good algorithms is by providing us with a mathematical explication of 'good' in the context of complexity theory. In doing so it provides us with a mathematical framework for our inquiries, within which we can express precise questions, and generate precise answers.

> ...if only to motivate the search for good, practical algorithms, it is important to realise that it is mathematically sensible even to question their existence. For one thing the task can then be described in terms of concrete conjectures. (Edmonds 1965, p. 451).

And yet, while the principle is generally a good guide, it is we who must ultimately decide, in every case upon which we bring it to bear, whether or not to follow its advice.

> It would be unfortunate for any rigid criterion to inhibit the practical development of algorithms which are either not known or known not to conform nicely to the criterion. Many of the best algorithmic ideas known today would suffer by such theoretical pedantry. ...And, on the other hand, many important algorithmic ideas in electrical switching theory are obviously not "good" in our sense. (Edmonds 1965, p. 451).

Just as the polynomial principle is a practical principle, so is complexity theory a practical science, in the sense that its fundamental aim is to inform us with regards to the practical difficulty of computing different classes of problems—i.e. with regards to the things we would like to do—on our various machines. Principles such as the polynomial and even the invariance principle (insofar as it serves as a methodological principle for carving out equivalence classes of machine models) illuminate the structure of this space of possible tasks. But they, and the structure with it, are ultimately guides which should be set aside whenever they cease to be useful. To some extent this is true in every science—the principles of Newtonian mechanics, say, must give way to the principles of modern spacetime theories. But principles such as the polynomial principle, and the theory of complexity that is built upon it, do not claim for themselves universal validity as Newtonian mechanics at one time did—nor do they even claim to have a well-defined sphere of application (how large must an exponent be before a polynomial-time algorithm is no longer considered to be 'really' efficient?). The polynomial principle, and complexity theory with it, are intrinsically practical in nature; they claim to be useful only 'most of the time' and for most of our practical purposes. This is the theory's core.

This said, if, with the development of the theory, the polynomial principle somehow ceased to be useful even in merely the majority of cases of practical interest, this would certainly require a substantial revision of much of the theory's framework, for so much of the conceptual structure of complexity theory is built upon the assumption of the polynomial principle. And yet even in this case the essential nature and subject matter—the metaphysical foundation, if you will—of the theory—a theory of what we are capable of doing in practice—would not change.

11.7 Conclusion

Cobham (1965) took complexity theory to be a science concerned with three general groups of questions: those related (i) to "specific computational systems", (ii) to "categories of computing machines", and (iii) to those questions which "are independent of any particular method of computation" (pp. 24–5). The third subdivision will always remain a part of complexity theory. In fact, machine-independent results can be had in the theory—though no one would argue that these provide a foundation for it—if one uses a very general and amorphous notion of a complexity measure (Seiferas 1990). Indeed studies such as these suggest that further reflection may be needed on precisely what is meant by the notion of 'intrinsic complexity'. But model-independence is not all of the theory; nor is it a foundation for the other two groups of questions mentioned by Cobham. Computational complexity theory is, at its core, a practical science. As a mathematical theory, it employs idealised concepts and methods, and appeals to formal principles which are justified insofar as they are useful in providing us with practical advice regarding the problems we aim to solve. Our solutions to such problems are implemented on particular machine models. Computational complexity theory studies the various notions of efficiency associated with these different models, and studies how these notions relate to one another. "Quantum computing", to quote Fortnow once again, "does not break the computational complexity paradigm but rather fits nicely within it." Indeed, far from breaking the complexity-theoretic paradigm, quantum computing serves to *remind* us of the point of it all.

Acknowledgements This project was supported financially by the Foundational Questions Institute (FQXi) and by the Rotman Institute of Philosophy. I am grateful to Scott Aaronson, Walter Dean, William Demopoulos, Armond Duwell, Laura Felline, Sona Ghosh, Amit Hagar, Gregory Lavers, and Markus Müller for their comments on a previous draft of this chapter. This chapter also benefited from the comments and questions of audience members at the Montreal Inter-University Workshop on the History and Philosophy of Mathematics, and at the Perimeter Institute for Theoretical Physics.

References

Aaronson, S. (2013a). *Quantum computing since Democritus*. New York: Cambridge University Press.

Aaronson, S. (2013b). Why philosophers should care about computational complexity. In B. J. Copeland, C. J. Posy, & O. Shagrir (Eds.), *Computability: Turing, Gödel, Church, and beyond* (pp. 261–327). Cambridge, MA: MIT Press.

Aaronson, S. (2016). Complexity zoo. https://complexityzoo.uwaterloo.ca/Complexity_Zoo

Agrawal, M., Kayal, N., & Saxena, N. (2004). PRIMES is in P. *Annals of Mathematics, 160*, 781–793.

Aharonov, D., van Dam, W., Kempe, J., Landau, Z., Lloyd, S., & Regev, O. (2007). Adiabatic quantum computation is equivalent to standard quantum computation. *SIAM Journal on Computing, 37*, 166–194.

Andréka, H., Madarász, J. X., Németi, I., Németi, P., Székely, G. (2018). Relativistic computation. In M. E. Cuffaro & S. C. Fletcher (Eds.), *Physical perspectives on computation, computational perspectives on physics*. Cambridge: Cambridge University Press.

Arora, S., & Barak, B. (2009). *Computational complexity: A modern approach*. Cambridge: Cambridge University Press.

Bennett, C. H., Bernstein, E., Brassard, G., & Vazirani, U. (1997). Strengths and weaknesses of quantum computing. *SIAM Journal on Computing, 26*, 1510–1523.

Bernstein, E., & Vazirani, U. (1997). Quantum complexity theory. *SIAM Journal on Computing, 26*, 1411–1473.
Briegel, H. J., Browne, D. E., Dür, W., Raussendorf, R., & den Nest, M. V. (2009). Measurement-based quantum computation. *Nature Physics, 5*, 19–26.
Bub, J. (2010). Quantum computation: Where does the speed-up come from? In A. Bokulich & G. Jaeger (Eds.), *Philosophy of quantum information and entanglement* (pp. 231–246). Cambridge: Cambridge University Press.
Carnap, R. (1962). *Logical foundations of probability* (2nd ed.). Chicago: The University of Chicago Press.
Carnap, R. (1980 [1950]). Empiricism, semantics, and ontology. In H. Morick (Ed.), *Challenges to empiricism* (pp. 28–45). Indianapolis: Hackett Publishing Company.
Cerf, N. J., Grover, L. K., & Williams, C. P. (2000). Nested quantum search and structured problems. *Physical Review A, 61*, 032303.
Church, A. (1936). An unsolvable problem of elementary number theory. *American Journal of Mathematics, 58*, 345–363.
Cobham, A. (1965). The intrinsic computational difficulty of functions. In Y. Bar-Hillel (Ed.), Logic, methodology and philosophy of science: Proceedings of the 1964 international congress, North-Holland (pp. 24–30).
Copeland, B. J. (2015). The Church-Turing thesis. In E. N. Zalta (Ed.), *The Stanford encyclopedia of philosophy* (Summer 2015 ed.). https://plato.stanford.edu/archives/sum2015/entries/church-turing/.
Cuffaro, M. E. (2012). Many worlds, the cluster-state quantum computer, and the problem of the preferred basis. *Studies in History and Philosophy of Modern Physics, 43*, 35–42.
Cuffaro, M. E. (2017). On the significance of the Gottesman-Knill theorem. *The British Journal for the Philosophy of Science, 68*, 91–121.
Cuffaro, M. E. (2018). Reconsidering no-go-theorems from a practical perspective. *The British Journal for the Philosophy of Science, 69*, 633–655.
Dawson, J. W., Jr. (2007). Classical logic's coming of age. In D. Jacquette (Ed.), *Philosophy of logic* (pp. 497–522). Amsterdam: Elsevier.
Dean, W. (2016a). Algorithms and the mathematical foundations of computer science. In L. Horsten & P. Welch (Eds.), *Gödel's disjunction: The scope and limits of mathematical knowledge* (pp. 19–66). Oxford: Oxford University Press.
Dean, W. (2016b). Computational complexity theory. In E. N. Zalta (Ed.), *The Stanford encyclopedia of philosophy* (Winter 2016 ed.). https://plato.stanford.edu/archives/win2016/entries/computational-complexity/.
Dean, W. (2016c). Squeezing feasibility. In A. Beckmann, L. Bienvenu, & N. Jonoska (Eds.), *Pursuit of the Universal: Proceedings of the 12th conference on computability in Europe* (pp. 78–88). Cham: Springer International Publishing.
Deutsch, D. (1985). Quantum theory, the Church-Turing principle and the universal quantum computer. *Proceedings of the Royal Society of London Series A Mathematical and Physical Sciences, 400*, 97–117.
Deutsch, D. (1989). Quantum computational networks. *Proceedings of the Royal Society of London Series A Mathematical and Physical Sciences, 425*, 73–90.
Duwell, A. (2007). The many-worlds interpretation and quantum computation. *Philosophy of Science, 74*, 1007–1018.
Duwell, A. (2018). How to make orthogonal positions parallel: Revisiting the quantum parallelism thesis. In M. E. Cuffaro & S. C. Fletcher (Eds.), *Physical perspectives on computation, computational perspectives on physics*. Cambridge: Cambridge University Press.
Edmonds, J. (1965). Paths, trees, and flowers. *Canadian Journal of Mathematics, 17*, 449–467.
Farhi, E., Goldstone, J., Gutmann, S., & Sipser, M. (2000). *Quantum computation by adiabatic evolution* (Technical report MIT-CTP-2936), MIT. arXiv:quant-ph/0001106.
Fortnow, L. (1994). The role of relativization in complexity theory. *Bulletin of the European Association for Theoretical Computer Science, 52*, 229–244.

Fortnow, L. (2003). One complexity theorist's view of quantum computing. *Theoretical Computer Science, 292*, 597–610.

Fortnow, L. (2006). The efficient Church-Turing thesis. http://blog.computationalcomplexity.org/2006/12/efficient-church-turing-thesis.html. Posted: Thursday, December 07, 2006. Retrieved: Monday, April 26, 2010.

Gödel, K. (1956). Private letter to John von Neumann, March 20, 1956. Translated by Wensinger. In Sipser (1992).

Goldin, D., & Wegner, P. (2008). The interactive nature of computing: Refuting the strong Church-Turing thesis. *Minds & Machines, 18*, 17–38.

Goldreich, O. (2008). *Computational complexity: A conceptual perspective*. Cambridge: Cambridge University Press.

Grover, L. K. (1996). A fast quantum mechanical algorithm for database search. In *Proceedings of the Twenty-Eighth Annual ACM Symposium on Theory of Computing, STOC'96* (pp. 212–219). New York: Association for Computing Machinery.

Hagar, A. (2007). Quantum algorithms: Philosophical lessons. *Minds & Machines, 17*, 233–247.

Hagar, A., & Korolev, A. (2007). Quantum hypercomputation—hype or computation? *Philosophy of Science, 74*, 347–363.

Hartmanis, J. (1993). Gödel, von Neumann and the P=?NP problem. In G. Rozenberg & A. Salomaa (Eds.), Current trends in theoretical computer science (pp. 445–450). River Edge: World Scientific.

Hartmanis, J., & Stearns, R. E. (1965). On the computational complexity of algorithms. *Transactions of the American Mathematical Society, 117*, 285–306.

Hewitt-Horsman, C. (2009). An introduction to many worlds in quantum computation. *Foundations of Physics, 39*, 869–902.

Hilbert, D., & Ackermann, W. (1928). *Principles of mathematical logic*. Berlin: Springer.

Howard, M., Wallman, J., Veitch, V., & Emerson, J. (2014). Contextuality supplies the 'magic' for quantum computation. *Nature, 510*, 351–355.

Lenstra, A. K., Lenstra, H. W., Jr., Manasse, M. S., & Pollard, J. M. (1990). The number field sieve. In *Proceedings of the Twenty-Second Annual ACM Symposium on Theory of Computing, STOC'90* (pp. 564–572). New York: Association for Computing Machinery.

Müller, M. (2010). Stationary algorithmic probability. *Theoretical Computer Science, 411*, 113–130.

Nakhnikian, G., & Salmon, W. C. (1957). Exists as a predicate. *Philosophical Review, 66*, 535–542.

Nielsen, M. A., & Chuang, I. L. (2000). *Quantum computation and quantum information*. Cambridge: Cambridge University Press.

Nishimura, H., & Ozawa, M. (2009). Perfect computational equivalence between quantum Turing machines and finitely generated uniform quantum circuit families. *Quantum Information Processing, 8*, 13–24.

Papadimitriou, C. H. (1994). *Computational complexity*. New York: Addison-Wesley.

Piccinini, G. (2011). The physical Church-Turing thesis: Modest or bold? *The British Journal for the Philosophy of Science, 62*, 733–769.

Pitowsky, I. (1990). The physical Church thesis and physical computational complexity. *Iyyun: The Jerusalem Philosophical Quarterly, 39*, 81–99.

Pitowsky, I. (2002). Quantum speed-up of computations. *Philosophy of Science, 69*, S168–S177.

Raussendorf, R., & Briegel, H. J. (2002). Computational model underlying the one-way quantum computer. *Quantum Information and Computation, 2*, 443–486.

Rivest, R. L., Shamir, A., & Adleman, L. (1978). A method for obtaining digital signatures and public-key cryptosystems. *Communications of the ACM, 21*, 120–126.

Seiferas, J. I. (1990). Machine-independent complexity theory. In van Leeuwen J (Ed.), *Handbook of theoretical computer science, volume A: Algorithms and complexity* (pp. 165–186). Cambridge, MA: MIT Press/Elsevier.

Shor, P. W. (1994). Algorithms for quantum computation: Discrete logarithms and factoring. In *1994 Proceedings of 35th Annual Symposium on Foundations of Computer Science* (pp. 124–134).

Shor, P. W. (1997). Polynomial-time algorithms for prime factorization and discrete logarithms on a quantum computer. *SIAM Journal on Computing, 26*, 1484–1509.
Simon, D. R. (1994). On the power of quantum computation. In *1994 Proceedings of the 35th Annual Symposium on Foundations of Computer Science* (pp. 116–123). Los Alamitos: IEEE Press.
Sipser, M. (1992). The history and status of the P versus NP question. In *Proceedings of the Twenty-Fourth Annual ACM Symposium on Theory of Computing, STOC'92* (pp. 603–618). New York: Association for Computing Machinery.
Spekkens, R. W. (2007). Evidence for the epistemic view of quantum states: A toy theory. *Physical Review A, 75*, 032110.
Steane, A. M. (2003). A quantum computer only needs one universe. *Studies in History and Philosophy of Modern Physics, 34*, 469–478.
Timpson, C. G. (2013). *Quantum information theory & the foundations of quantum mechanics*. Oxford: Oxford University Press.
Turing, A. M. (1936–7). On computable numbers, with an application to the Entscheidungsproblem. *Proceedings of the London Mathematical Society Second Series, s2–42*, 230–265.
Turing, A. M. (1938). On computable numbers, with an application to the Entscheidungsproblem. A correction. *Proceedings of the London Mathematical Society Second Series, s2–43*, 544–546.
Turing, A. M. (1950). Computing machinery and intelligence. *Mind, 59*, 433–460.
van Emde Boas, P. (1990). Machine models and simulations. In J. van Leeuwen (Ed.), *Handbook of theoretical computer science, volume A: Algorithms and complexity* (pp. 1–66). Cambridge, MA: MIT Press/Elsevier.
Veldhorst, M., Yang, C. H., Hwang, J. C. C., Huang, W., Dehollain, J. P., Muhonen, J. T., Simmons, S., Laucht, A., Hudson, F. E., Itoh, K. M., Morello, A., & Dzurak, A. S. (2015). A two-qubit logic gate in silicon. *Nature, 526*, 410–414.

Part IV
Mathematics in Technology

Chapter 12
Mathematical Models of Technological and Social Complexity

Ronald Kline

Abstract This chapter recounts part of the history of mathematical modeling in the social sciences in the United States and England in the 1950s and 1960s. It contrasts the modeling practices of MIT engineer Jay Forrester, who developed the field of System Dynamics, with that of English cybernetician Stafford Beer, and American social scientist Herbert Simon, in regard to the contested issues of prediction and control. The analysis deals with the topic of mathematics and technology in three senses: the technological origins of mathematical modeling in cybernetics and System Dynamics in the fields of control and communications engineering; the use of digital computers to create models in System Dynamics; and the conception of scientific models, themselves, as technologies. The chapter argues that the different interpretations of Forrester, Beer, and Simon about how models should serve as technologies help explain differences in their models and modeling practices and criticisms of Forrester's ambitious attempts to model the world.

In 1948, Warren Weaver, chief of the natural sciences division of the Rockefeller Foundation in the United States, issued a challenge to scientists in an article entitled "Science and Complexity." An applied mathematician who had headed the government's research and development program on gunfire control systems in World War II and now consulted for the RAND Corporation on the military science of operations research, Weaver said that scientists in the first half of the twentieth century had learned to solve the problem of "disorganized complexity" by using statistical mechanics and probability theory to analyze random events occurring among a large number entities, such as the movements of gas molecules in thermodynamics. The challenge in the second-half of the twentieth century was to solve the problem of "organized complexity" in biology and the social sciences. That problem could not be analyzed with pre-war methods, Weaver maintained,

R. Kline (✉)
Science and Technology Studies Department, Cornell University, Ithaca, NY, USA
e-mail: rrk1@cornell.edu

S. O. Hansson (ed.), *Technology and Mathematics*, Philosophy of Engineering and Technology 30, https://doi.org/10.1007/978-3-319-93779-3_12

because structured systems such as living organisms and social organizations exhibited purposeful rather than random behavior. Weaver predicted that scientists would utilize two innovations coming out of World War II (the digital computer and the interdisciplinary approach of operations research) to solve the "complex, but essentially organic, problems of the biological and social sciences."[1]

In this chapter, I discuss two new scientific disciplines that contemporaries thought were well-suited to solve the problem of organized complexity in the social sciences – cybernetics and System Dynamics. I contrast the work of engineer Jay Forrester, who developed the field of System Dynamics at the Massachusetts Institute of Technology (MIT), with that of American social scientist Herbert Simon and English cybernetician Stafford Beer. I discuss why contemporaries thought cybernetics and System Dynamics could meet Weaver's challenge, how Forrester and his competitors mathematically modeled social systems using theories from control-systems engineering, and criticisms about the validity of Forrester's models. I focus on the contested issues of prediction and control in modeling practice in order to draw out wider issues about the mathematization of the social sciences in this period.

My analysis deals with the topic of mathematics and technology in three senses: the technological origins of mathematical modeling in cybernetics and System Dynamics; the use of digital computers to create models in System Dynamics; and the conception of scientific models, themselves, as technologies. For the latter theme, I draw on recent work in the history and philosophy of science that analyzes the construction and use of scientific models as technologies.[2] Although Forrester and his contemporaries used similar principles from control-systems engineering, they created much different sorts of models. I argue that their different interpretations of how models should serve as technologies help explain differences in their models and modeling practices, criticisms of Forrester, and Forrester's ambitious attempts to model first the industrial firm, then the city, and finally the world.

---

[1] Warren Weaver, "Science and Complexity," *American Scientist*, 36 (1948): 536–544, on 542. On his work in World War II and later at RAND, see David A. Mindell, *Between Human and Machine: Feedback, Control, and Computing before Cybernetics* (Baltimore: Johns Hopkins University Press, 2002), Chap. 7; and Martin Collins, *Cold War Laboratory: RAND, the Air Force, and the American State, 1945–1950* (Washington, DC: Smithsonian Institution press, 2002), Chap. 4.

[2] Margaret Morrison and Mary S. Morgan, "Models as Mediating Instruments," in *Models as Mediators: Perspectives on Natural and Social Science*, edited by Morgan and Morrison (Cambridge: Cambridge University Press, 1999): 10–37.

12.1 The Technological Basis of Cybernetics and System Dynamics

The technological basis of modeling in cybernetics and System Dynamics came from the engineering fields of control systems and communications. From the late nineteenth century to the 1930s, inventors and engineers devised numerous types of small regulating units to control the operation of much larger, often unruly machines. These control systems steered ships, stabilized the motion of ships and airplanes by means of gyroscopes, enabled airplanes to fly with a minimal amount of human control, and aimed large artillery at distant targets on land, sea, and in the air. The control systems worked on the principle of the servomechanism, in which the outputs of the machine to be regulated (e.g., its position, direction, or motion) were fed back in a closed loop and compared to the desired settings in order to generate error signals that would eventually move the machine to the desired goal (see Fig. 12.1). In the field of communications, engineer Harold Black and his colleagues at the American Telegraph and Telephone Company (AT&T) invented a negative-feedback amplifier in the 1920s that stabilized the transmission of long-distance telephone signals by using feedback to cancel out the distortions that came from temperamental vacuum tubes used in the repeater amplifiers.

Cybernetics and System Dynamics drew on the mathematical theories that engineers and scientists developed between the wars to analyze and improve the design of these feedback control systems. At MIT, Harold Hazen created a general theory of servomechanisms while working on the differential analyzer, an influential mechanical analog computer used to solve complex differential equations. At AT&T's Bell Telephone Laboratory, Harry Nyquist and Henrik Bode created mathematical theories and graphical techniques to design stable negative feedback amplifiers for the Bell System. During World War II engineers combined the fields of control and communication by recognizing that any servomechanism operated like a negative feedback amplifier and thus could be analyzed using the Nyquist Diagram and the Bode Plot. This merger of control and communication theory occurred when the National Defense Research Committee's (NDRC) division of Fire Control, headed by Warren Weaver, let contracts to MIT's Servomechanisms Laboratory and to Bell Labs to design anti-aircraft fire-control systems.[3]

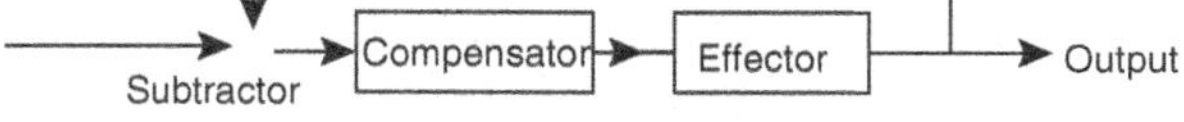

Fig. 12.1 Control-system diagram. Wiener, *Cybernetics*, 1948, p. 132

[3]Mindell, *Between Human and Machine*, Chaps. 3–5, 7–9, 11; and Stuart Bennett, *A History of Control Engineering, 1800–1930* (London: Peter Peregrinus, 1979), Chap. 4.

It was while working on a sophisticated anti-aircraft project for the NDRC that MIT mathematician Norbert Wiener created a theory of prediction and control, as well as a theory of information, that served as the basis for founding the new science of cybernetics in 1948. And it was while working at MIT's Servomechanisms Laboratory during the early Cold War that Jay Forrester learned the theory of feedback control and communication that would form the basis of System Dynamics.

12.2 Mathematical Modeling of Social Systems in the Cold War

The work of Forrester and other modelers was part of a larger movement to quantify the social sciences in the Cold War by replacing qualitative and physical models with equation-based mathematical models.[4] Although several scientists had created mathematical models of the economy in the 1920s and 1930s that consisted of linked differential equations, difference equations, or inferential statistics – most notably Dutch economist Jan Tinbergen – this endeavor did not gain ground outside of economics in the U.S. until after World War II.[5] In 1956, the first volume of *General Systems*, the journal of General System Theory, a field founded by biologist Ludwig von Bertalanffy, reprinted a 1951 article by economist Kenneth Arrow at Stanford University that surveyed mathematical models in the social sciences. The editors of the journal explained that physicists and, increasingly biologists, did not have to be convinced of the utility of mathematical models; social scientists did. Furthermore, the universal claims of mathematics were "put to a severe test in the evaluation of the role of mathematical models in social science."[6]

The social scientists and engineers who worked in cybernetics and Systems Dynamics did not have to be convinced because mathematical modeling had been central to those fields since their founding after the war. Both interdisciplines were part of a larger systems movement that grew to prominence in the 1960s with the rise of systems analysis, systems engineering, operations research, game theory, and General Systems Theory.[7] All of these fields, as well as economic input-output

[4]See Max Black, *Models and Metaphors: Studies in Language and Philosophy* (Ithaca: Cornell University Press, 1962), 223–226.

[5]George P. Richardson, *Feedback Thought in Social Science and Systems Theory* (Philadelphia: University of Pennsylvania Press, 1991, Chap. 2; and Adrienne van de Bogaard, "Past Measurement and Future Prediction," in *Models as Mediators*, edited by Morgan and Morrison, Chap. 10.

[6]Ludwig von Bertalanffy and Anatol Rapoport, "Preface," *General Systems: Yearbook of the Society for the Advancement of General Systems Theory*, 1 (1956): v; and Kenneth Arrow, "Mathematical Models in the Social Sciences," ibid., 29–47, which was reprinted from *The Policy Sciences: Recent Developments in Scope and Method*, edited by Daniel Lerner and Harold D. Lasswell (Stanford: Stanford University Press, 1952), 129–154.

[7]Ronald R. Kline, *The Cybernetics Moment, Or Why We Call Our Age the Age of Information* (Baltimore: Johns Hopkins University Press, 2015), 190–195.

analysis and linear programming, were related to what Philip Mirowski has called the "cyborg sciences" of the Cold War and Hunter Heyck has called the "age of system" in the social sciences.[8]

The central analogy of cybernetics, popularized by Wiener, who gave the field its name, was that both animals and machines could be analyzed using mathematical theories from control and communications engineering. A favorite analogy was to compare the central nervous system in humans – consisting of the brain, nerve net, effectors, and sensors – with control systems run by digital computers, popularly known as "electronic brains." Cyberneticians claimed they could model living organisms, intelligent machines, and society as systems structured by causal information-feedback loops, because such systems exhibited purpose, learning, and adaptation to the environment.[9] Early researchers in cybernetics created physical and mathematical models of human behavior and the brain.[10] Although Wiener had stated that the mathematics of cybernetics could not be applied to the social sciences because of the lack of sufficient runs of quality data in those fields,[11] several social scientists in the United States applied cybernetics in their work. These included such prominent figures as Talcott Parsons in sociology, Herbert Simon in management science, Karl Deutsch in political science, George Miller in psychology, Roman Jakobson in linguistics, and Gregory Bateson in anthropology.[12]

Sociologist Walter Buckley at the University of California, Santa Barbara, noted in 1967 that cybernetic and general system models were not simply a fashionable analogy based on the latest technology. "There is a difference between analogizing and discerning fundamental similarities of structure," Buckley stated. "The newer system view is building on the insight that the key to substantive differences in systems lies in the way they are *organized*, in the particular mechanisms and dynamics of the interrelationships among the parts with their environment." Thus, contemporaries believed that cybernetics and systems theory could solve the problem of organized complexity because purposeful systems (living and non-

[8]Philip Mirowski, *Machine Dreams: Economics Becomes a Cyborg Science* (Cambridge: New York: Cambridge University Press, 2002), Chap. 1; and Hunter Heyck, *The Age of System: Understanding the Development of Modern Social Science* (Baltimore: Johns Hopkins University Press, 2015).

[9]Steve J. Heims, *The Cybernetics Group* (Cambridge, MA: MIT Press, 1991); Geoffrey C. Bowker, "How to be Universal: Some Cybernetic Strategies, 1943–1970," *Social Studies of Science* 23 (1993): 107–127; and Kline, *Cybernetics Moment*, Chap. 3.

[10]See., e.g., Andrew Pickering, "Ross Ashby: Psychiatry, Synthetic Brains, and Cybernetics," in *The Cybernetic Brain: Sketches of Another Future* (Chicago: University of Chicago Press, 2010), Chap. 4; and Lily E. Kay, "From Logical Neurons to Poetic Embodiments of Mind: Warren S. McCulloch's Project in Neuroscience," *Science in Context*, 14 (2001): 591–614.

[11]Norbert Wiener, *Cybernetics: Or Control and Communication in the Animal and the Machine* (Cambridge, MA, and New York: Technology Press and John Wiley, 1948), 33–34, 189–191.

[12]Richardson, *Feedback Thought in Social Science and Systems Theory*, Chaps. 3–5; and Kline, *Cybernetics Moment*, Chap. 5.

living) were governed by actual, casual information-feedback loops assumed by cybernetics.[13]

12.3 Prediction and Control

Those who employed cybernetics and systems theory to model social processes often parted ways when it came to the issues of prediction and control, even though they relied on the same control-system theory and practices described above. By the 1930s, engineers had worked out how to use Laplace transforms to solve the linear differential equations that described such a system. Prediction and control are intimately related in such an analysis. By finding a general solution to the equations and plotting the results in a Nyquist diagram, engineers could *predict* the performance of the system under various inputs and determine its stability, while automatically *controlling* the system in the desired manner. More complex systems containing multiple information-feedback loops, non-linearities, and positive as well as negative feedback were considered to be mathematically intractable.[14]

To illustrate the application of this theory to the social sciences, consider the work of Herbert Simon. A polymath social scientist at the newly established Graduate School of Industrial Administration at the Carnegie Institute of Technology (now Carnegie-Mellon University), Simon was a strong advocate of mathematical modeling in the social sciences.[15] In 1952, he published an extensive paper that applied servomechanism theory to the problem of optimizing production control in manufacturing. Saying his method came under the rubric of "cybernetics," Simon gave a tutorial for social scientists on the mathematics of servo theory, which was common in electrical engineering. He explained how to use Laplace transforms to solve the linear differential equations of control systems by converting them to complex-number algebraic equations. He also showed how to use the Nyquist criteria to decide whether or not the system was stable by plotting the roots of the algebraic equations in the frequency domain. Simon acknowledged that the results of his mathematical analysis of the production problem were known intuitively by experienced managers, but he concluded that servomechanism theory provided a rigorous and precise methodology on which to base decision rules.[16] As noted

---

[13]Walter F. Buckley, *Sociology and Modern Systems Theory* (Englewood Cliffs, NJ: Prentice Hall, 1967), 3 (quotation), 38–39. On systems theory and organized complexity, see Ludwig von Bertalanffy, *General System Theory: Foundations, Development, Applications* (New York: George Braziller, 1968), 34, 68.

[14]See, e.g., Gordon S. Brown and Donald P. Campbell, *Principles of Servomechanisms* (New York: Wiley, 1948).

[15]Hunter Crowther-Heyck, *Herbert A. Simon: The Bounds of Reason in Modern America* (Baltimore: Johns Hopkins University Press, 2005).

[16]Herbert Simon, "On the Application of Servomechanism Theory in the Studies of Production Control," *Econometrica*, 20 (1952): 247–268, on 258.

by his biographer, Simon combined the science of control (cybernetics) with the science of choice (decision theory) to come up with his celebrated theory of bounded rationality.[17]

Prediction and control entered into Simon's model in several ways. The general solution of the differential equations precisely predicts future states of the system. Determining whether a system is stable or not is another form of prediction. The model represents how a system is controlled, and the modeler can indirectly control a system by recommending changes to it based on experiments done with the model. At this time, economists often associated prediction of future states with what they called "descriptive models" and indirect control with what they called "prescriptive models."[18] In both senses, these models are themselves technologies.

A third form of control can be seen in the work of Stafford Beer. An operations-research consultant in the British steel industry, Beer turned from the mathematical techniques of OR, such as linear programming, to cybernetics in the 1950s. Beer laid out the new approach in *Cybernetics and Management* (1959), and in several subsequent books, culminating in the ill-fated scheme to build a cybernetic system to control the economy of socialist Chile in the early 1970s.[19] Beer's proposal for a Cybernetic Factory, which is analogous to a biological model, illustrates his approach. The main idea behind the system (Fig. 12.2) is that the cybernetician designs a control system that homeostatically couples the Company to its Environment. The control system consists of the boxes in the bottom half of the diagram. The main feedback loop, running from the output of the Company's Homomorphic Model to the Company, controls the Company's operations. This is not a traditional, negative feedback control loop. The homomorphic models enable the two-part system (Company and Environment) to achieve ultrastability, a concept Beer adapted from British cybernetician W. Ross Ashby. Ultrastability ensures the survival of the Company, which is the main goal of Beer's cybernetic management.[20] Although Beer, like Simon, relied on information-feedback loops, his purpose was to directly control systems not to create models to make predictions. He did this by designing automatically adaptable software models of the system and its environment – the Homomorphic Models – and embedding them physically into an actual control system.

---

[17]Crowther-Heyck, *Herbert A. Simon*, Chap. 9.

[18]Mary S. Morgan, "Simulation: The Birth of a Technology to Create 'Evidence' in Economics," *Revue d'Histoire des Sciences*, 57 (2004): 341–377, on n. 16, 348, and 365–366.

[19]Pickering, *Cybernetic Brain*, Chap. 6; and Eden Medina, *Cybernetic Revolutionaries: Technology and Politics in Allende's Chile* (Cambridge, MA: MIT Press, 2011).

[20]Stafford Beer, *Cybernetics and Management* (New York: John Wiley, 1959), Chap. 1; W. Ross Ashby, *Design for a Brain: The Origin of Adaptive Behaviour* (New York: John Wiley, 1952); and Pickering, *Cybernetic Brain*, Chaps. 4 and 6.

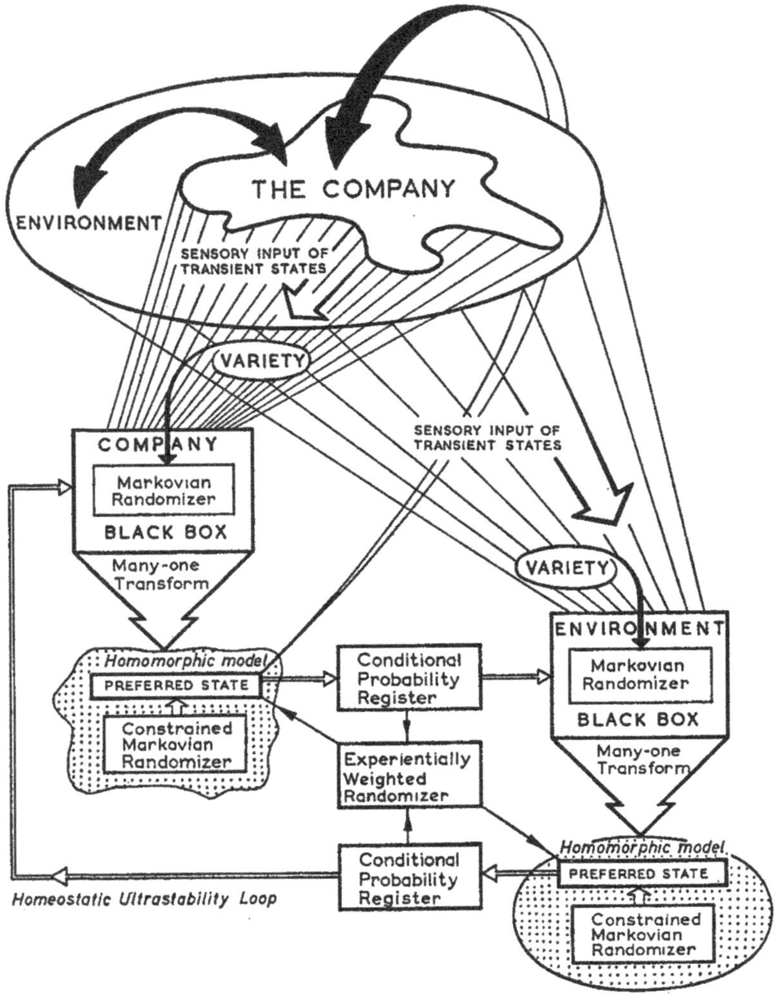

Fig. 12.2 Cybernetic factory. Beer, *Cybernetic Management*, 1959, p. 150

12.4 Jay Forrester and System Dynamics

These multiple meanings of prediction and control are evident in the work of Jay Forrester. Forrester modeled much more complex systems than did Simon and other social scientists, and tried to model the complexity Beer thought could not be represented. Forrester developed his method of modeling from his experience designing control systems and digital computers at MIT during World War II and the early Cold War. He worked on an analog flight simulator at Gordon Brown's Servomechanisms Laboratory during the war, then turned to digital computing after the war and led the group that developed the Whirlwind computer and magnetic-core memory. Whirlwind formed the basis for SAGE, the real-time, computer-controlled anti-aircraft radar system that was deployed across the northern United States during the 1950s.[21] Upon moving to the newly established Sloan School of Industrial Management at MIT in the mid-1950s, Forrester brought his experience with control systems and digital computers to bear on the solution of management problems. He embarked on a 5-year program – supported by the Ford Foundation, the Digital Equipment Corporation, and MIT's Computation Center – to develop his first modeling program, "Industrial Dynamics," whose main purpose was to train managers at the Sloan School. Forrester expanded his modeling technique to analyze the growth and decay of cities in *Urban Dynamics* (1969), then the world's population and natural resources in *World Dynamics* (1971). In all of these projects, which eventually came under the rubric of "System Dynamics," Forrester designed models to be learning tools, that is, technologies for policy makers.[22]

Forrester explained the purpose and practices of his modeling technique, which remained remarkably constant over the years, in his first book on the subject, *Industrial Dynamics* (1961). At the Sloan School, Forrester established a "management laboratory," in which he and his group created mathematical models to serve as "tools for 'enterprise engineering,' that is for the design of an industrial organization to meet better the desired objectives." Enterprise engineering consisted of four steps. First, Forrester and his group interviewed managers to identify the goals of the organization, problem areas to be investigated, and factors and decision policies to include in the model. Second, they created an information-feedback model on a digital computer to simulate the observed behavior of the organization. Third, they revised the computer model until it gave an "acceptable" representation of the organization's behavior (i.e., one agreed to by the modelers and their clients). Fourth, managers used these results to modify the organization to improve its performance.[23] Forrester and his followers defined the boundaries of their closed-system models so that the "behavior modes of interest [to their clients] are generated within the boundaries of the defined system."[24]

[21]Mindell, *Between Human and Machine*, Chap. 8; Paul N. Edwards, *Closed World: Computers and the Politics of Discourse in Cold War America* (Cambridge, MA: MIT Press, 1996); Chap. 3; and Thomas P. Hughes, *Rescuing Prometheus: Four Monumental Projects that Changed the Modern World* (New York: Pantheon, 1998), Chap. 2.

[22]Brian P. Bloomfield, *Modeling the World: The Social Construction of Systems Analysis* (London: Basil Blackwell), 1986; and William Thomas and Lambert Williams, "The Epistemologies of Non-Forecasting Simulations, Part I: Industrial Dynamics and Management Pedagogy at MIT," *Science in Context*, 22 (2009): 245–270.

[23]Jay W. Forrester, *Industrial Dynamics* (Cambridge, MA: MIT Press, 1961), Chaps. 1–2, quotation on 56.

[24]Jay W. Forrester, "Industrial Dynamics – After the First Decade," *Management Science*, 14 (1968): 398–415, on 406.

After a decade developing his program at MIT, modeling firms, and training modelers, Forrester could boast that industrial dynamics had spread to many U.S. companies, including Kodak, RCA, IBM, and the Coca-Cola company, and overseas to the Philips Lamp Works in the Netherlands. In 1968, Forrester maintained that industrial dynamics was not merely a modeling technique, but a profession. "Like the recognized professions," Forrester said, "there is an underlying body of principle and theory to be learned, applications to be studied to illustrate the principles, cases to build a background on which to draw, and an internship to develop the art of applying theory."[25]

In *Industrial Dynamics*, Forrester argued that six networks – handling flows of materials, orders, money, personnel, capital equipment, and information – could be interconnected to model any economic or company activity. Forrester modeled each network with the basic structure shown in the diagram in the top right section of Fig. 12.3. Levels accumulate flows of that network's activity (e.g., the flow of materials in production). Decision functions (shown by a valve symbol) regulate the flow rates between the levels (shown by a solid line) based on the information fed to them from the levels (shown by a dotted line). Forrester noted that this was a continuous rather than a discrete mathematical representation (i.e., an analog rather than a digital form of modeling, even though it was done on a digital computer). The symbols, in fact, resembled those used to set up equations on the analog differential analyzer invented by Vannevar Bush at MIT in the 1930s.[26] In mathematical terms, levels indicate integration, rates differentiation.

Consider the simplified model of a production-distribution system shown in the top left section of Fig. 12.3, which includes only three interconnected networks: materials, orders, and information.[27] To put this non-linear system into tractable mathematical form, Forrester wrote difference equations that related levels to rates, which the digital computer calculated at discrete time intervals. A level equation for this system is shown in the bottom section of Fig. 12.3. The equation states a simple accounting relationship: present retail inventory equals the previous retail inventory, plus the difference between the inflow shipment rate and the outflow shipment rate during the previous time interval, the difference in rates being multiplied by the computing time interval. As Forrester said, "In short, what we have equals what we had plus what we received less what we sent away."[28]

[25] Jay W. Forrester, "Industrial Dynamics – A Response to Ansoff and Slevin," *Management Science*, 14 (1968): 601–618, on 617. For an early list of industrial-dynamics clients, most of whom were modeled by consultants trained at MIT, see Edward B. Roberts, "New Directions in Industrial Dynamics," *Industrial Management Review*, vol. 6, no. 1 (Fall 1964): 5–14, on 11. On the experience of the Sprague Electric Company, which was the first firm to have its operations modeled by Forrester's group, see Bruce R. Carlson, "An Industrialist Views Industrial Dynamics," ibid., 15–20.

[26] Forrester, *Industrial Dynamics*, 67–71.

[27] Forrester added the other networks in modeling companies for his clients.

[28] *Industrial Dynamics*, 76. An example of a rate equation is OUT.KL = STORE.K/DELAY. See p. 78.

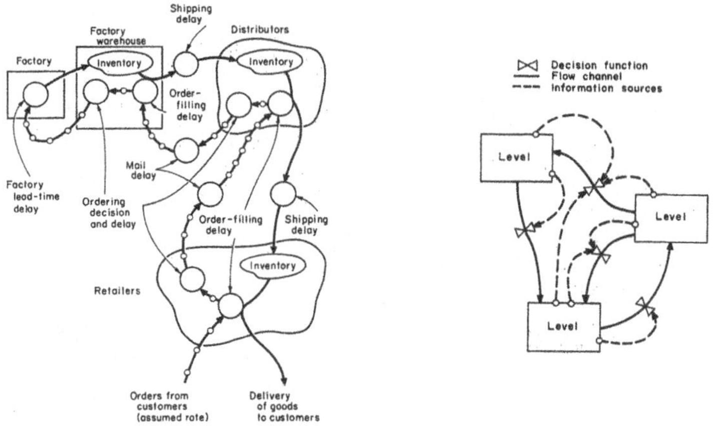

Example of a Difference Equation

```
IAR.K = IAR.J + (DT)(SRR.JK - SSR.JK)

IAR.K --- Inventory at Retail, level at present time K          [items]
IAR.J --- Inventory At Retail, level at previous time J         [items]
DT ------ computing time interval
SRR.JK -- Shipments Received at Retail, rate during time JK     [items/week]
SSR.JK -- Shipments Sent from Retail, rate during time JK       [items/week]

JK = DT.
```

Fig. 12.3 Principles of industrial dynamics. Forrester, *Industrial Dynamics*, 1961, pp. 67, 139

The industrial dynamics software, written in the language of the DYNAMO software compiler, created by Forrester's colleagues Phyllis Fox and Alexander Pugh to run on the IBM 700 series of digital computers, solved the level equations first, then the rate equations in each time interval DT. The difference equations were solved independently because the information-feedback channels were uncoupled from the rest of the model during the computing time interval. As Forrester explained, "The model traces the course of the system through time as the environment (levels) leads to decisions and actions (rates) that in turn affect the environment."[29]

Forrester's method differs substantially from the techniques employed by Simon and other modelers, who solved simultaneous, linked differential equations to obtain a general solution describing system behavior in a precise manner.[30] Forrester's use of the technology of the digital computer enabled him to model multiple-loop, nonlinear systems, containing both positive and negative feedback, whose differential equations were intractable. The computer simulation for actual systems often ran to more than 100 equations. Forrester stated that "When we no longer insist that

[29]Ibid., 75.

[30]Richardson, *Feedback Thought in Social Science and Systems Theory*, 153–155.

we must obtain a general solution that describes, in one neat package, all possible behavior characteristics of the system, the difference in difficulty between linear and nonlinear systems vanishes."[31]

In effect, Forrester traded the ability to model realistic, non-linear systems for the ability to make precise predictions of less realistic, linear systems. Forrester explained that he did not use the digital computer to do precise numerical integration of the difference equations as was common in scientific calculation.[32] These "elaborate numerical methods," Forrester argued, were not justified in industrial dynamics. "We are not working for high numerical accuracy. The information-feedback character of the systems themselves make the solutions insensitive to round-off and truncation errors." To complicate the model further, Forrester added an element of random noise to the rates, simulating actual conditions. The solution interval, DT, needed to be small to enable an accurate characterization of the overall performance of the system, but not so small that it led to excessive use of expensive computer time, which was a concern at the time, even at MIT. He empirically adjusted the solution interval "to observe whether or not the solutions are sensitive to the simplified numerical methods that are being used."[33]

Forrester addressed the issues of prediction and control under the topic of model validity. In line with the pedagogical goals of his program at the Sloan School,[34] Forrester developed industrial models as a technology to train managers how to redesign (i.e., control) industrial systems. Forrester thus related prediction and control to policy making (in this case management policy). Forrester argued that a model was valid if it predicted the general results that would "ensue from a *change* in organizational form or policy," especially the "*direction* of the *major changes* in system *performance*." It should also indicate the "approximate *extent* of the system improvements that will follow." The model could not predict precisely the future state of a system. Forrester thought such a prediction was "possible only to the extent that the correctly known laws of behavior predominate over the unexplained noise."[35]

Two response curves generated by the model of the production-distribution system discussed earlier illustrate Forrester's approach. Figure 12.4 shows the

---

[31] Forrester, *Industrial Dynamics*, 51 (quotation), Appendix B (list of equations for models). For a similar statement, see Forrester, *Urban Dynamics* (Cambridge, MA: MIT Press, 1969), 108.

[32] He explained the relationship between difference equations and differential equations, for example, as $IAR = IAR_{t=0} + \int (SRR\text{-}SSR)dt$. See Forrester, *Industrial Dynamics*, n, 8, p, 76. Numerical integration by the digital computer has a long history; see Thomas Haigh, Mark Priestly, and Crispin Rope, *ENIAC in Action: Making and Remaking the Modern Computer* (Cambridge, MA: MIT Press, 2016).

[33] Forrester, *Industrial Dynamics*, 80. DT was usually determined by the exponential delay (p. 79). He also says that statistical analysis probably cannot model non-linear, noisy, information-feedback systems. See p. 130.

[34] On this point, see, especially, Thomas and Williams, "The Epistemologies of Non-Forecasting Simulations, Part I."

[35] Forrester, *Industrial Dynamics*, 116, 124, his emphasis.

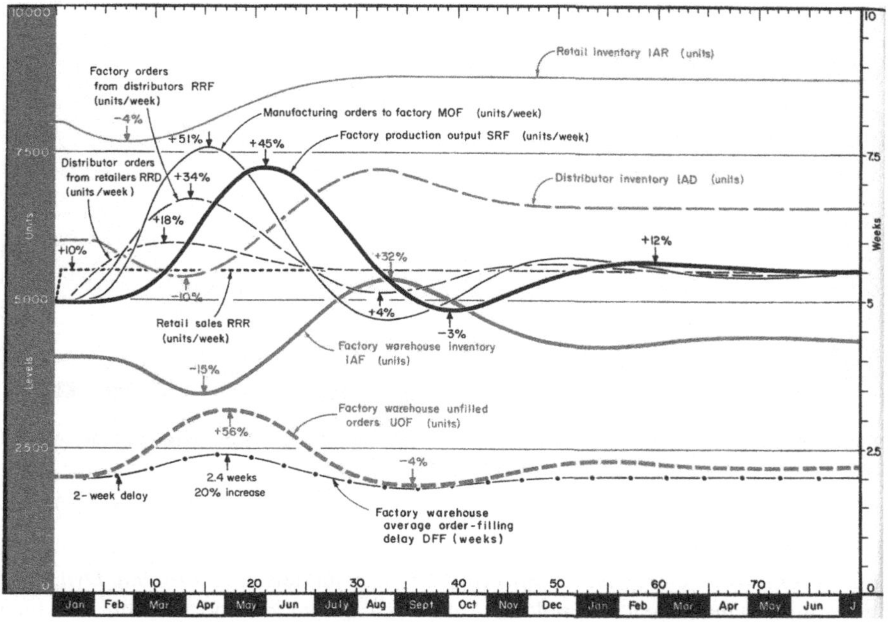

Fig. 12.4 Response of production-distribution system to a sudden 10% increase in retail sales. Forrester, *Industrial Dynamics*, 1961, p. 24

response of the model to a sudden and sustained increase of 10% in retail sales. The values for retail inventory, distribution inventory, factory production, factory inventory, and so forth fluctuate, then reach stability at higher values over a year after the step increase.[36] Figure 12.5 shows the response of the same production-distribution system to a 10% unexpected rise and fall in retail sales over a 1-year period. The periodic disturbance produces large swings in system values, showing that the "system, by virtue of its policies, organization, and delays, tends to amplify those retail sales changes to which the system is sensitive."[37] The effects of other disturbances are suppressed.

These curves illustrate how Forrester engaged in a policy form of prediction and control. He experimented with the model as a technology to investigate (predict) how the system responded in general to standard inputs. The next step was to "determine ways to improve management control for company success." Forrester altered the design of the system by changing the model's feedback-loop structure and policies (rates), such as the way orders were handled and inventory managed, to see the effects on system response. The mathematical models allowed experimental

[36]In this case, modeling showed that factory warehouse orders were not due to an industry increase in business volume, but to a transient.

[37]*Industrial Dynamics*, 28.

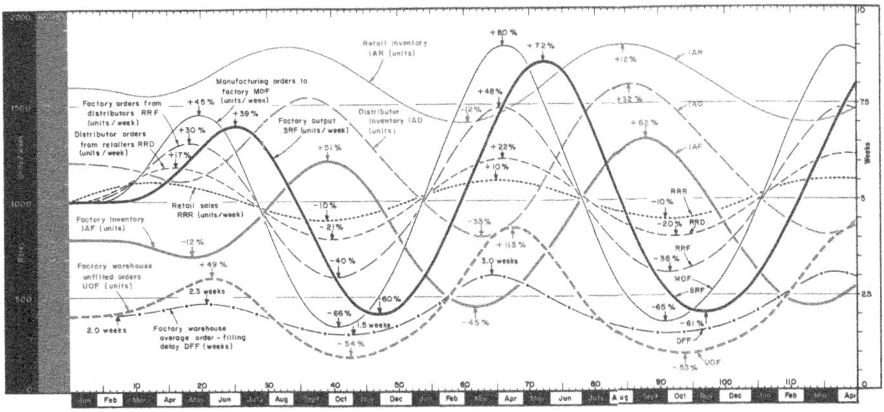

Fig. 12.5 Response of production-distribution system to 10% unexpected rise and fall in retail sales. Forrester, *Industrial Dynamics*, 1961, pp. 26–27

computer simulations to be conducted in Forrester's "management laboratory" in his program of "enterprise engineering."[38] The models in *Industrial Dynamics* were thus management technologies of policy prediction and control, in which representation served the interventionist purpose of redesigning industrial systems.

Forrester's engineering experience informed his modeling of social complexity. One of the foundations of industrial dynamics was that the "experimental model approach to the design of complex engineering and military systems can be applied to social systems."[39] The response curves generated by the model to various inputs answer the kinds of questions regarding prediction and control that would be posed by a control-system engineer: Is the system stable? How does it respond to standard types of inputs? How can the responses be improved to avoid disastrous results such as system oscillation and run away? These considerations outweighed those of precisely predicting future system states in the culture of the control engineer, as they did for Forrester when he trained managers in Industrial Dynamics at the Sloan School.

This technological approach is also present in Forrester's analysis of the complexity of social systems. In contrast to Beer, Forester believed that realistic, multiple-loop, feedback systems exhibited a complexity that had definable behaviors. He discussed these characteristics in *Urban Dynamics* (1969) under seven headings: Counterintuitive Behavior; Insensitivity to Parameter Changes; Resistance to Policy Changes; Control through Influence Points; Corrective Programs Counteracted by the System; Long-term vs. Short-Term Response; and Drift to Low Performance. Forrester argued that the interaction of non-linear feedback loops

[38]Ibid., 31, 43, 56.

[39]Ibid., vi. The other three foundations were the theory of information-feedback systems, military decision making, and the digital computer.

led to many of these behaviors.[40] As pointed out by critics, it was probably no coincidence that the conservative nature of Forrester's model of urban systems – in which government programs such as public housing produce ineffective or counter-productive results – matched Forrester's conservative worldview.[41]

12.5 Criticism of Forrester's Models

Forrester's models came under heavy criticism from social scientists in the 1960s and 1970s. Here, I focus on critiques relating to the contested issues of prediction and control. Management scientists criticized Forrester for his unbending views on prediction and model validity. Liberal social scientists criticized his models and those of Stafford Beer as technocratic technologies of authoritarian control.[42] The fact that Forrester was an outsider to the culture of mathematical modeling in the postwar social sciences helps explain these criticisms and his responses to them.

The most substantial critique of industrial dynamics came from management scientists Igor Ansoff and Dennis Slevin at the Graduate School of Industrial Administration at the Carnegie Institute of Technology, a rival institution to the Sloan School at MIT. In 1968 Ansoff and Slevin published a lengthy, supposedly non-partisan evaluation of industrial dynamics in *Management Science*, the main journal of that new field, as part of a series of expository papers commissioned by the Office of Naval Research and the Army Research Office. The paper appeared alongside an account by Forrester of the first decade of industrial dynamics, a response by Forrester to Ansoff and Slevin, and their brief reply. Writing from the point of view of management science, Ansoff and Slevin criticized many aspects of Forrester's modeling practices. They objected to the requirement to quantify all variables, which left out management judgment, that the DYNAMO complier dictated modeling practices, Forrester's ignorance of previous models of industrial systems, that the costs of industrial dynamics outweighed its benefits, and the inability of Forrester's models to predict system outcomes. The last complaint led Ansoff and Slevin to question the validity of Forrester's models, and, by implication, the validity of industrial dynamics itself.[43]

Ansoff and Slevin related the issue of prediction to model validity at several points. To them, ascertaining the validity of a model meant devising a "test

---

[40] Forrester, *Urban Dynamics*, Chap. 6. In contrast, Beer categorized systems in terms of two dimensions: complexity (as being "Simple," "Complex," or "Exceedingly Complex"); and determinism (as being "Deterministic" or "Probabilistic"). See Beer, *Cybernetics and Management*, 12, 18.

[41] See, for example, S. I. Schwartz and T. C. Foin, "A Critical Review of the Social Systems Models of Jay Forrester," *Human Ecology*, 1 (1972): 161–173, on 166; and Bloomfield, *Modeling the World*, 40–47.

[42] See, e.g., Robert Lilienfeld, *The Rise of Systems Theory: An Ideological Analysis* (New York: John Wiley, 1978); and Medina, *Cybernetic Revolutionaries*, Chap. 6.

[43] H. Igor Ansoff and Dennis P. Slevin, "An Appreciation of Industrial Dynamics," *Management Science*, 14 (1968): 383–397.

which establishes, first, that a model is capable of describing (and predicting) the behavior of the system with satisfactory accuracy; second, and more important for a management scientist, that changes in the model which produce desired improvements will produce closely similar improvements when applied to the real world systems." They thought Forrester's practice of developing models by adjusting the computer simulation to achieve dynamic characteristics – such as stability, oscillation, and growth – that were acceptable to modelers and their clients, was "largely subjective." They thought such tinkering was even more subjective than other forms of modeling in "prescriptive management science." For an exemplary model of management decision-making, Ansoff and Slevin pointed to the work of Herbert Simon, their colleague at Carnegie Tech, whose 1952 paper on servomechanism theory generated testable generalities for a production-control system. "What one gets instead [from Forrester] are prescriptions of how to construct models for individual situations; but no unifying insights are apparent."[44]

Although Simon had also based his model on control-system theory (the theory of servomechanisms), Ansoff and Slevin attributed some of Forrester's problems to that very field of engineering. They observed that Forrester's preference for modeling behavioral characteristics "has a strong resemblance to the typical typological techniques used in servomechanism design," e.g., Nyquist diagrams. More generally, they found the unreflective, imperial program of Forrester, an interloper from engineering, to be galling. They ended their essay, ironically titled "An Appreciation of Industrial Dynamics," by quoting Forrester to the effect that industrial dynamics should replace the failed program of mathematical modeling in management science. "This was written in 1961," Ansoff and Slevin concluded, "after a fifteen year period which many people, disagreeing with Forrester, would describe as a period of revolutionary advances in management science."[45]

This sort of reaction to Forrester left such an impression that Herbert Simon voiced a similar complaint a quarter of a century later in his autobiography, *Models of My Life* (1991). Simon wrote about the advice he had given the U.S. President's Science Advisory Committee (PSAC) about the Club of Rome's report *Limits to Growth* (1972), which relied on Forrester's model of world dynamics. Simon recalled that "My reaction was one of annoyance at this brash engineer who thought he knew how to predict social phenomena. In the discussion, I pointed out a number of naïve features of the Club of Rome model, but the matter ended, more or less, with that."[46] Simon was more caustic in private. In early 1972, he wrote PSAC that "My objection, of course, is not to system studies, but to the cavalier way that Forrester does them, and his complete ignorance of the relevant theoretical and empirical literature." He thought that Dennis Meadows, a co-author of *Limits to Growth* whom he had met, was "less doctrinaire about what he is doing than is Forrester, but apparently lives as a satellite to the latter." That spring, Simon wrote

[44]Ibid., quotations on 387, 388, 390, 395.

[45]Ibid., quotations on 386, 396.

[46]Herbert Simon, *Models of My Life* (New York: Basic Books, 1991), 307.

the director of the Brookhaven National Laboratory about using linear programming to model energy policies: "Meanwhile the hullaballoo about the rather silly Club of Rome model of everything-in-the-world has had at least the good effect (so far) of stirring up some positive attitudes toward models. I don't know why it should take a bad model to convince people that modelling is a good thing, but I will not look this gift horse in the teeth."[47]

Many reviewers of Forrester's *World Dynamics* (1971), on which the Club of Rome model was based, were just as critical. Martin Shubick at the Department of Administrative Sciences at Yale, voiced many of the same criticisms Ansoff and Slevin had made, including satirizing Forrester's ignorance of the social sciences. In regard to Forrester's method of testing a model's validity via consensus, Shubick harshly said, "Most behavioral scientists even when they want to be 'relevant' are not completely satisfied with a criterion of validation that amounts to no more than acceptance by top decision makers or use by those in power. Such a criterion can fast lead to a Lysenko style of science. And it appears to be the one that Forrester accepts." Simon told the PSAC that he agreed with Shubick's review. One of Forrester's main defenders was Denis Gabor at Imperial College, a physicist who had turned to cybernetics to model social systems and thus rigorize the supposedly "soft" social sciences.[48]

Yet some critics apparently did not understand that Forrester had gone beyond Simon and other modelers to use servo theory to model *nonlinear* social systems. Shubick missed this point, as did Simon. In fact, Simon had to apologize in 1989 to Donella Meadows, another co-author of *Limits to Growth*, that he had mistakenly described Forrester's model as being linear in a manuscript Meadows was reviewing for *OR Forum*. Simon corrected the error, but he did not change his statement that it was not news that the Club of Rome model, like earlier ones of prey-predator relationships, would "explode" and show "large limit cycles of booms and busts of population and other variables." Simon concluded that this result "could have been inferred from textbook treatments of dynamic systems without any computation at all."[49]

Forrester responded to these criticisms by defending the tests of model validity he gave in *Industrial Dynamics*, charging that critics (usually, non-engineers) did

[47]Herbert Simon to David Beckler, Jan. 27, 1972; and Simon to Kenneth Hoffman, May 3, 1972, both in Herbert Simon Papers, Carnegie-Mellon University, box 51, Consulting, PSAC correspondence, available on-line at http://diva.library.cmu.edu/Simon

[48]Martin Shubick, "Modeling on a Grand Scale," *Science*, n.s., 174 (1971): 1014–1015, on 1014; Simon to Beckler (note 47 above), and Denis Gabor, "World Modeling," *Science*, n.s., 176 (1972): 109. On Gabor's approval of cybernetic modeling of social systems, see, e.g., Gabor, "Cybernetics and the Future of Our Industrial Civilization," *Journal of Cybernetics*, 1, no. 2 (April–June 1971): 1–4.

[49]Donella Meadows to Herbert Simon, August 26, 1989; Simon to Meadows, Sep. 6, 1989, both in Simon Papers, box 79, Publications, "Prediction and Prescription of Systems Modeling"; and Simon, "Prediction and Prescription of Systems Modeling," *OR Forum*, 38, no. 1 (Jan.-Feb. 1990): 7–14, on 9. On Shubick not understanding the non-linearity of Forrester's approach, see Harold Hemond, et al. "World Modeling," *Science*, n.s., 176 (1972): 109.

not understand his modeling practices, and promising that further development of his program would answer all charges.[50] Forrester's responses had become so predictable that a reviewer of his collected papers in a British operations-research journal in 1975 remarked, "Throughout the twenty years of I.D. and S.D. [Industrial Dynamics and System Dynamics], it is repeatedly stated that the subject is just beginning, that much research remains to be done, and that there are as yet few people sufficiently trained and competent to understand the conceptual and theoretical background necessary to apply the work. None of the published S. D. work to date – least of all 'World Dynamics' or 'Limits to Growth' – can be said to substantiate these claims for intellectual profundity." The reviewer thought Forrester's papers were "an attempt to place an MIT-exclusive brand name on a product which is an already widely available commodity."[51]

12.6 Discussion

What do my examples from cybernetics and System Dynamics tell us about the relationship between technology, mathematics, and modeling in the social sciences during the Cold War? In attempting to solve the problem of organized complexity, Simon, Beer, and Forrester drew on a successful technological theory (the theory of servomechanisms) to mathematically model complex social systems in a variety of ways using dynamic information-feedback loops in order to predict their behavior (Simon) or control them, either directly (Beer), or by prescribing improvements to the system (Forrester).

They debated at length the question of how well their adaptations of servomechanism theory modeled social systems. The question of whether or not it was permissible to borrow a highly mathematical model from engineering was not of concern to them because it had become the norm in such areas as Operations Research.[52] The question was how well the models worked in practice, how good a technology they were.

In England, Beer staked his claim on his ability to control extremely complex systems using an unconventional control theory derived from Ross Ashby. He succeeded in the field of operations research by designing systems that worked on the basis of performative control, rather than prediction.[53]

In the United States, social scientists in the related field of management science critiqued engineer Jay Forrester for creating a modeling technique at MIT's

[50]See, e.g., Forrester, "Industrial Dynamics – A Response to Ansoff and Slevin"; and Forrester, "World Modeling," *Science*, n.s., 176 (1972): 109–110.

[51]Mark Cantley, "[Review of] 'Collected Papers of Jay W. Forrester'...," *Operations Research Quarterly*, 28 (1977): 111–113, on 112–113.

[52]William Thomas, *Rational Action: The Sciences of Policy in Britain and America, 1940–1960* (Cambridge, MA: MIT Press, 2015).

[53]Pickering, *Cybernetic Brain*, Chap. 16.

business school outside the culture of social science. Ansoff, Slevin, and Shubick criticized Forrester for not knowing the social-science literature on mathematical modeling. Simon called him a "brash engineer" who had the audacity to model social systems. Forrester's reliance on servomechanism theory was not the issue because they viewed Simon's work in that area as a paragon of modeling practice. Instead, they criticized Forrester for slavishly abiding by the culture of control-system engineering to privilege behavioral characteristics over precise results – for valuing indirect control over prediction, prescription over description. Ironically, they upheld a representative ideal of scientific research more so than did the highly-respected physicist Denis Gabor. The conflict between these engineering and social-science cultures apparently did not encourage Forrester to read the social-system modelers, nor did they read Forrester very carefully either.

Ever the evangelist, Forrester acted in a manner that his critics called hubris and his disciples called leadership. He and his followers worked hard to establish System Dynamics as an autonomous field. They imitated the famous MIT summer studies at the Sloan School, made a fetish out of computer simulation, relied on corporate sponsors rather than peer-review scientific agencies for support, and established their own journal and professional society.[54] These efforts further insulated System Dynamics from mathematical modeling in the social sciences, which set the stage for the severe criticisms of the Club of Rome report.

In the end, it was Herbert Simon's theory of rational choice, derived from cybernetics, that prevailed in the social sciences. By combining the science of choice (decision theory) with the science of control (servomechanisms theory), Simon created a hybrid that was not so closely tied to the closed-loop feedback systems of engineering and technology.

[54] Forrester, "The Beginning of System Dynamics," Banquet Talk at the International Meeting of the System Dynamics Society, Stuttgart, Germany, July 13, 1989, available at http://leml.asu.edu/jingle/Web_Pages/EcoMod_Website/Readings/SD+STELLA/Forrester-Begin'g-SD_1989.pdf

Chapter 13
The Rise and Fall of the Anti-Mathematical Movement

Sven Ove Hansson

Abstract Ever since the beginnings of modern engineering education at the end of the eighteenth century, mathematics has had a prominent place in its curricula. In the 1890s, a zealous "anti-mathematical" movement emerged among teachers in technological disciplines at German university colleges. The aim of this movement was to reduce the mathematical syllabus and reorient it towards more applied topics. Its members believed that this would improve engineering education, but many of them also had more ideological motives. They distrusted modern, rigorous mathematics, and demanded a more intuitive approach. For instance, they preferred to base calculus on infinitesimals rather than the modern ("epsilon delta") definitions in terms of limits. Some of them even demanded that practically oriented engineers should replace mathematicians as teachers of the (reduced) mathematics courses for engineers. The anti-mathematical movement was short-lived, and hardly survived into the next century. However calls for more intuitive and less formal mathematics reappeared in another, more sinister context, namely the Nazi campaign for an intuitive "German" form of mathematics that would replace the more abstract and rigorous "Jewish" mathematics.

13.1 Introduction

Technological work has always required calculations. Alloys, mortars, and paints have to be mixed in the right proportions, the sizes of building elements and machine parts have to fit in with the construction as a whole, and in most crafts the required amounts of raw materials have to be determined before the work begins. But the use of more advanced mathematics, in particular mathematical analysis, to solve technological problems did not get off the ground until the eighteenth century (Klemm 1966). The French military engineer Bernard Forest de Bélidor (1698–

S. O. Hansson (✉)
Division of Philosophy, Royal Institute of Technology (KTH), Stockholm, Sweden
e-mail: soh@kth.se

S. O. Hansson (ed.), *Technology and Mathematics*, Philosophy of Engineering and Technology 30, https://doi.org/10.1007/978-3-319-93779-3_13

1761) published a famous four-volume book, *L'architecture hydraulique* (1737, 1739, 1750, and 1753) that represents the first extensive use of integral calculus to solve engineering problems. In 1773, the physicist Charles-Augustin de Coulomb (1736–1806), who is now best known for his work on electricity, published his *Essai sur une application des règles de maximis et de minimis à quelques problèmes de Statique relatifs à l'Architecture* in which he applied mathematical analysis in innovative ways to what is now called structural mechanics. In 1775, the Swedish ship builder Fredrik Henrik af Chapman published a treatise on naval architecture that made use of Thomas Simpson's method for the approximation of integrals (Harris 2001). The technological use of mathematics has continued to develop ever since.

The new profession of engineering was established in the late eighteenth and early nineteenth centuries. From the beginning, applied mathematics was one of its hallmarks. Mathematics has retained its central role in the education of engineers, but its role has sometimes been subject to heated controversies in engineering schools. In the 1890s a movement that called itself anti-mathematical flourished among German professors in the engineering disciplines. This chapter traces the activities and concerns of that rather short-lived movement. A particularly interesting aspect is its denunciation of abstract methods in mathematics and its promotion of *Anschauung* (apperception) at the expense of mathematical rigour. In the 1920s and 1930s this ideal was relaunched for entirely different purposes in the Nazi "German mathematics" movement that will also be briefly discussed. But let us first have a look at how it all started.

13.2 The French Connection

The word "engineer" derives from the Latin *ingenium*, which was used in the classical period for a person's talent or inventiveness, but could also refer to a clever device or construction. In the Middle Ages, *ingenium* was a general term for catapults and other war machines for sieges. A constructor or master builder of such devices was called *ingeniarius* or *ingeniator* (Bachrach 2006; Langins 2004).

In the eighteenth century, "engineer" was still a military category. Engineering officers worked with war machines, but they also drew maps and built fortifications, roads and bridges. Several European countries had schools for engineering officers where these skills were taught along with considerable doses of mathematics (Langins 2004). Outside of the military, advanced technological tasks were still performed by master craftsmen without any theoretical education. It was not until 1794 that the first civilian school for engineering was founded in Paris under the name *École polytechnique* (Grattan-Guinness 2005). It was led by Gaspard Monge (1746–1818), an able mathematician and a Jacobin politician. He was determined to use mathematics and the natural sciences, including mechanics, as the foundation of engineering education (Hensel 1989a, p. 7). About a third of the curriculum hours were devoted to mathematics (Purkert and Hensel 1986, pp. 27 and 30–35).

Monge himself developed a new discipline, descriptive geometry. Largely based on perspective drawing, it provided a mathematical basis for technical drawing. It was put to use in machine construction in the École polytechnique, and it spread rapidly to other countries as an important part of the mathematical education of engineers (Lawrence 2003; Klemm 1966).

In addition to the practical usefulness of mathematics, the emphasis on mathematical knowledge was well in line with the meritocratic, anti-aristocratic ideology of the young republic. Mathematical proficiency was an objectively verifiable standard that provided a non-arbitrary and decidedly non-aristocratic criterion for selection and promotion, and it was therefore perceived as democratic. This approach was largely modeled from the education of artillery engineers, which had a strong mathematical component in addition to extensive technical training (Alder 1999).

The École polytechnique became the paragon of polytechnical schools in other countries in Europe and also in the USA. A sizable number of polytechnical schools were founded in the 1820s and 1830s in the German-speaking countries, and a similar development took place in other parts of Europe (Purkert 1990, p. 180; Schubring 1990, p. 273; Scharlau 1990). The new schools all followed the example of the École polytechnique in providing their students with a high level of mathematical and natural science education. Initially, most of them fell far behind the École polytechnique, but they tried to catch up. Beginning in the 1860s, they modelled their education after the established universities (Hensel 1989a, pp. 6–7; Grayson 1993).

13.3 Heightened Mathematical Ambitions

The use of mathematical methods for various practical engineering tasks increased throughout the nineteenth century. One prominent example is the use of Karl Culmann's graphic statics in the construction of the Eiffel Tower (Gerhardt et al. 2003). In consequence, treatises and textbooks were published on the application of mathematics to technological topics such as optics, structural mechanics, building construction, machine construction, shipbuilding, and engineering thermodynamics (Klemm 1966). The number of mathematical teaching positions in the technological colleges increased rapidly, and they provided a large part of the new academic positions in mathematics[1] (Schubring 1990, p. 273; Scharlau 1990, 264–279; Hensel 1989b). Several prominent mathematicians started their academic career in technological colleges. One of the foremost among them was Richard Dedekind

[1]The advanced polytechnical schools in the German-speaking countries were called "Technische Hochschulen". Most of them were renamed "Technische Universitäten" in the 1960s–1980s. In this chapter, these terms are translated as "technological college" respectively "technological university".

(1831–1916), who taught first at what is now the ETH in Zurich and then at what became the Braunschweig University of Technology. He is still known for his path-breaking studies on real numbers, set theory and abstract algebra, but his strict methods were sometimes considered impractical for engineers (Purkert 1990, p. 188).

Around the middle of the nineteenth century, professors in mechanical engineering increasingly emphasized new and more stringent mathematical approaches to their discipline. This put higher mathematical demands on their students. Consequently, mathematics teaching expanded on the curricula, and more advanced mathematics was introduced. However, the heightened mathematical ambitions were not always easy to implement. Many of the students had a rather weak mathematical background from their previous education.

In 1865, the influential Association of German Engineers (Verein Deutscher Ingenieure, VDI) adopted a new policy for the education of engineers. It was based on a committee report that emphasized the difference between the German engineering schools with their "proclivity for an extensive scientific education" and the "more immediate and empirical introduction" to the engineering profession in the corresponding English institutions.[2] The commission was aware that the English system had proponents among German engineers, but their own opinion was favourable to extensive studies of mathematics and the natural sciences, which they described as the "foundations" of technology. In contrast, the historical, aesthetic, and economic disciplines had more limited roles as "auxiliary sciences"[3] (Anon. 1865, pp. 706, 716, 721). Based on this report, the VDI adopted a resolution that recommended "the teaching of mathematics and the natural sciences to an extent and intensity not inferior to the universities".[4] It was also emphasized that these sciences should be studied "for their own sake, not just as a preparation to make it possible to study the special courses"[5] (Hensel 1989a, pp. 14–15). This should be read against the background that at this time, the technological colleges were striving to achieve the same status as the traditional universities. The VDI's policy seems to have contributed to the continued recruitment of prominent mathematicians to technological colleges in the 1870s and 1880s. These recruitments were based primarily on excellence in pure mathematics (Hensel 1989a, pp. 16–21 and 240–243).

[2]"Neigung zu einer umfassenden wissenschaftlichen Ausbildung", "mehr unmittelbare und empirische Einführung".

[3]"Grundlage", "Hülfswissenschaften".

[4]"die Mathematik und die Naturwissenschaften in einer den Universitäten nicht nachstehenden Ausdehnung und Intensität gelehrt werden sollen".

[5]"diese Wissenschaften an der polytechnischen Schule um ihrer selbst willen, nicht nur als Vorbereitung für die Fachcurse studiren zu können".

13.4 The Counterreaction

The expansion of mathematical teaching and research at technological universities was largely driven by professors in mechanical engineering who were engaged in the introduction of new, more mathematically advanced models in their disciplines. One of them was Franz Grashof (1826–1893). He was president of the VDI, and instrumental in developing its pro-mathematical policy of 1865. But many representatives of other technological subjects had a different opinion. They did not see the need for more mathematics, but they were worried that the new mathematics courses would infringe on their own subjects. This led to a growing tension that sometimes gave rise to open conflicts, in particular in decisions on recruitments and appointments. In the technological college in Munich such conflicts broke out already in the 1870s. The director of the school, Karl Maximilian von Bauernfeind (1818–1894), used his influence to recruit practically oriented mathematicians who put little emphasis on the more abstract and foundational issues in mathematics. He was actively opposed by the young mathematician Walther von Dyck (1856–1934), who wanted to recruit mathematicians with excellent research qualifications (Hashagen 1998, p. 174).

Contacts with educators in other countries fuelled the conflicts over the role of mathematics. Those favouring an extended mathematical curriculum looked to France and in particular the École polytechnique, whereas their opponents turned their eyes to the more practically oriented education of engineers in Britain with its strong focus on on-the-job training (Hensel 1989a, p. 6). Increasingly, their focus shifted to America, whose engineering education was quite similar to that in Britain. The world exhibition in Philadelphia in 1876 led to a vivid discussion in Germany about technological education (Manegold 1970, pp. 146–147).

However, it was another world exhibition, namely that in Chicago in 1893, that triggered an intensified and often outright hostile discussion about the teaching of mathematics in technological colleges (Manegold 1970, pp. 146–147). Once again, the VDI was the main forum of the discussions. The organization had organized German participation in the exhibition, and afterwards it also provided forums for discussions on what could be learned from the transatlantic visit. Considerable concerns were vented about Germany's competitiveness in comparison to the US, both in terms of actual engineering achievements and the education of new generations of engineers (Hensel 1989a, pp. 54 and 56–58). In consequence of these discussions the VDI decided to develop a new policy for higher technological education. A report published in 1894 by Alois Riedler (1850–1936), professor in mechanical engineering, had a central role in the society's deliberations. In this report, Riedler described in detail the educational laboratory facilities of the best American engineering schools. These laboratories were much superior to anything seen in Germany. The educational methods were equally impressive. "Value is attached not only to the use of instruments and equipment, but primarily to methods of scientific investigation and independent work as means to learning"[6] (Riedler 1894, p. 512). He put much emphasis on the difference between such scientific laboratories and workshops for learning practical work methods. Such workshops "have no place in higher education"[7] (p. 635). Riedler recognized that the American engineering schools were largely modeled after the English ones (p. 612), but the latter had much less resources and had not reached the same level as their American counterparts (p. 630). In order to catch up with the Americans, the German educational institutions would need resources for building laboratories, but they also had to make room for laboratory work in their curricula. The major obstacle to increased laboratory work was in his view "an excess of mathematical education" (p. 632) in the German schools that could not be found on the other side of the Atlantic[8]:

> The mischief that subjects such as physics, mechanics, etc. that should be exclusively devoted to knowledge about natural science, are represented and treated as mathematical, cannot be found there... In our schools the interest and efforts of the students are consumed by an overabundance of a farreaching and onesided "theoretical" education. There, the students' interest is stimulated by the superior means of education in laboratories and independent work in these laboratories.[9] (Riedler 1894, p. 632)

In an additional article, published in 1895, Riedler accused the mathematics teachers at technological colleges of "an unmeasurable overestimation of analytical methods" (Riedler 1895, p. 954). He saw it as imperative to remove "the theoretical speculations of modern university mathematics" from the syllabus (p. 955).[10] Since mathematicians could not be trusted to implement these changes, they would have to be replaced by teachers with another background:

> The technological colleges should themselves educate the teachers in mechanics, physics, and mathematics, since only those who know the needs and goals of the special sciences from working themselves in these sciences, can satisfy the demands of these teaching tasks.[11] (Riedler 1895, p. 955)

---

[6]"Es wird nicht nur auf Handhabung der Instrumente und Apparate sondern vor allem auf wissenschaftliche Untersuchungsmethoden und auf das selbschaffende Arbeiten als Mittel des Lernens Wert gelegt."

[7]"gehören überhaupt nicht an die Hochschulen".

[8]"ein Übermaß von mathematischer Ausbildung".

[9]"Der Unfug, dass Fächer, wie Physik, Mechanik usw., die ausschließlich der naturwissenschaftlichen Erkenntnis gewidmet sein sollen, als mathematische ausgegeben und behandelt werden, besteht dort nicht... Bei uns werden das Interesse und die Kraft der Schüler durch das Übermaß eines weitläufigen und einseitigen 'theoretischen' Unterrichts verbraucht, dort wird das Interesse durch das hervorragende Mittel der Unterweisung in Laboratorien und durch selbständige Arbeit in diesen angeregt."

[10]"eine maßlose Überschätzung der analytischen Methoden", "die theoretischen Spekulationen der modernen Universitäts-Mathematik".

[11]"Die technischen Hochschulen müssten die Lehrer der Mechanik, Physik und Mathematik selbst ausbilden, den nur derjenige, der Bedürfnisse und Ziele der Fachwissenschaften aus eigener fachwissenschaftlicher Arbeit kennt, vermag dem genannten Lehrerberufe zu genügen."

13.5 The Anti-mathematical Movement

Riedler's attacks on the teaching of abstract mathematics found resonance among many of his colleagues. As the mathematician Felix Klein wrote a few years later, "what had long been slumbering under the surface broke out with elemental force: the conflict between the engineers and the mathematicians on the amount and nature of the preparative mathematical education that is necessary for an engineer"[12] (Klein 1898, p. 1092). For instance, in a speech in 1894 at a meeting in the VDI, Adolf Ernst (1845–1907), who was professor in the Stuttgart technological college, went even further than Riedler and attacked not only the teaching methods but also the validity and relevance of modern mathematics:

> It is a fact that a too extensive mathematical apparatus is used to develop a whole series of hypotheses whose conditions are not satisfied and whose conclusions therefore lead to false results... The overemphasis on purely theoretical studies and lectures also [leads] to an overestimation of a prioristic thinking and to a highly detrimental underestimation of the value that perceptive ability has for our discipline, since our professional practice always deals with concrete rather than abstract cases.[13] (Ernst 1894, p. 1352)

According to Ernst, mathematics was just an "auxiliary science".[14] It was taught "far in excess of the limits to what is necessary" (p. 1354), and it had to be substantially reduced in order to make room for more practical approaches to engineering.[15] Ernst's speech was seen as the starting-point of the "anti-mathematical movement" that now unfolded[16] (Hensel 1989a, pp. 58–60).

Riedler's report was vigorously discussed in the local and regional branches of the VDI. Most of them were in favour of a reduction of higher mathematics in engineering schools (Anon. 1895d, p. 1214). However, this standpoint was far from unanimous. For instance, the branch in Aachen adopted a statement according to which they could "by no means endorse a limitation of education in mathematics", in particular considering the "uneven and often inadequate" mathematical skills of

[12]"[W]as lange unter der Oberfläche geschlummert hatte, das brach mit elementarer Gewalt hervor: der Gegensatz zwischen den Ingenieuren und den Mathematikern inbezug auf das Maß und die Art der für den Ingenieur erforderlichen mathematischen Vorbildung".

[13]"Sodann ist es Tatsache, dass eine ganze Reihe von Hypothesen mit weitschweifigen mathematischen Apparat verarbeitet wird, deren Voraussetzungen nicht zutreffen, und deren Schlussfolgerungen deshalb auch zu falschen Ergebnissen führen... [D]as Übergewicht der rein theoretischen Studien und Vorlesungen [führt] auch zu einer Überschätzung des a prioristischen Denkens und zu einer höchsts nachteiligen Unterschätzung des Wertes, der dem reinen Beobachtungsvermögen gerade für unser Fach unbedingt zuzuerkennen ist, weil wir es in der Ausübung unseres Berufes stets mit konkreten, nie mit abstrakten Fällen zu thun haben."

[14]"Hilfswissenschaft".

[15]"weit über die Grenzen des Notwendigen".

[16]According to Hensel, this movement was in its own time called the "antimathematische Bewegung" (Hensel 1989a p. 1). The earliest use of the term that I have been able to find is in a speech that Felix Klein held in 1904 (Klein 1905, p. 37).

the students[17] (Anon. 1895b, p. 753). Similarly, the branch in Franken-Oberpfalz warned against a reduction of the mathematical curriculum, emphasizing that "the mathematical sciences as a whole contribute primarily to strengthening the engineering student's abilities in logical thinking"[18] (Anon. 1895a, pp. 721–722).

Based on Riedler's report and the recommendations of the local and regional branches, the 1895 Congress of the VDI, meeting in Aachen, adopted a policy in favour of the creation of educational laboratories. The policy specifically endorsed reductions in mathematics teaching as a means to make room for the new experimental studies. The use of abstract methods in mathematics should be reduced, and the focus of the (reduced) mathematics curriculum should be on the mathematical tools that were necessary for the technological disciplines.

> Therefore education in the auxiliary sciences should be kept within the limits of what is necessary for understanding the engineering sciences. It is in particular desirable that the mathematical education, while not being restricted in the achievement of these goals, is restricted in the use of abstract methods. Through vivid connections with the application areas the students will be led faster and more safely to sufficient mastery of the mathematical tools.[19] (Anon. 1895c, p. 1095)

This was followed up in a report from the board of the VDI in which the teaching of mathematics in technological colleges was explicitly criticized. According to the report, the education as a whole had become too abstract; "to put it briefly, it has become an end in itself and has neglected the constant contact with the practical tasks that it should serve"[20] (Anon. 1895d, p. 1214).

13.6 The Nature of the Controversy

The anti-mathematical movement combined several concerns, and its participants seem to have had in part different motives. There were at least three lines of conflict. First, there was competition for space in the curriculum. For some time, mathematics had expanded, and teachers in the more practically oriented subjects felt a need to defend their own disciplines. Although theoretically uninteresting, this was probably a major component in the conflict.

---

[17]"eine Beschränkung des mathematischen Unterrichtes durchaus nicht gutheißen können", "ungleichmäßigen und häufig ungeeignete Vorbildung".

[18]"die Gesamtheit der mathematischen Wissenschaften in erster Linie dazu beiträgt, die logische Denkkraft des Studirenden der Technik zu schärfen".

[19]"Deshalb muss dieser Unterricht in den Hilfswissenschaften das zum Verständnis der Ingenieurwissenschaften erforderliche Maß einhalten; insbesondere ist es wünschenswert, den mathematischen Unterricht nicht in diesen Zielen, aber in der Benutzung abstrakter Methoden zu beschränken und durch lebendige Beziehung zu den Anwendungsgebieten die Studirenden schneller und sicherer als bisher zu ausreichender Beherrschung der mathematischen Hilfsmittel zu führen."

[20]"der Unterricht im ganzen zu sehr abstrakt gestaltet worden; er ist mit einem Worte zu sehr Selbstzweck geworden und hat die stete Berührung mit den praktischen Aufgaben, denen er dienen soll, vernachlässigt."

A second line of conflict concerned the nature of technological science. The technological colleges had begun as schools for craftsmen. They fought a long battle to achieve academic status, a battle that was finally to be won in the twentieth century, when they received the right to confer doctorate degrees and most of them changed their names to "technological universities". However, although the professors in technological colleges agreed on the goal to achieve academic status, they were divided on how this should be done. There were two competing strategies. The original strategy was closely connected with a view of technological science as applied mathematics and natural science. Formulas from mechanics could be used to characterize the movements of machine parts, and electromagnetic theory could be used to design electrical machines and appliances. In this way, technological science could be based on mathematics, physics, and chemistry. Consequently, the obvious way to obtain academic status was to excel in these foundational sciences. If the students of engineering schools learned as much physics and mathematics as those of the established universities, then what reason could there be to deny the technological schools the status of universities? Above, we saw this strategy at play in the VDI's policy document from 1865.

But there was also another opinion on the nature of technological science, namely that it consisted primarily in the use of scientific methods in direct investigations of the subject matter of technology, namely machines and other constructions by engineers. According to this approach, the empirical basis of technological science consisted in experiments with machines and other technological objects. Many of the teachers in engineering disciplines conducted this type of research. They built machines and machine parts and tested the functionality of alternative constructions in order to optimize the construction (Faulkner 1994; Kaiser 1995). In many cases this was the only way to solve the problems of practical engineering, for the simple reason that available physics-based theories either did not cover all aspects of the problems to be investigated or required calculations that were too large to be performed (Hendricks et al. 2000). German researchers who promoted this form of science were much encouraged by what they saw in American laboratories. Unsurprisingly, they regarded mathematics and physics as auxiliary rather than foundational sciences for the study of technology. With this approach to technological science, it was not necessary to compete with the universities in physics and mathematics in order to justify an academic status. Instead, that status could be based on a type of research that was unique to the technological colleges, namely empirical studies of technology.

There is an interesting parallel with developments in medical faculties in the late nineteenth century. Although these faculties were already parts of the university system, they did not have the high status that the natural sciences were increasingly favoured with. Here as well, there were two competing strategies for achieving a higher status. One was to develop medicine as an application of the natural sciences. Through laboratory studies of sick and healthy organs, the causes of diseases could be discovered and remedies developed. Claude Bernard (1813–1878) was a leading proponent of this strategy. The other strategy was based on treatment experiments in the clinic, i.e. what we today call clinical trials. By systematic evaluations of the outcomes of different treatment methods, the most beneficial methods could be identified (Booth 1993; Wilkinson 1993; Feinstein 1996; Hansson 2014). Today these approaches are seen as complementary, but in the late nineteenth century they were considered to be in conflict, in much the same way as the two strategies of technological educators just referred to.

The third line of conflict concerns two different views of mathematics. This was a dividing line with interesting philosophical implications. The major target of the anti-mathematical movement was the teaching of "higher mathematics", by which was meant differential and integral calculus and analytic geometry (Hensel 1989a, p. 25). The anti-mathematical activists were particularly hostile to the new, more stringent methods that mathematicians introduced into their teaching in this field (Purkert 1990, pp. 179, 188). Previously, the calculation of integrals was based on infinitesimals (hypothetical objects that are larger than zero but smaller than any positive number). Infinitesimals had been used successfully for many years, but mathematicians had discovered cases in which they give the wrong answers. They were therefore replaced by new more stringent methods that were based on limits. This transformation was largely based on work by Karl Weierstraß (1815–1897), who showed how geometrical reasoning about infinitesimals could be replaced by more precise reasoning expressed in formulas. This was commonly called the "arithmetization" of analysis, but that designation is somewhat misleading since the new method was based on a much more sophisticated manipulation of formulas than that of common arithmetic. Felix Klein provided an excellent explanation in a popular lecture:

> A glance at the more modern textbooks of the differential and integral calculus suffices to show the great change in method; where formerly a diagram served as proof, we now find continual discussions of quantities which become smaller than, or which can be taken to be smaller than, any given quantity. The continuity of a variable, and what it implies, or does not imply, are discussed, and a question is brought forward whether we can, properly speaking, differentiate or integrate a function at all. This is the Weierstrassian method in mathematics, the "Weierstrass'sche Strenge", as it is called. (Klein 1896b, p. 242)

Many teachers in engineering subjects were highly critical of the new methods. In 1896 the journal of the VDI contained a book review by Gustav Holzmüller (1844–1914) in which he denounced the "purely abstract theory of sizes that completely refrains from geometrical illustrations", and proposed a return to the old geometrical methods in which "spatial means of apperception, especially geometrical presentations" were used[21] (Holzmüller 1896, p. 108).

Holzmüller used the word "Anschauung"; in translations I have followed the tradition and rendered it as "apperception". The term is strongly connected with Immanuel Kant's epistemology. In his *Kritik der reinen Vernunft* (*Critique of Pure Reason*), Kant distinguished between two forms of apperception, namely empirical apperceptions that are provided by the sense organs and pure apperceptions that are

[21]"eine rein abstrakte Größenlehre, die auf geometrische Veranschaulichungen vollständig verzichtet", "räumliche Anschauungsmittel, besonders die geometrische Darstellung".

a priori, i.e. independent of sensory experiences. These pure apperceptions referred to time and space, which we can, according to Kant, conceptualize independently of our perceptions. Holzmüller and others proposed that students should develop their abilities to apperceive geometrical objects in space. This, in his view, was the right road to mathematical knowledge, rather than the precisely defined operations on formulas that followers of Weierstraß recommended as a means to achieve sufficient stringency.

In 1896, Alois Riedler published a new series of articles in the journal of the VDI, in which he defended the old approach to calculus. He accused modern, abstract mathematics of a "one-sided lack of apperception" that led to "fear of reality and escape from it"[22] (Riedler 1896a, p. 305). Instead, all teaching should "stand on the foundation of apperception and, as its highest task, strive for *apperceptive* logical thinking *without formulas*, that is thinking and operating *with apperceptive concepts*"[23] (p. 305). He did not trust academic mathematicians to perform such mathematics teaching "without formulas", and therefore proposed to put an end to their teaching at the technological colleges. Instead, engineering students should be taught a reduced mathematics curriculum by teachers who were themselves engineers (pp. 342–343). He polemicized against the mathematician Felix Klein who had spoken in conciliatory terms about the important tasks that mathematicians had in making their subject relevant and useful for students of engineering. Mathematicians did not have any such task at all, said Riedler. Instead, the students should encounter "*mathematics* as an indispensable tool in the hands of those who are educated in technology or at least the natural sciences"[24] (Riedler 1896b, p. 990). His choice of Klein as the main target of this attack on mathematicians may have been injudicious; among the leading mathematicians of his time Klein was one of those who most emphasized the prudent use of geometrical intuitions (Klein 1896b, p. 246).

13.7 The End of the Movement

With this escalation of the rhetoric against them, it is no surprise that the mathematicians at the German technological colleges felt obliged to respond. In December 1896, all the 33 professors in mathematical subjects made a joint statement. (One of the signatories was Richard Dedekind who taught at the technological college in Braunschweig.) They pointed out that mathematics was a foundational science,

[22]"anschauungslose Einseitigkeit", "Furcht und Flucht vor der Wirklichkeit".

[23]"auf dem Boden der Anschauung stehen und gerade das *anschauliche* logische Denken *ohne Formeln*, das Denken und Operiren in *anschaulichen Begriffen*, als die höchste Aufgabe erstreben".

[24]"der *Mathematik* als unerlässlichem Werkzeug in den Händen technisch oder mindesten naturwissenschaftlich Gebildeter".

not an auxiliary one.[25] They also made it clear that a reduction in the time spent on mathematics was impossible due to "the difficulty and the size of the material that is necessary to put forward, given the previous education that the students currently receive in the secondary schools".[26] The teachers should have a complete mathematical eduction, and it was "out of the question that a technician can hold mathematical lectures even for beginners".[27] But on the other hand, they emphasized that as mathematicians at technological colleges, they had a particular obligation to pay close attention to the technological uses of mathematics. They also conceded that "extensive references to apperceptive methods" were pedagogically useful[28] (von Braunmühl et al. 1897).

This was followed by a sharp retort from the anti-mathematical movement, signed by 57 teachers in engineering disciplines. They said:

> In the education of engineers, mathematics does not have the importance of an essential foundation, but that of a tool. The contrary standpoint of the mathematicians explains the errors that are made in the mathematical education at [technological] colleges... The education in higher mathematics currently exceeds the actual needs, is a too heavy load in the first four terms and should therefore be reduced in favour of a better preparatory technological education during these terms. At the same time it should be strengthened technologically through as much applications as possible for instance in technological calculation exercises. Those parts of the mathematical sciences that are suitable for enhancing spatial conception and graphical representation of quantities deserve to be favoured... The current educational programme for mathematicians does not make them able to correctly judge the needs of technology, which they misconstrue to the benefit of mathematics. Therefore teachers with an education essentially based in technology should be found for the mathematical education.[29] (Arnold et al. 1897)

[25]"grundlegende Wissenschaft", "Hülfswissenschaft".

[26]"der Schwierigkeit und dem Unfang des nothwendig voranzutragenden Stoffes, wie bei der von den Mittelschulen gegenwärtig gegebenen Vorbildung der Schüler".

[27]"kann keine Rede davon sein, dass ein Techniker mathematische Vorlesungen auch nur für Anfänger halte!"

[28]"ausführliche Heranziehung anschauungsmässiger Methoden".

[29]"[D]ie Mathematik hat für die Ausbildung des Technikers nicht die Bedeutung einer wesentlichen Grundlage, sondern die eines Hülfsmittels; der entgegengesetzte Standpunkts der Mathematiker erklärt die Fehler, welche im mathematischen Unterrichte an den Hochschulen begangen werden... Der Unterricht in der höheren Mathematik geht heute über die thatsächlichen Bedürfnisse hinaus, belastet die ersten vier Semester zu schwer und ist daher zu Gunsten einer bessern technischen Vorbildung in diesen Semestern einzuschränken, zugleich aber durch möglichst weitgehende Anwendung – etwa in technischen Rechenübungen – technisch zu vertiefen. Bevorzugung verdienen diejenigen Theile der mathematischen Wissenschaften, welche geeignet sind, die Fähigkeit der räumlichen Vorstellung und der bildlichen Darstellung der Größen zu fördern.... Der heutige Ausbildungsweg der Mathematiker befähigt diese nicht zu richtiger Erkenntnis der Bedürfnisse der Technik, welche sie nach der mathematischen Seite überschätzen. Deshalb müssen für den mathematischen Unterricht Lehrer mit wesentlich technischer Grundlage ihrer Ausbildung gewonnen werden."

In the short run, the anti-mathematical movement had some success. In 1896, the mathematics curriculum was considerably reduced in the technological college in Berlin, where Alois Riedler was himself was professor. Such reductions were also made in several other technological colleges (Hensel 1989a, pp. 78–81). But some members of the movement were still not satisfied. In an article published in April 1899, Paul von Lossow (1865–1936) who was professor in mechanical engineering at the technological college in Munich, extended his attacks and denounced not only all mathematicians, but all teachers with a university background. Such people "seriously overestimate the impact of education on later achievements"[30] (von Lossow 1899, p. 361). His description of university teachers was far from friendly:

> After the end of their studies they remain stuck to the school, become assistants and later Privatdozenten – the worst possible career path for a teacher in the art of engineering. Such a man, who has spent his whole life in school and never broke free from the spell of one-sided theoretical speculations, can do an infinite amount of harm when he later, as a professor, exerts his influence for decades on hundreds of students.[31] (von Lossow 1899, p. 356)

von Lossow's article was the last major expression of the anti-mathematical movement. In the last years of the nineteenth century, the mathematicians managed to calm down the conflict by adjusting their teaching to the needs of engineers, for instance by giving examples from engineering a more prominent role in their lectures (Hensel 1989a, pp. 84–86; Purkert 1990, p. 192; Schubring 1990; Scharlau 1990, pp. 264–279). Felix Klein seems to have had an important role as a mediator in these developments. He said already in 1898 that "an actual, though not formal agreement" had been reached[32] (Klein 1898, p. 1092). On the one hand it was agreed that the teaching of mathematics should be better adapted to the needs of engineers, on the other hand that engineers needed a broad base in mathematics (Cf.: Klein 1896a, 1905). One of the reasons why the movement lost its momentum may have been that mathematics and other theoretical disciplines had an important role in the argumentation for conferring on the technological colleges the right to grant doctor's degrees (Manegold 1970, p. 157). The decline of the movement was so fast that the technological college in Munich decided already in 1904 to increase its mathematics curriculum, which had been cut down in the previous decade (Otte 1989, p. 177). In 1903 Arnold Sommerfeld (1868–1951), a physicist and mathematician at the technological college in Aachen, described the fight between theoreticians and engineers as a conflict that had been "still lively a few years ago" but had now been replaced by "an unhesitant appreciation of the different fields of research"[33] (Sommerfeld 1903, p. 773). In 1919 the mathematician Eugen Jahnke characterized the anti-mathematical movement as belonging entirely to the past (Jahnke 1919).

[30]"überschätzen den Einfluss der Schulung auf die späteren Leistungen des einzelnen arg".

[31]"Sie bleiben nach Abschlus ihrer Studienzeit an der Schule kleben, werden Assistenten und später Privatdozenten – der verkehrteste Werdegang für einen Lehrer der Ingenieurkunst. Wenn solch ein Mann, der sein ganzes Leben lang nicht aus der Schulstube und nicht aus der Banne einseitig theoretischer Spekulationen herausgekommen ist, später als Professor Jahrzehnte lang auf hunderte von Studirenden seinen Einfluss äußert, so kann er undendlich viel Unheil anrichten."

[32]"eine allerdings nicht formelle, wohl aber thatsächliche Uebereinstimmung".

[33]"noch vor weningen Jahren lebhaft", "eine bereitwillige Würdigung der verschiedenen Forschungsrichtungen".

13.8 Aftermath: A Nazi Movement Against Abstract Mathematics

The anti-mathematical movement's attacks on the abstract and strictly rule-bound methods of modern mathematics had a brief resurgence in a much more sinister context, namely attempts to align mathematics with Nazi ideology.

Although the new, more stringent, methods in mathematics had acquired a dominant role in the 1930s, some mathematicians defended a traditional approach that assigned a central role to intuition and apperception in validating mathematical statements. The most influential among them was the Dutch mathematician L.E.J. Brouwer (1881–1966). One of its most prominent German proponents was Ludwig Bieberbach (1886–1982). In his inaugural lecture in Basel in 1914 he took a formalistic view, but in the twenties he became a proponent of mathematical apperception, in particular in geometry (van Dalen 2013, p. 496; Mehrtens 1987, p. 166; Segal 2003, p. 348).

Like all other parts of German intellectual life, mathematics suffered great losses during the Nazi regime. Between 1933 and 1937, about 30% of the mathematicians at German universities lost their jobs due to racial or political persecution (Schappacher 1998, p. 127; Mehrtens and Kingsbury 1989, p. 49). One example is the statistician Emil Julius Gumbel (1891–1966) who was severely persecuted already in the 1920s and had to emigrate in 1932. He was the only mathematician on the Nazi regime's first list of persons who were deprived of their citizenship in 1934 (Remmert 2004; Mehrtens and Kingsbury 1989, p. 49). (Albert Einstein was on the same list.) Another was Emmy Noether (1882–1935), one the principal founders of modern abstract algebra. In spite of strong support from David Hilbert (1862–1943) and other prominent colleagues, her career was hampered first by Prussian antifemale legislation and then by Nazi persecutions that targeted her because she was a Jew and a socialist. She was expelled from the university and had to emigrate (Segal 2003 p. 15).

Many German mathematicians took a clear stand against these persecutions. For instance, David Hilbert, who was arguably the most influential mathematician of his time, wrote in 1928:

> [A]ll limits, especially national ones, are contrary to the nature of mathematics. It is a complete misunderstanding to construct differences or even incompatibilities according to peoples and races, and the reasons for which this has been done are very shabby ones. Mathematics knows no races... For mathematics, the whole cultural world is a single country. (Quoted in Siegmund-Schultze 2016)

However, there were also mathematicians who sided with the Nazis and tried to obtain support from the regime for their own strivings. One of them was Ludwig Bieberbach, who joined the Nazis in 1933 (Remmert 2004). In 1933 the prominent philosopher Carl G. Hempel (1905–1997) described "disturbing events set in motion by National Socialism: the eminent mathematician Ludwig Bieberbach strutting about in a Nazi uniform, greeting his classes with the Hitler salute, and talking of the racial facets of mathematics" (Hempel 1991, p. 9).

In two articles published in 1934, Bieberbach divided mathematicians into two major styles, which he attributed to different races. Basically, he associated axiomatic and formal work with Jewish and French national character, and a more intuitive or apperceptive approach with German character. However, he twisted the classification in order to avoid classifying axiomatically oriented Germans like David Hilbert and Richard Dedekind along with the Jews. Even Karl Weierstraß who explicitly criticized reliance on intuition was classified among the intuitively oriented mathematicians, for the simple reason that he was a German (Segal 1986, 2003, pp. 360–368).

Bieberbach took the lead in a movement for so-called German mathematics, centring around the journal *Deutsche Mathematik* ("German mathematics") that appeared from 1936 to 1943 (Schappacher 1998; Remmert 2004). To put it mildly, the contents of the journal did not do much to corroborate the supposed superiority of German mathematics.

In this period, mathematicians who emphasized formal rigour and deductive reasoning were mostly opponents of the Nazi regime, whereas many proponents of apperceptive and intuitive mathematics went in the other direction. However, there is certainly no necessary connection between intuition-based mathematics and this or any other political ideology. There was, and still is, a highly respectable intuitionist standpoint in mathematics. It sees mathematical intuitions as common to humankind, which is of course very different from the Nazi view that mathematical intuitions differ between the "races" (Segal 2003, pp. 33–34; Mehrtens 1987 p. 171).

There was a parallel Nazi movement in physics, the Deutsche Physik (German Physics). Its members were opponents of relativity theory due to its unintuitive nature. The term apperception (Anschaung) was used in this context as well. So-called "Jewish" physics was accused of being too abstract and lacking in apperception (Wazeck 2009). One of the leaders of Deutsche Physik was the Nobel laureate in physics Phillip Lenard (1862–1947), who joined the Nazi party already in the early 1920s. He rejected relativity theory due to its, as he saw it, non-apperceptive nature. On a physics conference in 1920 he debated this with Einstein, who retorted:

> I want to say that what appears apperceptive to the human, and what does not, has changed. The way of thinking about apperceptiveness is in a way a function of the time. I would say that physics is conceptual, not apperceptive.[34] (Quoted in Wazeck 2009, pp. 183–184)

Lenard went even further than Bieberbach in his criticism of abstract mathematics. In an article in 1936 he denounced in principle all mathematics from the last century or so, claiming that it had lost contact with the real world:

> Gradually, presumably from approximately Gauss' time on, and in connection with the penetration of Jews into authoritative scientific positions, however, mathematics has in continually increasing measure lost its feeling for natural research to the benefit of a

[34] "Ich möchte sagen, daß das, was der Mensch als anschaulich ansieht, und was nicht, gewechselt hat. Die Ansicht über Anschaulichkeit ist gewissermaßen eine Funktion der Zeit. Ich meine, die Physik ist begrifflich, nicht anschaulich."

development separated from the external world and playing itself out only in the heads of mathematicians, and so this science of the quantitative has become completely a humanities subject [Geisteswissenschaft]. Since the role of the quantitative in the world of the spirit is, however, only a subordinate one, so this mathematics is presumably to be designated as the most subordinate humanities subject... It is certainly not good to allow this humanities subject with all its newest branches any large space in the school curriculum. (Quoted in Segal 2003, p. 375)

Fortunately, the influence of Bieberbach, Lenard and their collaborators ended with the defeat of the Nazi regime.

13.9 Conclusion

The proper extent and form of mathematics in engineering education has not ceased to be contentious. The issues debated are much the same. Proposals are still being made to reduce the mathematical rigour, to focus more on applications from engineering subjects, and to let engineers rather than mathematicians teach the subject (Barry and Steele 1993; Cardella 2008; Flegg et al. 2012; Ahmad et al. 2001; Klingbeil et al. 2004). Hopefully, something can be learned from the history of the anti-mathematical movement of the 1890s. The old methods in calculus that were promoted by its adherents have since long been given up in mathematics education. This can be seen as an indication that Einstein was right in his answer to Lenard: What we consider to be intuitive changes with time. Demands for apperceptiveness, or immediate intuitive appeal, can be counterproductive since they tend to hamper progress in both fundamental and applied mathematics.

And we should not take it for granted that there is a conflict between applicability and rigour. The purpose of mathematical rigour is to make sure that one's conclusions are valid, and that is certainly a paramount concern in engineering.

References

Ahmad, J., Appleby, J., & Edwards, P. (2001). Engineering mathematics should be taught by engineers! In P. Egerton (Ed.), *Proceedings of the 27th Undergraduate Mathematics Teaching Conference*, (pp. 33–40). Birmingham, England, September 3–6, 2001. Downloaded from http://files.eric.ed.gov/fulltext/ED482954.pdf.

Alder, K. (1999). French engineers become professionals; or, how meritocracy made knowledge objective. In W. Clark, J. Golinski, & S. Schaffer (Eds.), *The sciences in enlightened Europe* (pp. 94–125). Chicago: University of Chicago Press.

Anon. (1865). Achte Hauptversammlung der Vereines, *Zeitschrift des Vereines deutscher Ingenieure, 9*, 691–734.

Anon. (1895a). Sitzungsberichte der Bezirksvereine. Fränkisch-Oberpfälzischer Bezirksverein. *Zeitschrift des Vereines deutscher Ingenieure, 39*, 721–722.

Anon (1895b). Sitzungsberichte der Bezirksvereine. Aachener Bezirksverein. *Zeitschrift des Vereines deutscher Ingenieure, 39*, 753.

Anon (1895c). Beschlüsse der 36. Hauptversammlung. *Zeitschrift des Vereines deutscher Ingenieure* 39:1095–1096.

Anon. (1895d). Bericht über die Ingenieurlaboratorien und die Gestaltung des Unterrichts and den technischen Hochschulen. *Zeitschrift des Vereines deutscher Ingenieure, 39*, 1212–1216.

Arnold, et al. (1897). Der mathematische Unterricht an den technischen Hochschulen. *Zeitschrift für Arkitektur und Ingenieurwesen, 3*, 609–611.

Bachrach, D. S. (2006). English artillery 1189–1307: The implications of terminology. *English Historical Review, 121*(494), 1408–1430.

Barry, M. D. J., & Steele, N. C. (1993). A core curriculum in mathematics for the European engineer: An overview. *International Journal of Mathematical Education in Science and Technology, 24*(2), 223–229.

Booth, C. C. (1993). Clinical research. In W. F. Bynum & R. Porter (Eds.), *Companion encyclopedia of the history of medicine* (pp. 205–229). London: Routledge.

Cardella, M. E. (2008). Which mathematics should we teach engineering students? An empirically grounded case for a broad notion of mathematical thinking. *Teaching Mathematics and its Applications, 27*(3), 150–159.

Ernst, A. (1894). Maschinenbaulaboratorien. *Zeitschrift des Vereines deutscher Ingenieure, 38*, 1351–1365.

Faulkner, W. (1994). Conceptualizing knowledge used in innovation: A second look at the science-technology distinction and industrial innovation. *Science, Technology and Human Values, 19*, 425–458.

Feinstein, A. (1996). Two centuries of conflict-collaboration between medicine and mathematics. *Journal of Clinical Epidemiology, 49*, 1339–1343.

Flegg, J., Mallet, D., & Lupton, M. (2012). Students' perceptions of the relevance of mathematics in engineering. *International Journal of Mathematical Education in Science and Technology, 43*(6), 717–732.

Gerhardt, R., Kurrer, K.-E., & Pichler, G. (2003). The methods of graphical statics and their relation to the structural form. In S. Huerta (Ed.), *Proceedings of the First International Congress on Construction History, Madrid, 20th–24th January 2003*. Madrid.

Grattan-Guinness, I. (2005). The Ecole Polytechnique, 1794–1850: Differences over educational purpose and teaching practice. *American Mathematical Monthly, 112*, 233–250.

Grayson, L. P. (1993). *The making of an engineer: An illustrated history of engineering education in the United States and Canada*. New York: Wiley.

Hansson, S. O. (2014). Why and for what are clinical trials the gold standard? *Scandinavian Journal of Public Health, 42*(Suppl 13), 41–48.

Harris, D. G. (2001). *Fredrik Henrik af Chapman: The first naval architect and his work* (revised ed.). Stockholm: Literatim.

Hashagen, U. (1998). Mathematik und Technik im letzten Drittel des 19. Jahrhunderts – eine bayerische Perspektive. In F. Naumann (Ed.), *Carl Julius von Bach (1847–1931). Pionier-Gestalter-Forscher-Lehrer-Visionär* (pp. 169–184). Stuttgart: Konrad Wittwer.

Hempel, C. G. (1991). Hans Reichenbach remembered. *Erkenntnis, 35*, 5–10.

Hendricks, V. F., Jakobsen, A., & Pedersen, S. A. (2000). Identification of matrices in science and engineeering. *Journal for General Philosophy of Science, 31*, 277–305.

Hensel, S. (1989a). Die Auseinandersetzungen um die mathematische Ausbildung der Ingenieure an den Technischen Hochschulen in Deutschland Ende des 19. Jahrhundert. In S. Hensel, K.-N. Ihmig, & M. Otte (Eds.), *Mathematik und Technik im 19. Jahrhundert in Deutschland. Soziale Auseinandersetzung und Philosophische Problematik* (pp. 1–111 and 215–305). Göttingen: Vandenhoeck & Ruprecht.

Hensel, S. (1989b). Zu einigen Aspekten der Berufung von Mathematikern an die Technischen Hochschulen Deutschlands im letzten Drittel des 19. Jahrhunderts. *Annals of Science, 46*(4), 387–416.

Holzmüller, G. (1896). Bücherschau. *Zeitschrift des Vereines deutscher Ingenieure, 40*, 108–109.

Jahnke, E. (1919). Die Mathematik als Grundlage der Technik. *Dinglers polytechnisches Journal, 100*(15), 161–165.

Kaiser, W. (1995). Die Entwicklung der Elektrotechnik in ihrer Wechselwirkung mit der Physik. In L. Schäfer & E. Ströker (Eds.), *Naturauffassungen in Philosophie, Wissenschaft, Technik, Band III; Aufklärung und späte Neuzeit* (pp. 71–120). München: Verlag Karl Alber Freiburg.
Klein, F. (1896a). Die Anforderungen der Ingenieure und die Ausbildung der mathematischen Lehramtskandidaten. *Zeitschrift des Vereines deutscher Ingenieure, 40*(1896), 987–990.
Klein, F. (1896b). The arithmetization of mathematics. *Bulletin of the American Mathematical Society, 2*(8), 241–249 [Translation by Isabel Maddison of a speech held in German in November 1895.].
Klein, F. (1898). Universität und technische Hochschule. *Zeitschrift des Vereines deutscher Ingenieure, 42*(1898), 1091–1094.
Klein, F. (1905). Bericht an die Breslauer Naturforscherversammlung über den Stand des mathematischen und physikalischen Unterrichts an den höheren Schulen. *Jahresbericht der Deutschen Mathematiker-Vereinigung, 14*, 33–47.
Klemm, F. (1966). Die Rolle der Mathematik in der Technik des 19. Jahrhunderts. *Technikgeschichte, 33*, 72–90.
Klingbeil, N. W., Mercer, R. E., Rattan, K. S., Raymer, M. L., & Reynolds, D. B. (2004). Rethinking engineering mathematics education: A model for increased retention, motivation and success in engineering. In *Proceedings of the 2004 American Society for Engineering Education Annual Conference & Exposition*. 2004. Downloaded from https://peer.asee.org/13166.
Langins, J. (2004). *Conserving the enlightenment: French military engineering from Vauban to the revolution*. Cambridge, MA: MIT Press.
Lawrence, S. (2003). History of descriptive geometry in England. In S. Huerta (Ed.), *Proceedings of the First International Congress on Construction History, Madrid, 20th–24th January 2003* (pp. 1269–1281). Madrid: Instituto Juan de Herrera.
Manegold, K.-H. (1970). *Universität, Technische Hochschule und Industrie: Ein Beitrag zur Emanzipation der Technik im 19. Jahrhundert unter besonderer Berücksichtigung der Bestrebungen Felix Kleins*. Berlin: Duncker & Humblot.
Mehrtens, H. (1987). The social system of mathematics and national socialism: a survey. *Sociological Inquiry, 57*(2), 159–182.
Mehrtens, H., & Kingsbury, V. M. (1989). The Gleichschaltung of mathematical societies in Nazi Germany. *Mathematical Intelligencer, 11*(3), 48–60.
Otte, M. (1989). Die Auseinandersetzungen zwischen Mathematik och Technik als Problem der historischen Rolle und des Typus von Wissenschaft. In S. Hensel, K.-N. Ihmig, & M. Otte (Eds.), *Mathematik und Technik im 19. Jahrhundert in Deutschland. Soziale Auseinandersetzung und Philosophische Problematik* (pp. 149–214). Göttingen: Vandenhoeck & Ruprecht.
Purkert, W. (1990). Infinitesimalrechnung für Ingenieure–Kontroversen im 19. Jahrhundert. In D. D. Spalt (Ed.), *Rechnen mit dem Unendlichen. Beiträge zur Entwicklung eines kontroversen Gegenstandes* (pp. 179–192). Basel: Birkhäuser.
Purkert, W., & Hensel, S. (1986). Zur Rolle der Mathematik bei der Entwicklung der Technikwissenschaften. *Dresdener Beiträge zur Geschichte der Technikwissenschaften, 11*, 117–128.
Remmert, V. R. (2004). Die Deutsche Mathematiker-Vereinigung im 'Dritten Reich': Krisenjahre und Konsolidierung. *Mitteilungen der Deutschen Mathematiker-Vereinigung, 12*(3), 159–177.
Riedler, A. (1894). Amerikanische technische Lehranstalten. *Zeitschrift des Vereines deutscher Ingenieure, 38*, 405–409 507–514, 608–615, 629–636.
Riedler, A. (1895). Zur Frage der Ingenieurerziehung. *Zeitschrift des Vereines deutscher Ingenieure, 39*, 951–959.
Riedler, A. (1896a). Die Ziele der Technischen Hochschulen. *Zeitschrift des Vereines deutscher Ingenieure, 40*, 301–309 337–346, 374–375.
Riedler, A. (1896b). Sehr geehrte Redaktion! *Zeitschrift des Vereines deutscher Ingenieure, 40*, 989–990.
Schappacher, N. (1998). The Nazi era: the Berlin way of politicizing mathematics. In H. Begehr, H. Koch, J. Kramer, N. Schappacher, & E.-J. Thiele (Eds.), *Mathematics in Berlin* (pp. 127–136). Basel: Birkhäuser.

Scharlau, W. (1990). *Mathematische Institute in Deutschland 1800–1945*. Braunschweig: Friedr. Vieweg & Sohn.

Schubring, G. (1990). Zur strukturellen Entwicklung der Mathematik an den deutschen Hochschulen 1800–1945. In W. Scharlau (Ed.), *Mathematische Institute in Deutschland 1800–1945* (pp. 264–279). Braunschweig: Friedr. Vieweg & Sohn 1990.

Segal, S. L. (1986). Mathematics and German politics: The national socialist experience. *Historia Mathematica, 13*(2), 118–135.

Segal, S. L. (2003). *Mathematicians under the Nazis*. Princeton: Princeton University Press.

Siegmund-Schultze, R.. (2016) 'Mathematics Knows No Races': A political speech that David Hilbert planned for the ICM in Bologna in 1928. *Mathematical Intelligencer, 38*(1):56–66.

Sommerfeld, A. (1903). Die naturwissenschaftligen Ergebnisse und die Ziele der modernen technischen Mechanik. *Physikalische Zeitschrift, 4*, 773–782.

van Dalen, D. (2013). *L.E.J. Brouwer – topologist, intuitionist, philosopher*. London: Springer.

von Braunmühl, A., et al. (1897). Über den mathematischen Unterricht an den technischen Hochschulen. *Zeitschrift für Architektur und Ingenieurwesen, 2*(22), 242–244.

von Lossow, P. (1899). Zur Frage der Ingenieurausbildung. *Zeitschrift des Vereines deutscher Ingenieure, 43*, 355–361.

Wazeck, M. (2009). *Einsteins Gegner. Die öffentliche Kontroverse um die Relativitätstheorie in den 1920er Jahren*. Frankfurt: Campus.

Wilkinson, L. (1993). Epidemiology. In W. Bynum & R. Porter (Eds.), *Companion encyclopaedia of the history of medicine* (Vol. 2, pp. 1262–1282). London: Routledge.

Chapter 14
Reflections on the Empirical Applicability of Mathematics

Tor Sandqvist

Abstract This paper addresses the not infrequently voiced view that the immense usefulness of mathematics in the physical sciences constitutes a deep philosophical mystery, with potentially far-reaching implications concerning the relationship between the inquiring mind and the material world. It grants the broadly Humean point that the very possibility of inductive projection from past to future, by whatever intellectual means, must be considered a remarkable and perhaps inexplicable fact, but calls into question the idea that the utility of mathematics in this regard is especially baffling. While the aims pursued in pure mathematics may differ radically from those of engineers and scientists, in their development of concepts and theories mathematicians are nevertheless beholden to the same fundamental standards of simplicity and similarity that must govern any reasonable inductive projection; and this fact, it is suggested, may go a considerable way towards explaining why many mathematical constructs lend themselves to empirical application.

14.1 Introduction

In his 1959 Richard Courant Lecture in Mathematical Sciences, later published under the title "The Unreasonable Effectiveness of Mathematics in the Natural Sciences" (Wigner 1960), the physicist Eugene Wigner gave voice to a deep perplexity over the way in which mathematical concepts and theories, originally developed in the pursuit of pure mathematics without any view to application, so often turn out to be perfectly suited to the purpose of describing and predicting physical phenomena. This seeming ability of mathematicians to presage the development of natural science is all the more baffling, Wigner argued, in light of the very different priorities of scientists and mathematicians. Whereas the prime objective of the former is to produce an accurate description of the physical world, following the data

T. Sandqvist (✉)
Department of Philosophy and History, Royal Institute of Technology (KTH), Stockholm, Sweden
e-mail: tor.sandqvist@abe.kth.se

S. O. Hansson (ed.), *Technology and Mathematics*, Philosophy of Engineering and Technology 30, https://doi.org/10.1007/978-3-319-93779-3_14

wherever they might lead, the activities of the latter are more akin to artistic creation. To be sure, as in any intellectual endeavour, mathematicians are constrained in their construction of definitions and proofs by the rigours of deductive logic; but in deciding which tracts of logical space to explore – what objects and operations to define, and what properties of these constructs to investigate – they are guided to a far greater extent by their sense of beauty and their appetite for intellectual adventure and competition than by any desire to understand or manipulate their physical surroundings. How is it, then, that ideas and theories resulting from such creative pursuits in the realm of the abstract end up as indispensable tools for studying and negotiating the material world? "The miracle", Wigner writes, "of the appropriateness of the language of mathematics for the formulation of the laws of physics is a wonderful gift which we neither understand nor deserve."

Stronger language still is employed in Mark Steiner's book-length exploration of the same topic, *The Applicability of Mathematics as a Philosophical Problem* (Steiner 1998). Citing a multitude of instances, Steiner ventures to show that eminent physicists in their mathematical development of theories have habitually engaged in blatantly "anthropocentric" reasoning, hypothesizing that observable reality will behave in accordance with certain equations that have been selected on the basis of purely formal "analogy" with successful earlier theories, unsupported by any substantively "physical" rationale. Such reasoning is anthropocentric, Steiner maintains, in that, in its faltering progression from one mathematically formulated hypothesis to another, it employs paths of association that owe their very existence to various accidents of human intellectual history. The fact that theories of such disreputable provenance have repeatedly wound up finding empirical vindication amounts, in Steiner's words, to a "challenge to naturalism" (pp. 75, 176) comparable to making a "substantial physical discovery based upon the statistical distribution of the letters of the Roman alphabet in Newton's *Principia*" (Steiner 1989, p. 454).

My ambition in this essay is to articulate a point of view where the empirical applicability of mathematics does not present itself as a significant philosophical riddle. My position is not so much that of a thinker who has grappled with a problem and finally solved it as that of someone who fails to perceive any real difficulty in the first place. If my (perhaps somewhat rambling) exposition can either guide the reader towards a similarly untroubled place or assist her in giving the problem a sufficiently sharp formulation to enable complacents like myself to see it, I shall consider my effort well spent. The leading thought, insofar as there is one, is that mathematicians and scientists are beholden to a common conceptual standard of simplicity and similarity, and that this fact may go a considerable way towards explaining why many mathematical constructs lend themselves to empirical application.

My discussion will mainly be focussed on the role of mathematics in enabling us to predict future events – such as an instrument reading, or the behaviour of a machine – on the basis of past observation. The "problem of induction" is sometimes construed as the general question of how to justify an inference from the premiss that a certain regularity has appeared in all observations of a certain phenomenon to date ("all ravens observed so far have been black") to the conclusion that this regularity will *always* obtain ("all ravens are black"). Such categorical conclusions, however, raise difficulties of a probabilistic nature (Chalmers 2013, p. 48) which are not encountered in more cautious predictions ("most of the ravens to be observed in the near future will be black"), and I will not be concerning myself with them here. Nor will my discussion be predicated on a conception of mathematically formulated laws of nature as providing in any sense *perfect* descriptions of physical reality. The general form of inductive inference to be considered in the present paper is something like the following: "in most cases observed thus far, events have played out approximately according to such-and-such a pattern; therefore, most near-future cases will conform approximately to this pattern as well." For pragmatic purposes, including typical technological applications, inferences of this kind (with suitable quantifications in place of "near", "most", and "approximately") are sufficient. Restricted to such cases, Wigner's problem becomes: how is it that mathematics is so effective in discovering and describing patterns capable of figuring in successful inductive inferences of this form?

14.2 Induction Without Mathematics

Empirical extrapolations from past to future are often made without any use of mathematics. It will be useful to have, as a backdrop for the discussion of our main issue, some (fictional) examples of such non-mathematical induction.

Scenario A. Astronomers have just discovered a curious fluctuation in the visual light emanating from a certain star some 500 light-years away. About once every hour, the star is seen to emit a pair of light pulses – brief increases in the intensity of radiation in a narrow frequency band – approximately 1 min apart. The first pulse is always either red or green, and the same is true of the second. At the time our story begins, the astronomers find themselves in the short interval between two of these pulses, their observations so far having turned out as follows. 'R' signifies a red pulse, 'G' a green one, and a comma the 1-h interval between pulse pairs.

RR, GG, RR, GG, RR, RR, GG, RR, GG, RR, RR, RR, GG, RR, GG, GG, GG, RR, GG, GG, RR, RR, RR, RR, GG, RR, GG, GG, R

Now, should we expect the next pulse to be a red or a green one? Well – we may not have much data to go on, nor any idea of what is causing the signals, but so far, the two pulses in each pair have always been of the same colour, so the obvious answer would seem to be that, since our last observation was one of a red pulse, in all likelihood the next one is going to be red also. That is, on the basis of the fact that all observations so far agree with the hypothesis stated below, we expect the next one to do so as well.

H1. The pulses in any pair are always of the same colour.

Now, to be sure, H1 is not the only rule consistent with the observations made to date. The same is true, for instance, of the following:

H2. The pulses are always of the same colour, except in such cases where the latest eight pairs form the pattern RR, RR, RR, RR, GG, RR, GG, GG; on such occasions the pulses in the immediately following pair take opposite colours.

On H2, the next pulse should be green, not red. Yet to expect the next observation to agree with H2, solely on the basis that all the previous ones have done so, seems ridiculous. Our instinct is to expect future observations to conform to the *simplest* patterns exhibited by past ones; in the circumstances envisaged, H2 is immediately disqualified on the grounds of its needless and arbitrary complexity.

This, however, is not to say that, *qua* empirical hypothesis, H2 is somehow beyond the pale come what may:

Scenario B. Several months have passed since our narrative from Scenario A left off. Our brave astronomers have continued to observe red-red and green-green pulse pairs arriving from the star in a seemingly random mix. From time to time, however, the sequence is interrupted by a pair of pulses of opposite colours. These exceptions to the general trend do not turn up at random; in fact, they have occurred exactly in the way described by H2. Despite a frenzy of speculation, no one has yet come up with a good explanation for the findings. The last few signals have been recorded as follows.

...GG, RR, GG, GG, RR, RR, RR, RR, GG, RR, GG, GG, R

In this scenario, expecting a green pulse, on the grounds that H2 has held without exception thus far, makes perfect sense. However mysterious the source of the signals, the ability of scientists to *notice* the regularity specified in H2, given that it occurs, should not be a cause for wonder; nor should their ability correctly to predict the colour of the next pulse. Identifying a curious pattern and making rational predictions on its basis does not depend on understanding why the pattern appears (though, of course, the formulation of a plausible explanation might well justify a greater degree of confidence in such predictions).

Scenario C. As in Scenario B, most, but not all, pairs of pulses arrive in matching colours. This time, however, the exceptions do not appear in accordance with H2 but instead as follows.

H3. The pulses in the n-th pair, counting from the first pair of pulses that were historically observed, are of opposite colours if and only if the n-th character in Jane Austen's *Sense and Sensibility* is a full stop.

At first, of course, the correlation is dismissed as a fluke of chance, and no reasonable person expects it to persist; but as one unmatched pair after another turns up in perfect conformity with H3, the statistical significance of the correspondence gradually reaches a level where the dismissive stance becomes untenable. When a neighbouring star is discovered to exhibit a similar correlation with *Pride and Prejudice*, a full-blown scientific crisis is precipitated. Is the observable universe just a giant mirror of human creativity, emitting cryptic reflections of its products centuries ahead of time? (Recall that the stars are 500 light-years away.) Or, conversely, is the artistic human mind somehow set up to foreshadow observations in the natural sciences? On either interpretation, the common, naturalistic view of Man and his causal place in the Cosmos is shaken to its core.

In each of the three scenarios we are dealing with an unexplained phenomenon; but the degree of weirdness of the phenomenon increases steeply from the first to the third. In Scenario A there is a simple regularity standing out against a background of noise: although the colour of any given pair cannot be predicted in advance, the second pulse is always similar to the first. In Scenario B it is still the case that the second pulse in a pair can always be predicted given a record of the last few observations. To be sure, the rule for doing so has an unsatisfactory, seemingly arbitrary element to it – but there is no indication that the phenomenon is in any way purposely adapted to human culture or cognition. In Scenario C, by contrast, an all-out anthropocentric world-view seems to be the only possible response to the data obtained.

In this essay I wish, firstly, to make an observation, and secondly, to formulate a conjecture.

Observation: The universe's propensity for exhibiting, amid its confusion of particular detail, certain general regularities that enable human and non-human animals to predict and shape the future on the basis of the past is indeed an as-yet unexplained, and perhaps ultimately inexplicable, matter of empirical fact. This unoriginal, broadly Humean point is fleshed out a bit in Sect. 14.3.

Conjecture: The fact that some of these regularities can be discovered and described with the aid of mathematics does not, in the final analysis, add any further mystery over and above the circumstance noted in the observation; contrary to what is suggested in some of the literature on the "unreasonable effectiveness" of mathematics, instances of empirically applied mathematics have more in common with Scenarios A and B than with Scenario C.

The conjecture – various aspects of which are developed in Sects. 14.4 through 14.7 – is labelled as such, rather than as a thesis or a contention, in recognition of the limitations of my own scientific erudition. Perhaps scholars with greater insight into the practice and history of science and mathematics will be in a position to reject my suggestions as predicated on inexperience and misunderstanding. If so, the reader is cordially invited to set me straight.[1]

14.3 The General Mystery of Inductive Projection

Consider all the possible ways of filling a rectangular grid, such as a computer screen, with black and white pixels. The vast majority of them exhibit no lawlike regularities at all; if we decide to select one by a random process such as flipping a coin for every pixel, we will be astonished if any sort of coherent pattern appears – for instance, if pixels whose vertical coordinates are multiples of 10 always turn out black, or if a picture of a galloping horse emerges on the screen.

[1]In the general thrust of its argument – acknowledging the existence of a problem concerning reasoning in general, while calling into question the idea of mathematical reasoning being particularly problematic – the present paper bears some resemblance to Sarukkai (2005).

To be sure, on the extremely rare occasion, such a pattern will actually appear by chance. Consider a situation where half of the pixels – say, the top half – have been filled in, and we are speculating about how the remaining half is going to turn out. Suppose that in the filled-in part of the screen we find a flawless depiction of the top half of a horse. The rational reaction would be to conclude that our coin-tossing is not in fact random, but in some mysterious way under the control of a furtive horse-painter, and that the most likely outcome for the lower half of the screen is a picture of the bottom half of a horse. But under the hypothesis that, contrary to appearance, the process *is* actually random, the likely continuation is still a lower half-screen of featureless noise – for the crushing majority of all possible screen configurations featuring a partial picture of a horse in their top halves still have nothing but noise in their bottom halves. To put it in Bayesian terms: if our initial credence function assigns equal probability to all possible configurations, then no possible top-half configuration gives any basis for projection to the bottom half.

Turning now from the fictional computer screen to the material world in which we find ourselves, there similarly seems to be no a priori reason why the latter should necessarily have been structured in such a way as to exhibit any kind of projectible regularity; why, as it were, does the universe contain anything but white noise? To be sure, if it did not, we would not be here to ask about it, so in this sense our very existence establishes *that* certain non-random patterns exist. But this observation goes nowhere towards explaining *why* they do.

Nor, as Hume noted, it is obvious why the existence of regularities in empirical events up to the present time should provide any sort of justification for expecting such regularities to persist in the future. Why, among those possible complete world-histories that begin in the way ours has, should we favour the infinitesimal minority that continue in similar fashion, as opposed to all those featuring nothing but white noise from this moment on? The response "Because that's what every sane person just has to do", while sufficient from a pragmatic point of view, does little to alleviate the philosophical puzzle.

So there are really two separate conundrums here: one of a broadly scientific/explanatory character, one purely epistemological. The scientific problem is how to explain the fact that the world thus far exhibits any discernible regularities at all; the epistemological riddle is how to justify the inference from the observation that the past exhibits regularities to the prediction that the future will, too. As far as I am aware, our position today with respect to these questions is little better than that of Hume in his day; to paraphrase Wigner, the possibility of empirical induction is a gift we neither deserve nor understand. In this way, insofar as mathematics-aided induction is a form of induction, its effectiveness is indeed "unreasonable". But now let us turn to the question whether the usefulness of mathematics in inductive projection adds any further mystery. My conjecture, to repeat, is that it does not.

14.4 Mathematics-Aided Induction

In Sect. 14.2, in the course of our discussion of hypotheses H1 and H2, we remarked that, in a situation such as Scenario A, where all of the available data are consistent with either one of the hypotheses, H1 will be preferred to H2 on account of its greater simplicity. It is hardly a controversial claim that similar considerations will also be in play in cases where the hypotheses under consideration are formulated in mathematical terms. For a highly simplified example, consider the fictional case of a team of researchers with a proto-Newtonian understanding of gravitation. They know that two 1-kg objects will attract one another with a force that depends on the distance between them, and are trying to determine the nature of the dependence. In all their observations to date, the force F, as measured in newtons, has been related to the distance r, as measured in meters, in accordance with the equation

$$F = (6.674 \cdot 10^{-11})/r^2. \tag{14.1}$$

While (let us suppose) no distance in the interval from 99 to 101 m has yet been investigated, the 100 m case is the next one up for trial, and on the basis of their observations so far our scientists are pretty confident that the result will be $(6.674 \cdot 10^{-11})/100^2 = 6.674 \cdot 10^{-15}$ newtons.

Now why is this? After all, all observations so far made have also been in agreement with the rule

$$F = \begin{cases} (6.674 \cdot 10^{-10})/r^2 & \text{if } 99 < \text{r} < 101, \\ (6.674 \cdot 10^{-11})/r^2 & \text{otherwise} \end{cases} \tag{14.2}$$

– would the scientists not be equally justified in concluding, on this basis, that the force observed at a 100 m distance will be $(6.674 \cdot 10^{-10})/100^2 = 6.674 \cdot 10^{-14}$ newtons? The obvious answer, just as in Scenario A, is that (14.2) will and should be rejected on the grounds of its gratuitous complexity.

Another alternative to (14.1) which might conceivably be entertained is this:

$$F = (6.674 \cdot 10^{-11})/r^{1.999}. \tag{14.3}$$

Let us suppose that the range and precision of the instruments used by our scientists are insufficient to distinguish between (14.1) and (14.3); the data obtained are no less closely approximated by the latter than by the former. Nevertheless, it is (14.1), not (14.3), that gets provisionally accepted – a decision that is subsequently borne out by measurements conducted with more sensitive instruments.

Why should this be? After all, we do not expect natural constants to assume integer values when expressed in antecedently adopted units of measurement – why take a different attitude towards exponents figuring in formulae like (14.1) or (14.3)? Isn't preferring an inverse-power-of-2 law of gravitation to an inverse-power-of-1.999 law tantamount to numerological mysticism? And yet, this is essentially what did happen historically: gravitational force was hypothesised to vary in inverse proportion to the square of distance well before instruments became sufficiently precise to pin down the value of the exponent with any great precision; and once more refined measurements became possible, the hypothesis was corroborated. In this sense, in the oft-recurring phrase (cf. Wigner 1960, p. 9; Dyson 1964, p. 129; Feynman 1967, p. 171) we "got more out" of our mathematically formulated law than was put into it by way of data. How is such a feat of prediction possible?

Again, while always acknowledging the general philosophical mystery of the possibility of predicting the future of the basis of the past, I would argue that this is just another case of preferring a simpler theory to a more complex one. Squaring a number is simpler than raising it to the power of 1.999 because the former operation, unlike the latter, can be reformulated in terms of ordinary multiplication, thus obviating the need to bring in the exponentiation function at all. In fact, in order to say that the force F is inversely proportional to the square of the distance r – i.e., that Fr^2 is constant – we do not even need to multiply any physical quantities together at all, but can confine ourselves to multiplication of physical quantities by positive integers, which is to say, to repeated addition of physical quantities: in a straightforward adaptation of the Eudoxian analysis of proportionality, $F_1 r_1{}^2 = F_2 r_2{}^2$ just in case it holds of all positive integers m and n that $(F_2 \cdot m) \cdot m < (F_1 \cdot n) \cdot n$ if and only if $r_1 \cdot m < r_2 \cdot n$. (If this equivalence seems less than obvious, note that the identity obtains just in case $\sqrt{F_1}/\sqrt{F_2} = r_2/r_1$, whereas the quantified biconditional holds good just in case it is true of every rational number m/n that $m/n < \sqrt{F_1}/\sqrt{F_2}$ if and only if $m/n < r_2/r_1$.)

Of course, even if we come to agree that an inverse-square law is in a non-arbitrary sense simpler than an inverse-power-of-1.999 law, there still remains the question why pursuit of simplicity should at all be conducive to the pursuit of accurate prediction. But this is just the general problem of induction again.

14.5 On the Genesis of Mathematical Concepts

It might be objected to the considerations of the previous section that they were predicated on an already settled mathematical terminology. Yes – the objection would go – *given* the basic concepts of mathematics, an inverse-square law of gravitation may be the simplest way of fitting theory to data; but the deeper issue at hand is why mathematical concepts should have any bearing on the physical world in the first place. How is it that constructs of mathematics, even when developed for purposes other than describing the empirical world, turn out, when combined with simplicity considerations, to be so useful for that purpose?

In large part, I would suggest, the answer lies in the fact that the development of mathematics always takes place under the influence of simplicity considerations similar to those guiding human concept-formation and inductive projection in general.

Consider again Eqs. (14.1) and (14.2). With respect to these rival hypotheses, our envisaged objector might point out that, while (14.1) is simpler than (14.2) when formulated in terms of the conventional operation of division, it is easy to find mathematical functions with respect to which the situation is reversed. For instance, define the binary function ‡ as follows:

$$x \ddagger y = \begin{cases} 10x/y & \text{if } 99^2 < y < 100^2, \\ x/y & \text{otherwise.} \end{cases}$$

Then (14.2) – the law we dismissed as being gratuitously complex – may be rewritten

$$F = (6.674 \cdot 10^{-11}) \ddagger r^2, \tag{14.2$'$}$$

whereas (14.1) – the simple and sensible one – comes out as

$$F = \begin{cases} (6.674 \cdot 10^{-12}) \ddagger r^2 & \text{if } 99 < r < 101, \\ (6.674 \cdot 10^{-11}) \ddagger r^2 & \text{otherwise.} \end{cases} \tag{14.1$'$}$$

In this sense, our opponent rightly observes, simplicity is relative to a terminology, and if I wish to maintain that the empirical hypothesis equivalently expressed in (14.1) and (14.1′) is in any real sense simpler than the one given by either one of (14.2) and (14.2′), I need to justify the choice of carrying out the comparison in terms of division rather than ‡.

As any reader who is familiar with Nelson Goodman's (1955) classic discussion of intuitively reasonable versus absurdly gerrymandered concepts will recognize, for the purpose of discrediting ‡ it will not be sufficient to point to its manifestly contrived definition in terms of division – for the converse characterization of division in terms of ‡ is no more attractive:

$$x/y = \begin{cases} (x \ddagger 10) \ddagger y & \text{if } 99^2 < y < 101^2, \\ x \ddagger y & \text{otherwise.} \end{cases}$$

Rather, the case for division has to be based on the clean-cut and uniform way in which it fits into its surrounding conceptual framework. To begin with, division being the inverse of multiplication ($a = b/c$ just in case $b = ac$), any argument for assigning a central role to multiplication will *ipso facto* do the same for division. Now multiplication by a positive integer k is just repeated addition:

$$a \cdot k = \overbrace{a + \cdots + a}^{k}, \tag{14.4}$$

and multiplication of arbitrary reals is just the linear (i.e., literally, the most straightforward) extension of this. Precisely put: given any real number a, the one-place function mapping each real x to ax is the only continuous function f such that (i) $f(k) = a \cdot k$, as specified by (14.4), for every positive integer k, and (ii) f maps equal intervals to equal intervals in the sense that $f(y_1) - f(x_1) = f(y_2) - f(x_2)$ whenever $y_1 - x_1 = y_2 - x_2$.

(To see that this is so, consider any continuous f satisfying (i) and (ii). Let n/m be any rational number, and μ the *ex hypothesi* constant amount by which $f(x)$ increases when x increases by $1/m$. $f(0) = 0$ since, by (i) and (ii), $a - f(0) = f(1) - f(0) = f(2) - f(1) = (a + a) - a = a$. For any integer l, therefore, $f(l/m) = l\mu$; in particular $a = f(1) = m\mu$ and so $f(n/m) = n\mu = an/m$. Thus $f(x) = ax$ for every rational x; by continuity the same must hold good for every real x whatever.)

In this way, the twin concepts of multiplication and division make for a natural continuation of the theory of addition and its inverse, subtraction; one would be hard put to portray ‡ in a similar light. As our hypothetical critic would have it, the apparent perverseness of hypothesis (14.2) is but an artifact of an arbitrary decision to formulate its content in terms of the traditional operation of division. What he fails to appreciate is that, given the conceptual context of addition and its inverse, that decision is supported by the same sort of consideration as informed our assessment of the relative merits of (14.2) and (14.1) in the first place; the absurdity of (14.2) and that of ‡ are two faces of the same coin.

As for addition itself, its utility in any empirical context only requires that individual quantities remain unaltered when aggregated: after emptying a sack of 800 grains of wheat into one containing 1000, we find ourselves in the possession of 1000 + 800 grains because none has been destroyed or created in the process; adjoining a 1.5-m plank to a 2.1-m one will create a body measuring 2.1 + 1.5 meters because the operation does not change the lengths of the individual planks; etc. Of course, a philosopher of Heraclitan inclination may find cause for wonder in the fact that this sort of constancy from one moment to the next ever occurs in the physical world – let alone with sufficient regularity to allow for the confident prediction of future events – but this only brings us back to the considerations of Sect. 14.3; no additional mystery is incurred by bringing a smattering of arithmetic into the picture.

Thus far, our discussion has confined itself to the most elementary concepts of mathematical analysis. But considerations of overall simplicity of theory are very much in operation at more advanced levels as well. An instructive case in point (and one accorded importance by both Wigner and Steiner) is the theory of complex numbers – i.e., the theory that results from positing, in addition to the real numbers, a number i such that $i^2 = -1$, while retaining the usual laws (commutative, distributive, etc.) of addition and multiplication. As suggested by the term 'complex', the resulting field of numbers forms a more complicated structure than that of the reals, topologically isomorphic to a plane rather than a line. But, as so often happens in mathematics, the stipulative incorporation of a richer domain of objects brings about a considerable streamlining on the level of theory. Whereas, on the real line, for every *non-negative* number a there exists an x such that $x^2 = a$ (i.e., a root to the polynomial $x^2 - a$), complex numbers allow us to drop the restriction and simply state that for every a whatsoever there exists such an x. More generally – the so-called "fundamental theorem of algebra" – every polynomial in one variable has at least one root. What is more, even when the coefficients and roots of such a polynomial are all real, the introduction of complex numbers often makes it possible to specify the roots in algebraically uniform ways where otherwise no such characterization exists – indeed this is how, in the sixteenth century, mathematicians' interest in complex numbers was originally piqued in the study of cubic equations. In the realm of transcendental functions, too, the introduction of complex numbers brings increased uniformity: for instance, rather than constructing exponential and trigonometric functions separately from scratch, we can now define the latter in terms of the former by identifying the sine and cosine of x with $(e^{ix} - e^{-ix})/2i$ and $(e^{ix} + e^{-ix})/2$, respectively. And so on.

This picture of the development of mathematical concepts is not altogether unlike that given by Wigner. On his account,

> mathematics is the science of skillful operations with concepts and rules invented for just this purpose. [...] Most more advanced mathematical concepts [...] were so devised that they are apt subjects on which the mathematician can demonstrate his ingenuity and sense of formal beauty. [...] [Mathematical concepts] are defined with a view of permitting ingenious operations which appeal to our aesthetic sense both as operations and also in their results of great generality and simplicity. [...]
>
> Certainly, nothing in our experience suggests the introduction of [complex numbers]. Indeed, if a mathematician is asked to justify his interest in complex numbers, he will point, with some indignation, to the many beautiful theorems in the theory of equations, of power series, and of analytic functions in general, which owe their origin to the introduction of complex numbers.

In addition to describing (as I have just done) qualities such as generality and simplicity as desiderata of mathematical constructs, Wigner stresses the role of these qualities as criteria of aesthetic beauty, picturing mathematicians as artists in creative pursuit of these values, untroubled by concerns of empirical adequacy. While I have no quarrel with this picture of the psychological forces driving mathematicians, the crucial observation for our present philosophical concerns is that simplicity and generality are precisely the core concepts at work in conventional accounts of empirical induction. Take the simplest theory consistent with your data; make the generalizing assumption that this theory applies equally well to cases yet to be tried; there is your prediction for the future. Now if the original impetus for the theory of real numbers, addition, and multiplication comes from pragmatic concerns with the natural world; if empirical induction is the practice of extrapolating from experience in the most straightforward and uniform way possible; and if the complex plane is the most straightforward and uniform way of generalizing and rounding out the theory of real numbers – then is it really a great cause for wonder that complex numbers have found empirical application?

Once again I hasten to add that I do not pretend to offer any solution to the Humean problem of why human standards of simplicity and uniformity should prove conducive to successful prediction in the first place. All I am suggesting is that, insofar as these same standards are at work both in the development of mathematical theory and in the scientific effort to understand the natural world, this fact goes a good way towards explaining why the two exhibit a considerable degree of confluence.

14.6 On the Genesis of Empirical Hypotheses

Thus far in our discussion of mathematically formulated physical theories, we have been considering the fact of their empirical adequacy on an abstract level, without any view to the question how such theories arise in the minds of working scientists. While it is tempting to dismiss questions of the latter sort as matters of psychology with little import for deeper philosophical issues, some authors have held that, on the contrary, this is where the most theoretically significant cases of "unreasonably effective" mathematics are to be found.

Steiner (1989) identifies a number of patterns of reasoning whereby scientists have arrived at empirical hypotheses, subsequently experimentally verified, through a process of what he characterizes as purely mathematical "analogy", as opposed to any consideration of (what the scientist takes to be) real physical happenings. For instance:

> Equation E has been derived under assumptions A. The equation has solutions for which A are no longer valid; nevertheless, one looks for these solutions in nature, just because they are solutions of the same equation. (pp. 456–57.)

As an example of this pattern of discovery, Steiner – and, following him, Mark Colyvan (2001) – discusses the process through which Maxwell first arrived at his celebrated equations of electromagnetism, now a staple of physical theory. Here, somewhat abbreviated, is Colyvan's account (pp. 267–68; emphases in original):

> Maxwell found that the accepted laws for electromagnetic phenomena prior to about 1864, namely Gauss's law for electricity, Gauss's law for magnetism, Faraday's law, and Ampère's law, jointly contravened the conservation of electric charge. Maxwell thus modified Ampère's law to include a *displacement current*, which was not an electric current in the usual sense [. . .], but a rate of change [. . .] of an electric field. This modification was made on the basis of formal mathematical analogy [with Newton's theory of gravitation, where energy and momentum are conserved], *not* on the basis of empirical evidence. [. . .]
>
> The interesting part of this story for the purposes of the present discussion [. . .] is that Maxwell's equations were formulated on the assumption that the charges in question moved with a constant velocity, and yet such was Maxwell's faith in the equations, he assumed that they would hold for *any* arbitrary system of electric fields, currents, and magnetic fields. In particular, he assumed that they would hold for charges with accelerated motion and for systems with zero conduction current.

And, Colyvan goes on to recount, this latter assumption allowed Maxwell to predict the existence of electromagnetic radiation – a prediction which was later confirmed experimentally.

Thus, we have here a case of a scientist who, pursuing mathematically formulated criteria of well-formedness of theory, was able to construct a new hypothesis which then turned out to be applicable, and empirically supported, under a wider range of conditions than those for which it was originally developed – and this without guidance (in the theory-development stage) from any empirical data gathered under these wider conditions, or from any well-developed idea of the nature of the underlying physical forces. Mathematics itself, as it were, seems to have done the better part of the work.

But, while intellectual feats like these are without question admirable, it is not clear to me that, in the last analysis, they are really of a different kind from the more humdrum projections from experience discussed in Sect. 14.4. Maxwell sought a modification of Ampère's law which would agree with the existing empirical data and yet allow for the conservation of electric charge. Having found it, he hypothesised that the new equation would hold up in a wider range of circumstances than those for which data were available – and it did. How does this story differ from that in Sect. 14.4, where our scientists, upon noticing that Eq. (14.1) had held in a range of cases not including that of 100 m, hypothesised that it would hold up in the 100 m case as well? To be sure, there is a difference of degree – in the Maxwell story, the new cases are more radically dissimilar to the old ones than in our little fiction. But is the difference of such a nature as to raise any new philosophical puzzle? I do not see how. In either of the episodes, it is a matter of inferring, from the observation that a certain regularity has held in a limited class of situations, that it will hold in a greater class.

Perhaps Steiner and Colyvan will wish to call attention to the fact that Maxwell, in order to get things to work out properly, needed to postulate the existence of a new physical process (the displacement current mentioned in the block quote above) for which there was no empirical evidence at the time, thus venturing out in pursuit of purely mathematical cohesion without anchoring his reasoning to any previously known, robustly physical bedrock. But a similar lack of anchoring was assumed to obtain in Scenario B, and may likewise be supposed to afflict the physicists in Sect. 14.4 without detracting from the plausibility of that story. To notice a regularity, be it of a combinatorial or an arithmetical nature, and rationally project it from observed cases to unobserved ones, does not require any theory as to the underlying cause of the regularity. Ask Newton: *Hypotheses non fingo.*

What about the fact that the original impetus for Maxwell's work came from his feeling that, just as Newtonian mechanics implies the conservation of energy and momentum, so the theory of electromagnetism ought to conserve electric charge – what Colyvan classifies as a case of reasoning by "mathematical analogy"? Well, what of it? Yes, it was a good hunch, and one that could not have been spelled out in a precise way without employing mathematical concepts. Now to Steiner's and Colyvan's way of thinking, if I have understood them correctly, the latter fact suggests that conservation of a quantity, be it momentum or charge, should be classified as a property of the "formalism" employed in the presentation of the theory – a property, that is, of roughly the same dignity as, say, the number of symbols figuring in an equation – rather than as a substantive feature of the theory itself. On such a view, no doubt, the fact that Maxwell struck gold must appear as a crazy fluke. But surely, to conceive of things this way is to draw the line between notation and content in the wrong place. Mathematics is a body of theory, not a notational toolkit. To say that two forces are both inversely proportional to distance squared is not to say that physicists have elected to use similar symbols when writing about them – it is to say that the forces have a structural property in common. Likewise when two phenomena are both seen to obey conservation laws – it is a more abstract kind of similarity, but surely the difference is one of degree, not kind.

As always, we must of course acknowledge the possibility-in-principle of standards of similarity so unlike our own that, for instance, a black cat will be seen to have more in common with a white bicycle (they are both black-if-and-only-if-feline) than with a white cat ("Both are *cats*?"), or a black bicycle. ("Both are *black*? What strange and arbitrary categories your thinking employs!") From such a perspective, to be sure, the structural similarities cited in the previous paragraph may count for nothing – but now we are back to the quandaries of Hume and Goodman once again. *Given* the classificational tendencies inherent to basic human sanity, should commonalities of mathematical structure be considered as being on a level with mere notational similarities? I cannot see any reason for thinking that they should.

The story of Maxwell's equations is but one of the multifarious historical examples that have been cited by those who consider the applicability of mathematics a philosophical enigma (and, in fairness to Steiner, it is not the one to which he attaches the greatest significance). I could not possibly discuss them all, but must leave it to the reader to decide to what extent other potentially perplexing instances of mathematics-aided theory formulation can be accounted for in similar ways. For my part, I remain unconvinced that we are faced with a real problem.

14.7 Mathematicity as a Matter of Perspective

My principal aim in this paper has been to argue that cases of applicable mathematics which have been painted by some authors, more or less explicitly, as similar to the outlandish Scenario C, may really have more in common with Scenarios A and B: however striking, on closer inspection they do not give much reason to think of mathematicians as mysteriously capable of foreshadowing empirical observation. Still, though, in many cases one is likely to feel that the mathematical laws in question have a rather arbitrary character to them – why this specific constellation of mathematical operations, as opposed to some other? In this final section I wish to take the paper's main argument one tentative step further by suggesting that such seeming Scenario B-style cases might actually in the final analysis be more akin to Scenario A.

The point can be brought out by means of yet another fictional example, for the presentation of which we shall first need to provide a bit of mathematical background. The functions of *hyperbolic sine and cosine* are defined as follows.

$$\sinh\theta = \frac{e^{\theta} - e^{-\theta}}{2}, \qquad \cosh\theta = \frac{e^{\theta} + e^{-\theta}}{2}.$$

The hyperbolic functions obey laws closely resembling those governing the ordinary trigonometric functions. For instance, whereas, for any ϕ, ψ, and θ,

$$\cos(\phi + \psi) = \cos\phi\cos\psi - \sin\phi\sin\psi,$$
$$\sin(\phi + \psi) = \sin\phi\cos\psi + \cos\phi\sin\psi,$$

and

$$\cos^2\theta + \sin^2\theta = 1,$$

for the hyperbolic functions (as is easily verified by reference to their definitions) we have

$$\cosh(\phi + \psi) = \cosh\phi\cosh\psi + \sinh\phi\sinh\psi,$$
$$\sinh(\phi + \psi) = \sinh\phi\cosh\psi + \cosh\phi\sinh\psi,$$

and

$$\cosh^2\theta - \sinh^2\theta = 1.$$

The analogy goes further, and is one of the reasons the functions have attracted the attention of mathematicians since their introduction in the eighteenth century.

Now imagine the following counterfactual scenario. Experimental physicists, as yet unaware of the principles of special relativity, are conducting empirical investigations into the addition of speeds in a common direction of movement. That is, they are studying situations in which an object B is moving with speed u relative to an object A, while another object C is moving, in the same direction, with speed v relative to B. (By the relative speed of Y and X we mean the speed of Y as measured by equipment that is stationary with respect to X, or vice versa; as a matter of empirical fact, the result will be the same either way.) Now let w be the speed of C relative to A. Naturally, our relativistically innocent scientists expect to observe that

$$w = u + v.$$

To their considerable surprise, however, the relative speeds are instead consistently found to obey the law

$$\theta_w = \theta_u + \theta_v, \tag{14.5}$$

where θ_u, θ_v, and θ_w are the numbers satisfying the equations

$$u \cosh \theta_u = c \sinh \theta_u,$$
$$v \cosh \theta_v = c \sinh \theta_v,$$
$$w \cosh \theta_w = c \sinh \theta_w,$$

c being the speed of light in a vacuum.

One can easily imagine the experimentalists scratching their heads over this seemingly arbitrary law. Of all possible moderately complicated functions of two variables, why should w be determined by u and v in precisely the way described by (14.5)? Why not equally well some entirely different monotonically increasing function? On the face of it, the situation looks similar to our Scenario B.

But now a clever theoretician points out that Eq. (14.5) can in fact be *derived* in a natural way from one simple (if counter-intuitive) postulate: that c, the speed of light, must be the same in all inertial frames of reference. The reasoning will be familiar to any reader who (unlike our fictional scientists) has studied basic relativity theory. Nevertheless, in order to get a firmer grip on the sense of "derivation" in play here – physicists have been known to deploy the term rather more freely than a mathematician typically would – let us recapitulate the argument.

Consider two frames of reference $\mathfrak{A}$ and $\mathfrak{B}$, stationary with respect to the above-mentioned objects A and B, respectively; for an unspecified event e, let x be its spatial coordinate in $\mathfrak{A}$ along the axis parallel to the movement of B, and t the time at which e occurs in $\mathfrak{A}$. Similarly, let x' and t' be e's spatial and temporal coordinates in $\mathfrak{B}$, x' increasing in the same direction as x. For simplicity, let the origins of the two coordinate systems coincide, so that when $x = 0$ and $t = 0$, then likewise $x' = 0$ and $t' = 0$. (Such coincidence can always be arranged by picking an arbitrary event and using it as origin of both frames.) We now turn our attention to the mathematical relation between x and t, on the one hand, and x' and t', on the other. In particular, how is the spatial coordinate in the one system determined by the spatial and temporal coordinates in the other; in other words, what are the functions f and f' such that

$$x = f(x', t') \qquad \text{and} \qquad x' = f'(x, t)?$$

Some constraints on f and f' may be laid down at once. Firstly, for reasons of symmetry, we must expect f and f' to be related in such a way that, for any numbers ζ, ξ and τ, if $\zeta = f'(\xi, \tau)$ then $-\zeta = f(-\xi, \tau)$, meaning that

$$-x' = f(-x, t). \tag{14.6}$$

For – if a bit of hand-waving be allowed – $\mathfrak{A}$ as seen from $\mathfrak{B}$ is exactly like $\mathfrak{B}$ as seen from $\mathfrak{A}$ except that whereas $\mathfrak{B}$ is moving through $\mathfrak{A}$ in a positive direction, the direction of movement of $\mathfrak{A}$ through $\mathfrak{B}$ is negative; and this in turn is exactly like having the direction of movement the same but the signs of spatial coordinates flipped.

Secondly, lest the character of the transformation differ arbitrarily from one time and place to another, we should expect equal intervals to transform into equal intervals. That is to say, if $x'_1 - x'_2 = x'_3 - x'_4$ and $t'_1 - t'_2 = t'_3 - t'_4$, it ought to hold that $x_1 - x_2 = x_3 - x_4$ and $t_1 - t_2 = t_3 - t_4$ (where x_i and t_i are the $\mathfrak{A}$-coordinates of an event with $\mathfrak{B}$-coordinates x'_i and t'_i). But this can only hold in general if the function f is linear; in other words, there must exist numbers γ and δ such that always $f(\xi, \tau) = \gamma\xi + \delta\tau$, i.e.

$$x = \gamma x' + \delta t' \tag{14.7}$$

and, by (14.6), $-x' = \gamma(-x) + \delta t$, i.e.

$$x' = \gamma x - \delta t. \tag{14.8}$$

Thirdly, the origin of $\mathfrak{B}$, by definition always located at position 0 on the x' axis, will be moving trough $\mathfrak{A}$ in accordance with the equation $x = ut$, so that $0 = x' = \gamma x - \delta t = \gamma ut - \delta t$, which is to say that $\delta = \gamma u$, allowing us to recast (14.7) and (14.8) as

$$x = \gamma(x' + ut'), \qquad x' = \gamma(x - ut). \tag{14.9}$$

Having thus established the general form to be expected of f and f', in order to pin down γ we now invoke the postulate of the invariance of c. By this postulate, a light pulse emitted at the common origin of $\mathfrak{A}$ and $\mathfrak{B}$ will travel in the positive direction according to the equations

$$x = ct, \qquad x' = ct'. \tag{14.10}$$

From (14.9) and (14.10) it follows by simple algebraic reasoning – the details of which need not concern us here – that either

$$\gamma = \frac{1}{\sqrt{1 - u^2/c^2}} \tag{14.11}$$

or

$$\gamma = -\frac{1}{\sqrt{1 - u^2/c^2}}. \tag{14.12}$$

While, from a purely logical point of view, one might perhaps allow that (14.11) should hold in some cases, (14.12) in others, the simplest, most uniform way of having the disjunction obtain universally is for one of the disjuncts to obtain universally, so as to render γ a continuous function of u. Universal validity of (14.12), however, would entail, in the limiting case where $u = 0$, that $\gamma = -1$ and consequently $x = -x'$, which is impossible since in this situation frames $\mathfrak{A}$ and $\mathfrak{B}$ are one and the same. So (14.11) it must be.

Thus we arrive at the conclusion that time and place in frame $\mathfrak{B}$, and therewith speed relative to B, are related to time and place in frame $\mathfrak{A}$, and therewith speed relative to A, in the way indicated by Eqs. (14.9) and (14.11). And from these relations, as it turns out, the puzzling velocity-addition Eq. (14.5) follows on purely mathematical grounds. Again, the intra-mathematical specifics are immaterial to our philosophical concerns; readers wishing to delve into the details may consult a suitable textbook exposition, for instance sections 12.5 and 14.3 of Shankar (1989).

At this point, the reader might not unreasonably take issue with our description of the foregoing line of reasoning as a derivation of (14.5) from nothing but the postulate that the speed of light be the same in all inertial frames of reference. In addition to this premiss, it will be objected, in several instances we appealed to extra-mathematical considerations of simplicity, symmetry, etc. Objection granted; but the point of the example is that none of these additional premisses is nearly as mathematically involved as (14.5) itself, and our invocation of them does nothing to undermine the following lesson: an empirical regularity which initially gives a mysterious impression of having been instituted for the mathematical gratification of experimental scientists may on closer inspection turn out be a necessary consequence of a set of mathematically much more pedestrian principles. What initially (in our fictional chronology) looked like a Scenario B-like situation has turned out, when regarded from the right point of view, to be more closely comparable to Scenario A.

Here, then, is my suggestion. Perhaps it holds as a general rule that, whenever scientists observe that the material world exhibits a lawlike regularity, describable in mathematical terms but seemingly arbitrary in its specifics, in fact the mathematical character of the law is an effect of the specific perspective from which they are observing it. From a different conceptual vantage point it may be possible to give the phenomena in question an equally full and precise description without employing, on the level of basic postulates, any sophisticated mathematical functions whatsoever.

I am not suggesting that finding such a vantage point will typically be an easy task. On the contrary, it may well be that in many cases the mathematics-free point of view requires fundamental conceptual categories so far removed from the natural workings of human cognition that we will never be able to attain it, and from whatever perspective we are capable of looking at them, the phenomena will take on a mathematically artful aspect. In this regard, the picture I am sketching differs radically from the above-discussed example from relativistic kinematics, in which the only real challenge on a conceptual level is to think of temporal intervals as relative to a frame of reference. But what I am suggesting is that the difference between the relativistic case, where a less mathematically involved point of view is readily available, and the legions of cases where no such alternative is known, may be one of degree rather than kind.

The suggestion is a vague one, and is not being advanced in a programmatic spirit. Certainly I do not have any splendid mathematics-free reformulations of physical theories on offer, nor am I admonishing physicists to do a better job of coming up with them. What I am attempting is perhaps best described as a shift of the burden of proof with regard to the philosophical issue of the empirical applicability of mathematics. Granted, our best physical theories make heavy use of mathematics, not only in teasing out the testable consequences of their fundamental postulates, but also in the formulations of these postulates themselves. But, pending any argument that the mathematical character of physical theories is an essential feature of the world they are describing, rather than a (possibly humanly unavoidable) artifact of the conceptual lens through which that world is being studied, perhaps a bit of caution is in order when pronouncing on the wider philosophical implications of applied mathematics.

References

Chalmers, A. (2013). *What is this thing called science?* St Lucia: University of Queensland Press.

Colyvan, M. (2001). The miracle of applied mathematics. *Synthese, 127*, 265–277.

Dyson, F. (1964). Mathematics in the physical sciences. *Scientific American, 211*(3), 128–147.

Feynman, R. (1967). *The character of physical law*. Cambridge, MA/London: The M.I.T. Press.

Goodman, N. (1955). *Fact, ficion, and forecast*. Cambridge, MA: Harvard University Press.

Sarukkai, S. (2005). Revisiting the 'unreasonable effectiveness' of mathematics. *Current Science, 88*(3), 415–423.

Shankar, R. (2014). *Fundamentals of physics: Mechanics, relativity, and thermodynamics*. New Haven/London: Yale University Press.

Steiner, M. (1989). The application of mathematics to natural science. *The Journal of Philosophy, 86*(9), 449–480.

Steiner, M. (1998). *The applicability of mathematics as a philosophical problem*. Cambridge, MA: Harvard University Press.

Wigner, E. (1960). The unreasonable effectiveness of mathematics in the natural sciences. *Communications on Pure and Applied Mathematics, 13*(1), 1–14.

Chapter 15
What the Applicability of Mathematics Says About Its Philosophy

Phil Wilson

Abstract What does the existence of applied mathematics say about the philosophy of mathematics? This is the question explored in this chapter, as we take as axiomatic the existence of a successful applied mathematics, and use that axiom to examine the various claims on the nature of mathematics which have been made since the time of Pythagoras. These claims – on the status of mathematical objects and how we can obtain reliable knowledge of them – are presented here in four "schools" of the philosophy of mathematics. The perspective and claims of each school and some of its subschools are presented, along with some historical development of the school's ideas. Each school is then examined under what we call the lens of the existence of applied mathematics: what does the existence of applied mathematics imply for the competing claims of these various schools? Although, unsurprisingly, this millennia-old debate is not resolved in the next few pages, some of the key issues are brought into sharp focus by the lens. We end with a summary and a tentative discussion of the physicist Max Tegmark's Mathematical Universe Hypothesis.

15.1 Introduction

We use mathematics to understand the world. This fact lies behind all of modern science and technology. Mathematics is the tool used by physicists, engineers, biologists, neuroscientists, chemists, astrophysicists and applied mathematicians to investigate, explain, and manipulate the world around us. The importance of mathematics to science cannot be overstated. It is the daily and ubiquitous tool of millions of scientists and engineers throughout the world and in all areas of science. The undeniable power of mathematics not only to predict but also to

P. Wilson (✉)
School of Mathematics and Statistics, University of Canterbury, Christchurch, New Zealand
e-mail: phillip.wilson@canterbury.ac.nz

S. O. Hansson (ed.), *Technology and Mathematics*, Philosophy of Engineering and Technology 30, https://doi.org/10.1007/978-3-319-93779-3_15

explain phenomena is what physics Nobel laureate Eugene Wigner dubbed the "unreasonable effectiveness of mathematics in the natural sciences" (Wigner 1960).

Yet the success of mathematics in explaining the world belies a great mystery: why is that possible? Why are our abstract thought and our manipulation of symbols able to successfully explain the workings of distant stars, the patterns of stripes on a tiger, and the weirdest behaviour of the smallest units of matter? Why is applying mathematics to the real world even possible?

This is a question in *the philosophy of mathematics*. The traditional approach to answering it is to first decide (hopefully on rational grounds) what to believe about the nature of mathematics and its objects of study, and then to explore what this philosophical standpoint says about the applicability of mathematics to the world. In this chapter, I take a different approach.

I take as given *the existence of applied mathematics*. On this foundational axiom, I ask the question *"what does the existence of applied mathematics say about the philosophy of mathematics?"* In this way, we treat the existence of applied mathematics as a lens through which to examine competing claims about the nature of mathematics. What then do we mean by the existence of applied mathematics, by the philosophy of mathematics, and what are the claims on the nature of mathematics?

15.1.1 Applied Mathematics

It is not easy to define applied mathematics. The authoritative *Princeton Companion to Applied Mathematics* (Higham 2015a) sidesteps this difficulty by instead describing what applied mathematics is based on what applied mathematicians do. This is a strategy, the *Companion* argues (Higham 2015b, p. 1), with some distinguished historical precedent (for example, Courant and Robbins 1941).

In this chapter I borrow a concise definition of applied mathematics from mathematician Garrett Birkhoff (1911–1996), who took inspiration from physicist Lord Rayleigh (1842–1919): "mathematics becomes 'applied' when it is used to solve real-world problems" (quoted in Higham 2015b, p. 1). The breadth of this definition, which includes "everything from counting change to climate change" (Wilson 2014, p. 176), is important. It means that we can use the shorthand "applied mathematics" for any application of mathematics to understanding the real world, and the name "applied mathematician" for any person doing so. This usage of "applied mathematics" and "applied mathematician" means we avoid any confusion over how a particular example or person might be categorised according to contingent academic disciplines in the workplace.

For our purposes, then, applied mathematics, is simply *mathematics* which is *applied*. An applied mathematician is anyone who applies mathematics.

In a book like this we can take it for granted that the existence of applied mathematics is undisputed. Its chapters present case after case of the overwhelming success and importance of the application of mathematics to the world around us.

Applied mathematics not only predicts the outcome of experiment, it also provides understanding and explanation of the forces, fields, and principles at work. Indeed, "Mathematics ... has become the definition of explanation in the physical sciences" (Barrow 2000). This is what I mean by *the existence of applied mathematics*, a useful phrase which I will abbreviate to TEAM.

Here I take mathematics, science, and technology seriously, in that I believe they have something important and objective to say about the world. While there are cultural and social concerns with the institutional forms of transmission of mathematics, I firmly reject the "woefully inadequate explanation" (Barrow 2000) that mathematics is merely a social construct. This postmodern fallacy has been hilariously exposed by Sokal (1996, 2008) and others. As a mathematician and scientist, I also reject the notion, fashionable among some famous physicists, that philosophy has nothing useful to say about science; see for example Weinberg (1992), or Krauss (2012). This chapter is evidence against that view.

15.1.2 The Four Schools of the Philosophy of Mathematics

What is mathematics? What is the status of the objects it studies? How can we obtain reliable knowledge of them? These are the general types of questions which animate and define the philosophy of mathematics, and on which we will focus below. If you think this sounds vague, I agree with you. In *Philosophy of Mathematics: selected readings* (Benacerraf and Putnam 1983) compiled by the highly influential philosophers of mathematics Paul Benacerraf and Hilary Putnam, the editors write in their first sentence "It would be difficult to say just what comprises the philosophy of mathematics".

But we have to talk about something, so in what follows I present some of the main ideas from the long history of this vaguely-defined area of philosophy. This is not an exhaustive study of all of the schools of the philosophy of mathematics, neither will we see all of the main areas of study. Those in the know might find it shocking that I do not mention Descartes, Locke, Berkeley, or Wittgenstein, and spend scant time on Kant and Hume. Their ideas fill these pages through their influence on their contemporaries and those who came after them and on whose ideas I focus. And while I try to present some historical development, this can only ever be cursory in a single chapter covering over 2500 years from Pythagoras to the present. I am painfully aware of the Western bias in my presentation, with no mention of the great Indian, Chinese, and Arabic traditions. I hope that you are intrigued enough to follow the references. If you are eager to start right now, then Bostock (2009) gives a highly readable and comprehensive introduction, Benacerraf and Putnam (1983) contains selected key papers and readings, Horsten (2016) is an excellent starting point for an educational internet journey, and Mancosu (2008) is a survey of the modern perspective. But I hope you will read this chapter first.

The chapter divides the philosophy of mathematics into four schools, each of which has its own section. This division is broadly accepted and historically relevant, but not without controversy. I have also tried to present the arguments of smaller subschools of the philosophy of mathematics. Sometimes this has required discussing a subschool when a theme arises, even if historically it does not belong in that section. I hope that historians of the philosophy of mathematics, and the philosophers themselves, will forgive me.

Mostly I have tried to avoid jargon, but there are some important concepts that I have tried to develop as they arise. However, there are two words needed from the start: *ontology* and *epistemology*. Ontology concerns the nature of being. In terms of mathematics: what do we mean when we say that a mathematical object exists? Are mathematical objects pure and outside of space and time, as the platonist insists, or are they purely mental, as the intuitionist would argue, or the fairy tales of the fictionalist? Epistemology concerns the nature of knowledge, how we can come to have it, and what justifies our belief in it. Speaking loosely, we can say that if ontology concerns the nature of what we know, then epistemology concerns how we know it.

15.1.3 The Lens

I focus on what TEAM says about the philosophy of mathematics. It is important to distinguish this concern with what the applicability of mathematics says about the nature of mathematics from a concern (even a philosophical one) with the nature of the work done in applying mathematics. This latter question focusses on the praxis of applying mathematics: how applied mathematicians choose which problems to work on, how they turn a real-world problem into a mathematical one, what their aesthetic is, how they choose a solution method, how they communicate their work, and related questions. See for example Davis and Hersh (1981), Ruelle (2007), Mancosu (2008), and Higham (2015a).

In training our TEAM lens on the four main schools of the philosophy of mathematics, we bring into focus some aspects of old questions. This is complementary to a more modern focus on the so-called "philosophy of real mathematics" (Barrow-Green and Siegmund-Schultze 2015, p. 58). This "new wave" as outlined in the introduction to Mancosu (2008), currently avoids the daunting ontological question of why mathematics is applicable, and focusses instead on expanding the epistemological objects of study to include "fruitfulness, evidence, visualisation, diagrammatic reasoning, understanding, explanation" (Mancosu 2008, p. 1) and more besides. These everyday epistemological issues raised by working with mathematics are used to refine what is meant by applied mathematics, to study how applied mathematics and its objects of study relate to the rest of mathematics, and what mathematical value there is in applied mathematics. Indeed, Pincock (2009, p. 184) states "a strong case can be made that significant epistemic, semantic and metaphysical consequences result from reflecting on applied mathematics". The interested reader is referred to the excellent overviews collected in Mancosu (2008) and Bueno and Linnebo (2009).

I take TEAM as axiomatic in order to examine the claims of various schools of the philosophy of mathematics. This is distinct from those like Quine (1948) and Putnam (1971) who take TEAM as axiomatic in order to provide a justification for "faith" in mathematics. As outlined by Bostock (2009, pp. 275 ff), the Quine/Putnam position is that mathematics is similar to the physical sciences in the sense that both postulate the existence of objects which are not directly perceptible with human senses. In the case of mathematics, this includes the integers, while for the physical sciences, this includes atoms, to take an example in each field. The Quine/Putnam position is that mathematics as well as the physical sciences should be exposed to the "tribunal of experience". In particular, since our atomic theory leads to predictions which conform to our experience, we should accept the existence of atoms as real. Crucially, claim Quine and Putnam, since all our physical theories are mathematical in nature, and since those theories work, we must accept the existence of the mathematical entities on which those theories depend as also being real. The Quine/Putnam *indispensability argument* is that we must believe that mathematical objects exist because mathematics works. We will return to the indispensability argument, but I reiterate that we will use TEAM as an axiom for examining competing claims on the nature of mathematics, rather than using TEAM as an axiom for a new claim on the nature of mathematics.

The remainder of the chapter is structured as follows. We will examine each of the four schools in turn, introducing their main ideas, explaining their ontology and epistemology, and giving a brief overview of their history and structure. Within each school's section, we will use the TEAM lens to bring into focus the challenges faced by the school's followers as they attempt to explain the applicability of mathematics. We end with a discussion and conclusion.

15.2 Platonism

The platonist believes that mathematical objects are real and exist independently of humans in the same way that stars exist independently of us. Stars burn in all ignorance of us, and while their properties are discoverable by humans, they are independent of us. The same is true, says the platonist, of the existence and properties of numbers, and of all mathematical objects. Thus the platonist mathematician believes that we discover mathematics, rather than invent it.

The platonist position is that all abstract objects are real. An "abstract object" is one which is both entirely nonphysical and entirely nonmental. The triangle formed by the three beams over my head is an entirely physical object. When I hold it in my mind, and as you now attempt to picture it in yours, we have a mental object which is drawn from our experiences of the physical. But this mental object is still not yet a platonic object. For the platonist there exists in a third "realm" apart from the physical and mental ones the perfect, ideal form of a triangle, of which the imperfect triangles in our minds, and the still less imperfect ones in our physical world, are merely poor approximations.

The platonist does not believe that mathematical objects are drawn off or abstracted from the physical world; rather, that they exist in a realm of perfect, idealised forms outside of space and time. But what does "existence" mean in this statement? Existence usually refers to an object embedded in time and space, yet these platonic forms are taken to be outside of time and space. Their existence is of a different type to all other forms of existence of which we know. We can say that as I type this my laptop rests on an oak table in New Zealand early in 2017. We can say that our sun will be in the Milky Way galaxy next year, and that Caesar lay bleeding in Rome two millennia ago. The verbs "rest", "be", "bleed" in these statements are fancy ways of saying "is", and the locations and times in each example are not two pieces of information but one: a single point in the fabric of spacetime which Albert Einstein (1879–1955) wove for us a century ago. By contrast, platonic objects "are" in a "place" outside of spacetime.

Platonism is the oldest of our four schools, and for many mathematicians in history this perspective was taken to be natural and obvious – and this remains true for the typical mathematician or scientist today (Bostock 2009, p. 263). There is some evidence that Plato (427–347 BCE) held this view (Cooper 1997), possibly swayed towards the life of the mind and away from the life of the engaged citizen philosopher after his great mentor Socrates was condemned to death. Plato presented his theory of forms in his *Phaedo*, and developed it in his *Republic* (Cooper 1997), with its enduring image of a shackled humanity deluded by shadows cast by ideal forms on a cave wall. It is much less clear that the platonism we are discussing here was a view held by Plato, since in later life Plato saw mathematical forms as being intermediaries between ideal forms and perceptible objects in our world (Bostock 2009, p. 16). For this reason I do not capitalise the word "platonism".

Mathematical platonism is the position that mathematical objects have a reality or existence independent not only of space and time but also of the human mind. Within this statement are the three claims that (1) mathematical objects exist, (2) they are abstract (they sit outside of spacetime), and (3) they are independent of humans or other intelligent agents (Linnebo 2013). All three claims have been challenged by various schools, but the claim of independence sets platonism apart from the other schools, as we shall see. For the platonist, the concept of number, the concept of a group, the notion of infinity – all of these would exist without humans, and even, remarkably, without the physical universe. The platonist ontology is that mathematical objects are real, the realest things that exist.

But how can we know about them? Even mathematicians are physical beings containing mental processes and which are embedded in space and time, so how can they access this platonic realm, which sits outside of spacetime? The only platonist answer to this epistemological problem is that we know about these abstract objects *a priori* – that is, that they are innate, and independent of sensory evidence.

This is surely an unsatisfactory answer. To say that we know something *a priori* is merely to rename the fact that we do not know how we know it. It is dodging the issue – begging the question. If the innateness claim is taken to its extreme, the idea that every abstract concept that humanity might ever uncover is somehow hardwired from birth into a finite brain of finite storage capacity seems questionable to say the least. And where is the information encoded which is uploaded into the developing foetal brain? DNA has a finite, if colossal, storage capacity (Extance 2016).

The other option is that (at least) the human mind somehow has the capacity to access the platonic realm. But how can a physical, mental being access a realm outside of those two realms? Plato himself saw this epistemological problem as a grave issue, and in his later life he moved away from the viewpoint which bears his name, as we saw above.

This *problem of epistemological access* was precisely formulated by Benacerraf (1973). By breaking the problem into its constituent assumptions and deductions, Benacerraf gave philosophers of mathematics more precise targets at which to aim. There have been many responses to this challenge, as we shall see. But as summarised in Horsten (2016), the fundamental problem of a how a "flesh and blood" mathematician can access the platonic realm "is remarkably robust under variation of epistemological theory" – that is, "[t]he platonist therefore owes us a plausible account of how we (physically embodied humans) are able" to access the platonic realm.

Such an account is elusive, although attempts are being made; see (Balauger 2016, section 5) for an excellent summary. It is worth noting here that even ardent platonists such as Kurt Gödel (1906–1978) failed to avoid dodging the issue. Gödel is a central figure in the philosophy of mathematics. As we shall see, he was a platonist, who destroyed both logicism and formalism, and shackled the consistency of intuitionistic arithmetic to that of classical arithmetic (Ferreirós 2008, p. 151), where *consistency* means that contradictions cannot be derived. But returning to the issue of epistemological access, we see for example, in Gödel (1947, pp. 483–4) how he skips over it by stating "axioms force themselves on us as being true. I don't see why we should have less confidence in this kind of perception, i.e. in mathematical intuition, than in sense perception." But how do we come by such intuitions? Whether they are innate (following the great Immanuel Kant (1724–1804)) or acquired (following the equally great David Hume (1711–1776)) there remains the question of how mental events correlated with physical brains localised in spacetime are able to have them.

Platonism is a kind of *realism*. The realist believes that mathematical objects exist, and do so independently of the human mind. Gödel was certainly a platonic realist (Bostock 2009, p. 261). There are, however, non-platonic forms of realism, and the Quine/Putnam position outlined in the Introduction is one example. Quine and Putnam argue that mathematics is real because it underpins our physical theories – since they work, mathematics must be true. By "work" here I mean precisely what I meant when I defined applied mathematics in the Introduction, and the breadth of that definition is important. Since it really does cover everything from counting change to climate change, it is not just the use of mathematics in highfalutin scientific domains such as climate modelling or fundamental particle physics, but also includes the utility of basic arithmetic for counting sheep.

15.2.1 Platonism Under the Lens

Under even the closest scrutiny beneath the TEAM lens, the ontology of platonism remains as pure and perfect as its own ideal forms. Since the platonist believes that the physical world is an imperfect shadow of a realm of perfect ideal objects, and since in this worldview mathematics is itself a very sharp shadow cast by a more ideal form, it is no surprise that our mathematics becomes applicable to the physical world. This is not *evidence for* platonism, but the TEAM lens does not reveal any *evidence against* platonism based on its ontology.

However, as we have seen, cracks appear when we examine the epistemology of platonism – that is, when we ask how we are able to have knowledge of the platonic realm of ideal forms. The problem of epistemological access is such a serious one that it has prompted a rejection of platonism altogether, which we consider in the following three sections. Another approach has been to recast platonism in forms which avoid the epistemic access problem.

One example is *plenitudinous platonism*; see Balaguer (1998) and Linsky and Zalta (1995, 2006) for two different versions. The central idea is that any mathematical objects which *can* exist, *do* exist. Summarising how this approach may solve the problem of epistemological access, Linnebo (2013) says "If every consistent mathematical theory is true of some universe of mathematical objects, then mathematical knowledge will, in some sense, be easy to obtain: provided that our mathematical theories are consistent, they are guaranteed to be true of some universe of mathematical objects."

While plenitudinous platonism may solve the epistemic access problem (though this remains controversial), it does not yet explain why mathematics is able to be applied to the real world. Both platonism and plenitudinous platonism fail to explain why any part of mathematics should explain the physical world. Simply assuming that our mathematical objects (platonism), or objects in all forms of mathematics (plenitudinous platonism) have an independent existence does not in any way explain why they are applicable to the real world around us. Something further is required, some explanation of why the platonic realm entails the physical realm. This is what the TEAM lens brings sharply into focus for the platonist and plenitudinous platonist arguments.

An idea similar to plenitudinous platonism and which goes some way to addressing epistemic concerns was developed by the mathematical physicist Max Tegmark (b. 1967). In a series of papers beginning with Tegmark (2008), and explained in layperson terms in Tegmark (2014), Tegmark shows that platonic realism about physical objects implies a radical platonic realism about mathematical objects. Tegmark argues that the hypothesis that physical objects have an independent existence implies his Mathematical Universe Hypothesis (MUH): "our physical world is an abstract mathematical structure" (Tegmark 2008, p. 101). He goes on to argue, echoing the plenitudinous platonists, that all mathematics which can exist does exist in some sense, that our physical world *is mathematics* (not simply *mathematical*), and that our minds are themselves self-aware substructures of this mathematical universe. In the MUH, ourselves, our universe, and the various multiverses which our physical theories imply are subsets of this grand mathematical ensemble. The MUH addresses (though was not motivated by) the same epistemic concerns which motivated the plenitudinous platonists. Tegmark's ideas have spawned much debate, and in the grand tradition he has both defended and amended his hypothesis. It is heartening to see a mathematical physicist engaging with philosophers and mathematicians precisely around the issues of this chapter. For a starting point of objections and Tegmark's responses to them, see Wikipedia (2017).

As for Quine/Putnam realism, Bostock (2009, p. 278) observes that when considering objections to the Quinean position it is important to be careful about what is meant by science and the applications of mathematics. He argues that adopting the kind of broad definition of applied mathematics that I have taken for this chapter will undermine some of the objections to the Quine/Putnam theory, such as those in Parsons (1979/1980) and Maddy (1990). However, surely we can conclude that the Quine/Putnam idea is attractive under the TEAM lens?

Not so, claims Bostock (2009, pp. 305–6). One problem is the tenuous nature of truth when it is defined in this quasi-instrumentalist and utilitarian way, when the only true mathematical things are those which currently support our physical theories. As the theories change, so does truth. Worse, it is possible to argue that fewer and fewer parts of classical mathematics are required for our scientific theories, leading, in the extreme, to the fictionalism of Hartry Field (b. 1946) in which absolutely no mathematical objects are necessary; see the discussion in the Formalism section. But even if a time-dependent notion of mathematical truth is accepted, Paseau (2007) observes that the Quine/Putnam theory leaves unspecified the ontological status of the objects it posits. Mathematical statements are true when they are useful, but the Quinean can only shrug when asked whether mathematical objects are platonic or have one of the other possible statuses given in the following sections.

A final comment concerns the issue of causal agency for Quine and Putnam. Their position argues that both quarks and real numbers are to be considered true in as much as they are required in our quantum mechanics. Yet the former is a name for something which has a causal role in the world, while the latter is the name for a temporarily useful fiction with no causal power.

15.3 Logicism

To the logicist, mathematics is logic in disguise. All of the varied fields of mathematics are simply the fecund outpourings produced when logic combines with interesting definitions (Bostock 2009, p. 114). Mathematics equals logic plus definitions.

In this way, logicists seek to *reduce* mathematics to something else: logic. This idea can trace its lineage to Aristotle (384–322 BCE), who invented logic and tried to formulate his mathematical arguments in logical terms. Aristotle rejected Plato's insistence on a higher realm of ideal objects. He did not reject abstraction, but saw it as a process of generalisation of examples in the world. To him, the concept of triangle generalised real-world triangles. While Plato believed that all Earthly triangles were poor shadows of an ideal triangle with an independent existence beyond space and time, Aristotle believed that the concept of a triangle was abstracted from our everyday experience of triangles in the world. All Aristotle's science and mathematics concerns these abstractions. His ontology is of generalised ideas in the human mind, and his epistemology is one of perception, even in mathematics (Bostock 2009, p. 16). Thus to Aristotle, and his *conceptualist* viewpoint just outlined, we invent rather than discover mathematics, which is why I described him as having *invented* logic.

Central to a reductionist view of mathematics is that it can be reduced to something more fundamental, that the definitions of mathematics are a type of name or shorthand for relationships between sets of the fundamental objects, and that the correspondence of those names with things in the real world is of little interest or relevance to mathematics. This type of reduction can be called *nominalism*, since it concerns names, and there are two types (Bostock 2009, p. 262). One is logicism, which reduces mathematics to logic, and states that mathematics is a collection of names applied to logical objects. In this view, mathematics is a set of truths derived (or discovered) by the use of logic. It is worth noting that in this nominalist account, the mathematical objects have no independent existence. The second type of nominalism is the fictionalism of Hartry Field, which we discuss below in the section on Formalism.

The logicist ontology is that mathematical objects are merely logical ones in disguise. This ontology neatly explains why the varied fields of mathematics are connected: they lie in correspondence with one another because their objects of study are at root the same logical objects (or collections of them), but with a different overlay of definitions. Moreover, the central practice of mathematicians, the proving of theorems, follows well-defined and closely prescribed logical rules which themselves guarantee the validity and truth of the outcomes. No matter the definitions of the objects, when logical operations are correctly applied to logical objects (disguised as mathematical ones) the outcome will certainly be true.

In the logicist worldview mathematicians take disguised logical objects and perform logical operations on them. Because of this derivation of new results by a logical analysis of existing concepts, it is tempting to refer to these truths as *analytic*, and thereby to invoke Kant, and in particular to set up an opposition with Kant's *synthetic* truths derived from experience. But to use these words here might be misleading, since Kant himself argued for the synthetic nature of some, if not all, mathematical truth (Bostock 2009, p. 50). To Kant, mathematical truths could not be wholly derived by the action of logic; some *a priori* "intuition" of the objects involved was required. In the context of logicism, an analytic truth means one which is derived by the action of logic on logical objects plus definitions. This is the usage employed by the key figure Gottlob Frege, as we shall see below.

To explore what it means to say that mathematics is logic plus definitions, we can ask: what is a number in the logicist worldview? Surely something so fundamental to mathematics, at the core of arithmetic, cannot be open to debate? Yet to the logicist, the idea of number is in some sense superfluous to the truths of arithmetic. Defining number in a mathematical way simply overlays mathematical definitions on logical objects. The overlay is done on multiple objects rather than single objects, since if the latter were true then the logicist worldview would be rather barren. Merely positing a one-to-one correspondence between mathematical objects and logical ones would be no more interesting than compiling a very accurate thesaurus. If I observe that every eggplant is an aubergine and that every aubergine is an eggplant, then I can merely use the two words interchangeably, and I have not learned anything new about eggplants. Or aubergines. Rather, in the logicist worldview, a mathematical definition is powerful because it encodes multiple logical objects and the relationships between them. The apparently simple task of defining number logically takes us from the budding of logicism in the garden of a man named Frege, through its flowering in the care of a man named Russell, to its wilting in the shadow cast by a man named Gödel.

The soil for Frege's garden was laid down by Richard Dedekind (1831–1916). Dedekind is known to undergraduate mathematicians for putting the real numbers on a solid basis. He defined them by means of "cuts": an irrational such as the square root of 2 cuts the rational numbers into two classes, or sets. One of these contains all of the rational numbers smaller than the square root of 2, while the other contains all of the rational numbers larger than the square root of 2. This gave Dedekind the hope that all of mathematics could be built on logic plus set theory, with sets conceived of as logical objects.

This dream was shared by Gottlob Frege (1848–1925), who is considered the founder of logicism. Bostock in his (2009, p. 115) says "Frege's first, and ... greatest contribution ... is that he invented modern logic." Extending Dedekind's ideas, Frege defined number in terms of classes of equinumerous classes. In this way, the number 2 is the name for all sets which have two elements. Although this smacks of circularity, it is formalised in a way which avoids it. However, Bertrand Russell (1872–1970) found a paradox nestled at the heart of logicism as conceived by Frege as a combination of set theory and logic. This is the famous Russell's paradox, which in words is the following. Consider a set which contains all sets which do not contain themselves as members. Does this set contain itself? If it does, then it does not, and if it does not, then it does.

A popular analogy is the following. Suppose there is a town in which every man either always shaves himself, or is always shaved by the barber. This seems to divide the men of the town into two neat classes; no man can be in both sets by definition. But what about the barber? If he is a man who always shaves himself then he cannot be, since he is also then a man shaved by the barber. And if he is a man who is always shaved by the barber, then he will always shave himself, which he cannot.

Thus even the definition of quite simple sets is problematic. The problem is surprisingly difficult to eliminate, leaving aside solutions such as a barber who does not shave or is a woman. So difficult, in fact, that Frege gave up on his own logicist dream. Russell did not. He developed with Alfred North Whitehead (1861–1947) a new theory of "types", which in essence are hierarchical sets. This "ramified" theory eliminated the type of paradoxes which bedevilled Frege's logicism. A set could no longer contain itself as a member. In the shaving story, it is as if the town now has a caste system, and a man can be shaved only by someone of a lower caste. Thus the barber can be shaved by someone of a lower caste, and can shave anyone in a higher caste, but no-one can shave themselves (the lowest caste grows beards).

Russell and Whitehead wrote the monumental *Principia Mathematica* (Russell and Whitehead 1910) to bring Frege's dream to fruition through their ramified theory of types. The power of the mantle of meaning which mathematics places over logic is revealed by the fact that it takes 378 pages of dense argument in the *Principia* to prove (logically) that one plus one equals two.

But despite these Herculean efforts, the dream of reducing mathematics to logic died when Gödel rocked the mathematical world in 1931 with the publication of his two *incompleteness theorems* (Gödel 1931; see also Smoryński 1977). The first theorem is bad enough news: it says that any system which aims to formalise arithmetic must necessarily be incomplete. *Incomplete* means that the system must contain true statements which cannot be proved. And Gödel showed that this is true for *any* system which aimed to formalise arithmetic, and, worse, for any system which contained arithmetic. Thus Gödel's theorem not only destroyed the approach based on a combination of logic and ramified types developed by Russell and Whitehead, but all possible approaches. This was a profound and philosophically disturbing shock to mathematicians, who until that moment believed that all true statements must be provable. Mathematics has not been the same since.

Even worse was to come from Gödel's second incompleteness theorem: it is impossible to prove the consistency of arithmetic using only the methods of argument from within arithmetic. Thus to prove even the most basic of mathematical areas consistent, that is to show that contradictions can never be derived within it, requires stepping outside of that area. But then the new area of mathematics used to establish consistency of the first area would itself require external techniques in order to establish its consistency, and so on.

Gödel showed that any system which aims to formalise an area of mathematics contains unprovable true statements, and that the consistency of the system can only be established by stepping outside of itself. Logicism (and not just logicism, as we shall see) seemed well and truly dead. But logicism lives on in modified forms; the idea of number as a powerful naming convention for a set of interconnected logical objects is closer to what is now called the neo-Fregean standpoint. The difference between Fregean logicism and neo-Fregean logicism revolves around "Hume's Principle", which asserts that two sets are of the same size if their members can be placed in a one-to-one correspondence. Neo-Fregeans aim to derive elementary arithmetic from Hume's Principal plus logic rather than Frege's axioms of set theory plus logic. Frege himself rejected this approach since he knew that Hume's Principle did not clearly define number per se. Indeed, he observed that with Hume's Principle alone it is impossible to say that Julius Caesar is not a number. But Neo-Fregeans attempt to avoid this problem by taking Hume's Principle to be the very definition of number; see for example (Bostock 2009, pp. 266 ff). Moroever, Russell's theory of types is now considered the start of *predicativism*. Both neo-Fregean logicism and predicativism seek to avoid paradox while retaining logic as fundamental. These ideas have been developed for example by Bostock (1980); see also his (2009, section 5.3).

If in some sense all mathematics can be reduced to logic, what is the ontology of logic? The logicist rejects the realist idea that mathematical objects have an independent existence in a platonic realm of ideal forms, and substitutes logic as a foundation for mathematics. But this merely shifts the ontological question on to logic, and here we see a divergence in the history of logicist thought. Its founding father, Frege, was a realist of sorts, since he believed that logic and its objects had a platonic existence (Bostock 2009, chapter 9). Although Russell's views were complex and evolved throughout his life, he also seemed to remain essentially a platonic realist when it came to mathematics. Other logicists choose to remain silent on ontology.

15.3.1 Logicism Under the Lens

What can the logicist say about the existence of applied mathematics? If at the heart of mathematics we find only logic, and if the familiar objects of mathematics are merely names under which hides a Rube Goldberg arrangement of logical objects, then why should mathematics have anything useful to say about the real world? The logicist is not allowed to answer that the universe is merely an embodiment of a higher platonic realm of logic. To do so makes them a platonist.

There does not seem to be much more to see of logicism under the TEAM lens. At its heart, there is either a dormant platonism in its classical form (which Gödel destroyed anyway), or an echoing ontological silence in the modern forms. Since these modern forms do not propose any ontology, it is hard to critique them via the existence of applied mathematics. However, even they seem to have an implied platonism at their heart, since the neo-Fregean adoption of Hume's principle brings with it a notion of infinity which is platonic in the extreme – see Bostock (2009, p. 270) for some of the controversy.

Perhaps one observation can be made using the TEAM lens. If even such a simple concept as number veils a hidden complexity of logical objects, maybe what mathematicians do is to select definitions which excel at encoding logical objects and their interrelations. Having done so, perhaps mathematics is then a process of selection and evolution. This *principle of fecundity* and an evolutionary perspective is sufficiently general that it may apply in a broad sense to other schools in the philosophy of mathematics. However, it has problems. For a start, what is the ontological status of the fecund objects upon which evolution acts? Secondly, there are epistemological problems with the claim (see for example Mohr 1977) that minds with the best model of reality are those which are selected as fittest evolutionarily. It is not clear that the objects of the human mind need faithfully represent the objects of the physical universe. Mental maps of reality survive not because they are faithful to reality, but because of the advantage they conferred to our ancestors in their struggles to survive and to mate. Moreover, while concepts such as number and causality have obvious correlates in the real world, our modern theories of physics involve concepts which have no obvious correlates in the real world, such as complex analysis or the common-sense defying nature of quantum mechanics.

15.4 Formalism

The formalist holds a radical ontological perspective: mathematical objects have no real existence, they are merely symbols. The mathematician shuffles and recombines these meaningless symbols according to the dictates of systems of postulates. No meaning is ever to be ascribed to the symbols or the statements in which they appear, nor is any kind of interpretation of these symbols or statements ever to be done. Some formalists may be content to remain agnostic on whether meaning can ever be ascribed to mathematical symbols and statements, preferring simply to insist that no meaning is necessary, that the symbols and their interrelations suffice. Others, more radically still, insist that no meaning can ever be given to mathematical symbols and statements, and the systems in which they are used.

These symbols are manipulated within systems of postulates and rules, the *formal systems* which give formalists their moniker. The formalist is in theory able to study any formal system, but usually certain restrictions are placed on what counts as a postulate, and what is an allowable rule. One of the main criteria for a formal system is the concept of consistency which we have already encountered.

A formal system is consistent when its axioms and rules do not allow the deduction of a contradiction. In the early days of the formalist school, its leader, David Hilbert (1862–1943) believed that consistency implied existence (Bostock 2009, p. 168). It is hard to discern what is meant by "existence" here, given the formalist insistence on the meaningless of mathematics – indeed, Hilbert himself seems somewhat agnostic on this point (Reid 1996). However, I take it to mean that any statement derived from the axioms and rules has (at the very least) the same ontological existence as the axioms themselves. Thus while mathematics may be seen as one among many formal systems, and while each can be studied in the same way, if the axioms of mathematics are shown to have a more significant existence then so do all other mathematical objects.

It is impossible to talk about formalism without talking about Hilbert. The school probably would not exist without him. Hilbert was a towering figure of nineteenth and twentieth century mathematics, and his name is attached to several important concepts and theories (Reid 1996). He is also famous for listing 23 open problems in mathematics in the published form of his address to the International Congress of Mathematicians in Paris in 1900 (Hilbert 1902). Many of Hilbert's problems are still unanswered and remain the focus of research today. Hilbert in 1920 began his so-called program to show that mathematics is a consistent formal system. As we have seen, Gödel would show a decade later that this is impossible.

Hilbert was already on the formalist track when in 1899 he published his *Grundlagen der Geometrie* (*The Foundations of Geometry*) (Hilbert 1899), in which he formulated axioms of Euclidean geometry and showed their consistency. Hilbert is not the only mathematician to axiomatize Euclid's geometry. The idea is to eliminate geometrical intuition from geometry and to replace that intuition with definitions and axioms about objects bearing geometrical names. From those postulates can be derived all the theorems of Euclid's geometry, but crucially and as a direct result of the formulation of geometry as a formal system, those theorems need no longer be taken as referring to geometrical objects in the real world. In fact, they need not even be taken as referring to any kind of abstract geometry, neither to the platonist's ideal geometry, not to the Aristotelian's geometry generalised from the real world. Although the postulates use words such as "line" and "point", these objects are only defined by the formal system, and are not supposed to be taken as referring to our everyday notion of lines and points. The words could just as easily be replaced by "lavender" and "porpoise" – but again, without any sense that there is any correspondence with lavender or porpoises in the real world. This is the start of the formalist dream.

It was no great surprise when Hilbert showed in his *Foundations of Geometry* that Euclidean geometry was consistent. At the time, the only area of mathematics over which there was any doubt as to its consistency was Georg Cantor's (1845–1918) theory of infinite numbers (Bostock 2009, p.168). To introduce this theory, we first need to consider the notion of countability.

A finite set is *countable* if it can be placed in one-to-one correspondence with a subset of the natural numbers. This is a formal definition of what it means to count the objects in the set. Counting means assigning each object a unique number, which puts them in a one-to-one correspondence with a subset of the natural numbers, say the subset of numbers from 1 to 10 if there are ten objects in the set. If the set is infinite, we call it countable if it can be placed in one-to-one correspondence with all of the natural numbers (not just a subset). (Some authors reserve countable for finite sets and call countable infinite sets *enumerable*.)

The concept of countability puts infinity within our grasp. If the elements in an infinite set can be paired with the counting numbers, then an incremental counting-type algorithmic process can be set up to "access" everything in the set. For every element in the set there is a unique positive whole number, and for every positive whole number there is a unique object. However, this immediately leads to apparent paradoxes. For example, the even natural numbers can be paired in an obvious way with the natural numbers, and are thus countable. This means that the size of the set of even natural numbers is the same as the size of the set of all natural numbers, despite the fact that the latter contains the former!

Cantor asked whether the set of all numbers is countable. This set of real numbers contains not just the natural numbers, but all integers, all rational numbers, and all irrational numbers. He assumed first that the reals are countable, in which case, by definition they can be listed alongside the natural numbers. The next step was Cantor's stroke of genius. He considered a real number whose decimal expansion differs from the first real number on the list in the first decimal place, from the second real number in the second decimal place, and so on for every decimal place. This number is therefore different from every number on the list, and so it is not on the list. Yet it is a real number, and so if the assumption of the countability of the reals were correct it is on the list. This contradiction implies that the assumption of countability was wrong, and Cantor concluded that the reals are uncountable. Stunningly, this means that there is a "bigger size" of infinity than the size of the set of natural numbers. Moreover, Cantor showed that there is an infinite succession of sizes of infinities, each bigger than the last, and he constructed a beautiful theory of these infinite numbers. Within this theory, his famous continuum hypothesis is that the second smallest size of infinity is the size of the set of real numbers (Bagaria 2008).

Hilbert so loved Cantor's theory that he desired that "[n]o one shall drive us out of the paradise which Cantor has created" (Hilbert 1926, p. 170), and so he was desperate to prove its consistency. He never did so, and Gödel incompleteness theorems showed its impossibility before Hilbert had even finished shoring up the foundations of arithmetic. As Hilbert waded through the mud he found in the formalist foundations, he repeatedly encountered the notion of infinity. Although he hoped to construct an edifice up to Cantor's theory, Hilbert did not want infinity in the formalist foundations on which he built. Hilbert could not prove the consistency of arithmetic based on a finitary formal system. This insistence that as a finite human in an apparently finite world we should use only "finitary" definitions and methods will recur in our final school of mathematical philosophy, intuitionism, to which Hilbert ironically was bitterly opposed.

The death blow for Hilbert and the formalist's dream came with Gödel's incompleteness theorems, as described in the Logicism section above. These theorems not only destroyed the logicist dream of a mathematics founded on (and in some sense no more than) logic, but simultaneously destroyed Hilbert's formalism. This is because the theorems showed that any formal system sophisticated enough to contain simple arithmetic would necessarily contain unprovable true statements, and whose consistency required an external system. There was no way out, and formalism was dead.

Consequently, it is unlikely that anyone would call themselves a formalist today (Bostock 2009, p. 195). The idea which died is that formal systems are primary in the sense that they are the object of study, and that any application of them to an area of mathematics is essentially meaningless. But formalism evolved and survived in the same way that dinosaurs both died out and are alive in the birds we see around us. One surviving form is *structuralism*. The idea behind it, as advanced by Dedekind (1888) and Benacerraf (1965) is that the common structures of particular areas of mathematics are the object of study; they are primary. Like the formalist, the structuralist believes that applications of the structures are secondary, and that it is the structures themselves which must be studied. For example, the natural numbers can be taken to be an example of a *progression*: a non-empty set of objects each of which has a successor, as formalised in Peano's axioms (Gowers 2008, pp. 258–9). Because natural numbers are an example of a progression, they are less interesting to the structuralist than the progression structure they model.

The idea of structure being fundamental seems to be attractive to some physicists, even if they do not necessarily acknowledge structuralism. Writing popular accounts of the power of mathematics in the physical sciences, people like John Barrow, David Deutsch, and Ian Stewart argue for the primacy of pattern or structure. For example, Deutsch (b. 1953), a mathematical physicist, argues that the human brain both embodies the mathematical relationships and causal structure of physical objects such as quasars, and that this embodiment becomes more accurate over time. This happens because our study of these objects aligns the structure of our brains with the structure of the objects themselves, with mathematics as the encoding language of structure (Deutsch 2011). What is the ontology of such structures? The question is somewhat avoided by structuralists, but in essence they must claim either a platonic existence for them, or one of the other positions detailed here. Thus any claims of the structuralist are subject to some of the same ontological and epistemological objections as the other schools herein.

Finally here, we consider not a variant of logicism but a subschool which has in common with logicism the denial of any meaning in the objects of mathematics. In the Logicism section I said that logicism could be considered to be one form of nominalism. Another is given in Field (1980); see also Bostock (1979). By this account, mathematics is a "fairytale world which has no genuine reality" (Bostock 2009, p. 262). In this fairytale world, numbers (and other mathematical concepts) are powerful names for a collection of underlying objects and structures. These names allow us to use, say, arithmetic rather than logic or set theory in our deductions. This use of arithmetic as a set of names and rules is conservative in the sense that we cannot prove anything in arithmetic that could not be proved by stripping away the arithmetical names and working with a more fundamental structure (such as logic). Thus the names are useful but not required, and no meaning is given to them. Moreover, even if it is a useful fiction to treat them as real, the things to which the names seem to point have no independent existence; they may be abstractions of some kind, but they are not real in the sense of having an independent platonic existence.

Of course, we sometimes choose names which correspond to things in the real world. We know about numbers when we count shirt buttons, which is a kind of instrumentalist view of the existence of numbers. Thus arithmetic can be taken to be about the countable things we encounter in the world, whose ontological status is either left vague or has a minimalist instrumentalist view. Any correctly derived arithmetical statements are true both of numbers as fictions and of real-world numbers. Arithmetical deductions which go beyond what can be encountered in the world are true, but only in some fictional sense.

15.4.1 Formalism Under the Lens

If mathematics is a game, why should it tell us anything about the world? To the pure formalist, mathematical objects have no "real" existence, and to do mathematics is simply to explore a formal system or systems. But no particular formal system should be privileged over any other – some may be more interesting than others, for sure, but none of them is taken to have any special ontological status. Why, then, does mathematics help to explain the world?

The only way out of this conundrum seems to be to take Hilbert's less hard line view in which mathematical objects have a special ontological status, and that the formal system or systems at the foundations of mathematics are therefore more special than others. Although this does fix one problem, it creates another: what does it mean for mathematical objects to have special ontological status? What is that ontological status? The options are presumably those held by one of the other schools of the philosophy of mathematics and therefore subject to the same criticisms under the TEAM lens (amongst others).

Putting those criticisms to one side, and playing devil's advocate, I could point out that some games do teach us about the world. For example, in 1970 Martin Gardner introduced the world to John Conway's "Game of Life" (Gardner 1970). Since that time, this simple game has become a field of study both in its own right and as a model for processes in biology, economics, physics, and computer science, as revealed by a quick search of Google Scholar. But although some features of the Game of Life are emergent and therefore could not be predicted, the simple rules of the game were chosen in order to mimic those of simple real-world systems. If we wish to claim that this is comparable to the far more complex game of mathematics mimicking the real world, then we would have to assert that the rules of mathematics were chosen in order to mimic those in the real world. Once again, we are forced to abandon the ontology of pure formalism, at least.

Other problems are visible under the TEAM lens. While it is easy to accept that, say, the rules of arithmetic have been chosen because they mimic real-world counting, it is harder to explain the important role that, say, complex analysis or Hilbert spaces play in our best theories of the universe. In geometry, it is "natural" to consider flat Euclidean geometry, and so the non-Euclidean geometry which arose in the last half of the nineteenth century was viewed initially with distaste and seen as something of a pointless game. Yet Einstein has taught us that our universe is non-Euclidean. How, then, are we to know which of our formal systems have special ontological status? Only those which are later shown to correspond to some aspect of the real world? But this is surely a poor ontological status which seems predicated both on time and on our ignorance. What if when our theories change we need an area of mathematics and so it becomes "real" – but then later find we no longer need it, at which it returns to being unreal? It seems that this is indistinguishable from the Quine/Putnam indispensability argument, and so arguments against that position are also valid here.

The structuralist might choose to argue that the structures of mathematics are chosen because they mimic some aspect of the real world. But does this not give a privileged ontological status to the real world, and the structures within it? What is their ontological status? At this point, the structuralist has passed the buck. The fictionalist seems Quinean when examined under the TEAM lens, for the only way to distinguish between the real and the fictional is to expose a truth to the crucible of the real world. The other option is to admit a platonic existence at the heart of your fictionalist worldview, as Field himself did when he sought to remove it in Field (1992).

15.5 Intuitionism

Intuitionism was the first and remains the largest "constructivist" schools of mathematics (Chabert 2008). Most of what I say in this section can be taken to be true of the other constructivist schools, which include (i) finitism, (ii) the Russian recursive mathematics of Shanin and Markov, (iii) Bishop's constructive analysis, and (iv) constructive set theory. It is always a pleasure to note that intuitionists claim constructivism as a subschool and constructivists claim intuitionism likewise, but I will mostly use the word "intuitionism" as an umbrella term in this section, and look forward to the deluge it provokes from constructivists.

The defining characteristic of intuitionism is that existence requires *construction*. The perspective of intuitionists, for example in Bridges (1999), is that believing that existence requires construction forces upon the mathematician the requirement to use a different logic. This logic is the intuitionistic logic which has at its heart a rejection of the *law of excluded middle* and a rejection of the *axiom of choice*. I will explain each of these points below. It is worth noting that, as in every area we discuss herein, the argument for intuitionism has at least two sides. For every Bridges arguing that construction implies intuitionistic logic, there is a Dummett arguing that this is untrue (see his 1977, and Bostock 2009, pp. 215 ff). But we continue, since all schools presented herein have adherents arguing their corner and antagonists arguing them into one.

All mathematicians distinguish between an *existence proof* and a *construction proof*. The former merely establishes whether a statement claiming the existence of some mathematical object is true or not. A construction proof, by contrast, gives steps which construct the properties of the object in question, and so gives in addition to a proof of truth some insight as to why. In the case in which the statement is not true, an actual counterexample is constructed. I now try to put a little flesh on these bones.

A common question in mathematics concerns the *existence* of a mathematical object. This is not the metaphysical notion of existence central to this chapter. When a mathematician asks whether a mathematical object exists, she is not worried about whether scientific methods can show it to be a real, physical thing in the world, nor is she usually bothered with the ontological status of that object. Instead, she is interested in whether the object exists in a mathematical sense.

For the majority of mathematicians, existence proofs suffice, even if construction proofs provide more information. Not so the intuitionists, who believe that existence is shown only when the object has been constructed. Construction here has a specific meaning, and once again this has nothing to do with building an object in the real world. Rather it has to do with providing a proof of a statement from which, at least in principle, an algorithm could be extracted which would compute the object in question, and any of its properties. Only when a constructive proof has been found is the object said to exist. For the intuitionist, "existence" means "construction".

For a real-world analogy, we can turn the weather. When I look up the weather records for my home town of Christchurch, New Zealand, I can see that in 2016 the maximum recorded temperature was 34 °C on 27th February, and the minimum recorded temperature was −5 °C on 11th August (WolframAlpha 2017). This means that with confidence I can claim that there was a moment between 27th February and 11th August when the temperature was precisely 0 °C. My assertion rests on two points: that for this time range the temperature starts at a positive value (34) and ends on a negative one (−5), and that temperature cannot instantaneously change. From these two observations, I know that there must exist a time, however short, when the thermometer read 0°, since it is impossible to go smoothly from 34 down to −5 without passing through 0. Of course, there were probably many such times, but the mathematician's interest in uniqueness is not our concern here, only existence. In our temperature analogy we have demonstrated the existence of a time at which the temperature was 0° in a way which would satisfy most mathematicians.

But the intuitionist weather-watcher would not be satisfied. She wants something more: she wants an actual moment at which the thermometer read 0. In our analogy, this means going through the weather station data until such a time is found. That is a "constructive" proof of the existence of a 0° time.

Our analogy has flaws, as all do. It could give the impression that intuitionistic mathematics is about data-sifting; this is untrue. Intuitionistic mathematics is mathematics, but with tighter constraints on what can be used in the logical arguments called proofs which establish truths. Indeed, Bridges argues in his (1999) that the intuitionistic mathematician is free to work with whatever mathematical objects she so desires. Another flaw is that although the analogy illustrates the difference between existence and construction, it does not have an analogy for intuitionistic logic.

I said above that intuitionistic logic has two features which distinguish it from classical logic, and both features involve a rejection. The first of these is a rejection of the law of excluded middle (LEM). For most mathematicians, something either is, or is not. A number is either rational, or irrational. It cannot be both; it is either. But the intuitionist will not say it is one or the other until it has been constructed. A classical mathematician may present the following argument. Object X can either have property P or not. If we assume for the sake of argument that it has property P, we can investigate the consequences of our assumption. Suppose that when we do that, we uncover a contradiction, an absurdity. Then (assuming we have done everything correctly) the only problem was our assumption that object X had property P. Thus it cannot have property P. This is the commonly used proof technique called *proof by contradiction*, and we saw an example of it above when we presented Cantor's diagonal argument.

Such a proof would not be considered valid in intuitionistic logic. The reason that it is invalid is that X has not been shown to have a particular property or not, but simply that by assuming the converse a contradiction has been found. At issue is not the assumption of whether or not X has property P. If the objects of study of which X is an example are such that they must either have property P or not, then it would be absurd to argue that they have neither, or, somehow, a superposition of both. The intuitionist does not argue this. Rather, the idea is of a radical redefinition of truth. To the intuitionist mathematician, a statement is true only when a constructive proof without recourse to the LEM has been given. A statement is false precisely when a counterexample has been given. Since truth now has this specific meaning, a statement is neither true nor false until such a constructive proof is furnished.

Although the truth of a statement becomes time-dependent, it is not the same time-dependency as in the Quine/Putnam indispensability argument. There, something is real only for as long as it is necessary for a successful theory of the real world; the status of mathematical objects are forever conditional. For the intuitionist, on the other hand, truth is defined to mean proof by construction. Thus an object is neither real nor not real until it is constructed, at which point it becomes and forever remains real (or becomes and remains forever not real when a counterexample is constructed).

To object that surely, say, the statement "the trillionth decimal digit of pi is zero" has been true or false since the dawn of time is to confuse the platonic notion of truth with the intuitionist one. The point is that although the trillionth decimal digit of pi has a value entirely independent of the free will of humans, that it is indeed dictated by something deeper than whatever human whimsy may want it to be, until its value is actually calculated the statement has no (intuitionist) truth value associated with it.

Although for the intuitionist mathematical objects have properties which can be rigorously defined or derived, they nevertheless have the ontological status of being purely mental objects. In this way, intuitionism is a form of the conceptualism which harks back to Aristotle (Bostock 2009, p. 44). By making mathematics mental, intuitionists avoid problems of epistemic access, since naturally we can access the objects of our own minds. There is an ontological issue associated with insisting that mathematical objects are purely mental. We must ask why they have properties independent of the individual mind which explores or creates them. Thus an obvious objection to this conceptualism is that these objects must rely on some deeper structure that at the very least is shared by other human minds. But that suggests that there is something more fundamental than the mathematics itself – and the intuitionist certainly cannot claim that something like logic, language, "structure", or a platonic realm of ideas is more fundamental.

Indeed, the founder of intuitionism, Luitzen Egbertus Jan Brouwer (1881–1966), echoing Kant and in agreement with the mathematicians Felix Klein (1849–1925) and Henri Poincaré (1854–1912), believed that the basic axioms of mathematics are intuited. In this he meant that they were known to our minds, but not that our intuition reveals anything which exists outside of the mind. He went further, claiming a stark independence of mathematics from both language and logic. If there was any relation there, it was that logic and language rested on mathematics, rather than the other way around. This was revolutionary, and put Brouwer directly in harm's way. His point of view, given in Brouwer (1907), was directly contrary to both logicism and to Hilbert's program of formalism as it developed in the 1920s. Hilbert's program was popular and Hilbert himself was powerful. Brouwer apparently did nothing other than disagree with Hilbert, yet Hilbert had Brouwer removed from the editorial board of the prestigious journal *Mathematische Annalen*, and sought to discredit him at every turn (van Dalen 2008, p. 800).

Having discussed construction, the law of excluded middle, and the redefinition of "truth", I now consider the other idea which intuitionists reject, the axiom of choice. Stated in words, it says that we can always select an element from each of a family of sets. This is uncontroversial for a finite family of finite sets, but becomes controversial otherwise, because an infinite number of choices can be made. For most mathematicians this is not a problem; to put it crudely, the fact that there are an infinite number of choices which can be made guarantees that one can be made. For an intuitionist, the mere fact that a choice can be made is not enough: the choice must be specified in order to count as a construction. Yet when a classical mathematician invokes the axiom of choice it is usually for very general cases in which specificity is impossible (or for which there is no perceived benefit in specifying the object).

To make this point clearer, suppose we have a countable number of sets, each of which is countable. Now suppose that we wish to form a superset containing all of the elements in all of the sets and to ask whether that new set is itself countable. This is easy for a classical mathematician. For each set, she first lists the elements, which we know can be done because every set in the family is countable. Then she runs the lists together in turn, and hey presto, the superset is listed out, and therefore countable. There is no "problem" with this proof for most mathematicians, but the intuitionist asks: how did she choose the ordering for each set, and for the family of sets? There is an infinite number of choices in each case, so the choice function is unspecified. The proof uses (in quite a disguised way) the axiom of choice. Whenever the axiom of choice is used, the proof is non-constructive.

Uncountable infinity is the heart of the rejections which define intuitionism. To be clear, if the axiom of choice is invoked either in a finite context or in one which is countable, then a choice function can be defined and the intuitionist is happy. The problem is in the uncountable case. Likewise, the law of excluded middle is connected with the notion of infinity; recently Bridges has argued that the continuum hypothesis implies LEM (Bridges 2016). Only potential infinities, namely those accessible through enumeration or by an algorithmic process are acceptable to the intuitionist.

But to return to our starting point that intuitionistic mathematics is mathematics done with intuitionistic logic, we note that it is sometimes possible to construct intuitionistic theories of mathematical objects which in classical mathematics require uncountable infinities. For example, Brouwer introduced the notion of *choice sequences* to create a theory of the continuum (that is, the real number line) which was apparently out of reach to intuitionists (Brouwer 1981). Brouwer never defined choice sequences carefully enough to avoid problems, but Bishop's constructive mathematics (Bishop 1967; Bishop and Bridges 1985) does contain an apparently sound theory of the reals which avoids uncountable infinities. This is an example of how something which in classical mathematics requires uncountable infinities can be given an intuitionistic theory which only uses countable processes.

15.5.1 Intuitionism Under the Lens

Intuitionism has never been popular with mathematicians, and few applied mathematicians insist on a constructive approach to their work. But is it possible to argue that intuitionistic logic's insistence on countability, apparently so true of our physical universe, is the reason for the success of mathematics in modelling the world?

Does the universe only appear to rely on countability, and so are there unavoidable instances of uncountable infinities, both in our theories of the world and in the universe itself? Since infinity is implied in our best theories of the very big and the very small, it is no wonder that when intuitionism is under the TEAM lens what comes into focus is quantum mechanics (QM) and general relativity (GR).

It may seem that on a large scale our universe is a finite (though huge) thing containing a finite number (though huge) of discrete things. But we do not know that to be a fact. At the other end of the scale, quantum mechanics suggests that the structure of spacetime is granular at the very smallest of time and length scales. However, that prediction has not yet been verified. It may be the result of our most successful and accurate theory of science, but we do not know it to be true. Could the universe be infinite in extent? Might spacetime be a continuum?

Continuous spacetime does not necessarily cause a fatal problem for intuitionism since Bishop's constructive mathematics has an intuitionistic theory of continua. A potentially deeper argument, given by Hellman (1993, 1997), that intuitionism must be wrong because QM requires a theory of unbounded operators which seems to defy intuitionism, has been refuted by Bridges (1995, 1999) on the grounds that such a theory is possible with an intuitionistic approach. These Hellman-type arguments have also been refuted in the context of GR: see Billinge's (2000) response to Hellman (1998). However, what of mathematical objects essential to our theories of the universe but for which no intuitionistic theory has yet been found? Does their necessity destroy intuitionism? Billinge (2000) says no, when she powerfully argues that just because we have not yet found a constructive proof of something does not mean that it cannot ever be found.

The intuitionist's belief that the objects of mathematics are purely mental avoids the plantonist's problem of epistemological access. But the TEAM lens shows us a deeper ontological problem: if the objects of mathematics are purely mental, why should they ever have any correspondence with the real world? Why should mathematics ever be useful?

15.6 Discussion and Conclusion

The platonist see mathematics as eternal and changeless, existing outside of spacetime. But how do we access such an ideal realm? How does this ideal realm cast the physical "shadows" in our world which mathematics explains? The logicist reduces mathematics to logic in disguise. But why should logic explain the world? Does logic have a platonic existence? The formalist is the ultimate reductionist, claiming that mathematics is naught but a game, a meaningless shuffling of semantically empty symbols. But why should the game of mathematics be able to explain the world? Why that game and not another? Finally, conceptualism returns with the intuitionists, who believe that only construction means truth. But while intuitionistic logic and an insistence on construction are not at odds with our best theories of the universe (our best applied mathematics), the intuitionist believes that all mathematical objects are mental constructions. Why should such mental constructions explain the world?

This last point is subtle, and slippery. Of course we expect that any idea which explains the world will be in our minds; that is where we experience ideas. The issue concerns how an idea can come to mimic and explain the outside world. This is a debate with a long history. In the middle stand two figures directly opposed to one another. Kant believed that our minds are primary, and thus that our applied mathematics works not because our minds come to mirror reality, but because reality must conform to the mind in order to be perceptible and comprehensible to us. By contrast, Hume was an empiricist, naturalist, and sceptic, who believed that our concepts came from experience of an independently-existing natural world, without imposing an ontology on that world. At the far end of the chronology is Plato, who believed that our mental realm can access a world of forms which projects the physical world. This raises more questions than it answers. Nevertheless, it seems to be the perspective of many theoretical physicists today, perhaps without considering its epistemic problems. The modern structuralist, by contrast, might argue that structure is fundamental, and so our mental world can be structured to mimic the external world. We have already observed in the Formalism section that such a perspective seems to pass the buck on the ontological status of structure. This structuralist approach seems attractive to physicists such as Deutsch, whom we encountered in our discussion of structuralism above, and who otherwise seems to be a realist in his worldview.

When physicists make pronouncements about mathematics they are usually motivated not by concern about what mathematics is or what its foundations are, but only by what sort of mathematics should or can be taken to be the foundation of *physics*. For example, the Nobel laureate in Physics Gerard 't Hooft (b. 1946) wants only finiteness in his theories of quantum mechanics (Musser 2013). It is not completely clear what he means by this, but it seems to be a kind of countability, since he mentions basing theory on the integers or finite sets (though the former are countably infinite). 't Hooft seems to be directly motivated by the granular discreteness of spacetime at the Planck scale predicted by QM. It would be wrong to suggest that he is rejecting classical mathematics and a platonic ontology in favour of, say, neo-Fregean logic, intuitionism, or a Hilbertean finitism, when he is only restricting himself to finite methods and objects for the mathematics of QM. He says nothing about the ontological status of other mathematics. Likewise, the physicist Lee Smolin (b. 1955) claims in his (2000) that topos theory is "required" for cosmology, and topos theory itself requires constructive set theory, a form of intuitionism. Once again, this is not a statement of ontological intent for the whole of mathematics, just for what mathematics can be applied to physics. In both cases, the question of epistemology is left open, as is the ontological status of the objects being studied. However, when applied mathematicians such as these physicists do not explicitly acknowledge their adopted philosophical position they may overlook some difficulties, especially when their position combines ideas from different philosophical schools. This seems particularly acute when the physical objects are considered real but the mathematics used to model them is considered to be entirely mental. Note that neither of these physicists claim that the mathematics which helps them is the only mathematics which is true; there is no evidence that they adhere to the Quine/Putnam indispensability argument.

The TEAM lens reveals other issues which we have not discussed above. For example, it is one thing to say that applied mathematics is possible, but we could also ask why *we* are able to do it. Why is the mathematics which seems to do so well at explaining the world accessible to our minds? We can imagine a universe in which rational, intelligent beings existed who were incapable of developing sufficiently advanced mathematics to understand that universe even though it were capable of being comprehended mathematically.

Also, what about beauty, or the role of aesthetics? This is a commonly-observed inspiration for both mathematicians and those who apply mathematics. The mathematician GH Hardy (1877–1947) said of mathematics "Beauty is the first test: there is no permanent place in the world for ugly mathematics" (Hardy 1940). Einstein is quoted in Farmelo (2002, p. xii) as saying "the only physical theories that we are willing to accept are the beautiful ones", while physicist colleague Hermann Weyl (1885–1955) said "My work has always tried to unite the true with the beautiful and when I had to choose one or the other, I usually chose the beautiful" (quoted in Stewart 2007, p. 278). But why should an aesthetic of mathematics help create new mathematics, and new applied mathematics? Are we simply wrong about beauty, especially when we use it as a selection criterion? Could ugly theories better explain the world, and even be more fecund mathematically? Perhaps we have been misled by mathematics because we are in the early days of science; are we even wrong about the power of mathematics to explain the world?

Another question we have overlooked as we peered down the TEAM lens concerns the meaning of deductive steps in an applied mathematical argument. More specifically, if I have a mathematical model of a physical process which I then analyse mathematically to arrive at a physically-verifiable result, need each of the intermediate logical steps also have physical meaning? This question has been considered by Nancy Cartwright, among others; see for example her (1984), in which she says "derivations do not provide maps of causal processes. A derivation may start with the basic equations that govern a phenomenon. It may be highly accurate, and extremely realistic. Yet it may not pass through the causes." This question, and the others raised above, deserve more attention.

15.6.1 Conclusion

We do not know the ontological or epistemological status of mathematical objects. We do not know why mathematics can be applied to the world around us. Though it was too much to hope that the TEAM lens would itself provide an *experimentum crucis* which would eliminate all but one philosophy of mathematics and therefore resolve a millennia-old debate, the TEAM lens has brought into focus the questions which must be clearly addressed when defending a particular philosophical standpoint.

I have attempted to summarise the systems of ideas which constitute these standpoints in four broad schools. Despite presenting them as separate, they are united in their concern with the ontological and epistemological questions, and in their focus on key ideas: what is number, what is a set, what is a proof, what is infinity, and more besides. As we saw, one person who has united them in a stunningly destructive way was Kurt Gödel.

Another figure may pull some of these strands together. Max Tegmark introduced the radically realist Mathematical Universe Hypothesis, which earns him a capital P on Platonist if anyone ever deserves it. The MUH is a tentative, new, and controversial idea, and my positive view of it may not be representative. But I do think it takes seriously these philosophical questions and that it represents an important attempt to think clearly about them, and possibly to unite some of the schools. For example, structuralists and fictionalists might observe that in the MUH all mathematical objects exist and all things which exist are mathematical, and so there is no need for any particular structure or fiction to be privileged. Even the debate between Kantian innateness and Humean empiricism may be erased: if the mind is a self-aware substructure of the mathematical universe, then there is no epistemic gap between the mind and the world. For platonists, the problem of epistemological access may be solved because the MUH is more than plenitudinous platonism, which addressed the epistemic concerns. However, it also potentially fixes the platonist ontological issues and leaves us with an inspiring thought: if everything is mathematics and mathematics is everything, then there is only one realm. We are self-aware substructures of mathematics.

References

Bagaria, J. (2008). Set theory. In T. Gowers (Ed.), *The Princeton companion to mathematics*. Princeton: Princeton University Press.

Balaguer, M. (1998). *Platonism and anti-Platonism in mathematics*. Oxford: Oxford University Press.

Balauger, M. (2016). Platonism in metaphysics. In E. N. Zalta (Ed.), *The Stanford encyclopedia of philosophy* (Spring 2016 Edition), viewed 12 Feb 2017. https://plato.stanford.edu/archives/spr2016/entries/platonism/

Barrow, J. D. (2000). *Between inner space and outer space*. Oxford: Oxford University Press.

Barrow-Green, J., & Siegmund-Schultze, R. (2015). The history of applied mathematics. In N. J. Higham (Ed.), *The Princeton companion to applied mathematics*. Princeton: Princeton University Press.

Benacerraf, P. (1965). What numbers could not be, *Philosophical Review, 74*, 47–73. Reprinted in Benacerraf, P., & Putnam, H. (Eds.). (1983). *Philosophy of mathematics: Selected readings*, second edition. Cambridge: Cambridge University Press.

Benacerraf, P. (1973). Mathematical truth. In P. Benacerraf & H. Putnam (Eds.), (1983) *Philosophy of mathematics: Selected readings* (2nd ed.). Cambridge: Cambridge University Press.

Benacerraf, P., & Putnam, H. (Eds.). (1983). *Philosophy of mathematics: Selected readings* (2nd ed.). Cambridge: Cambridge University Press.

Billinge, H. (2000). Applied constructive mathematics: On Hellman's 'Mathematical constructivism in spacetime'. *The British Journal for the Philosophy of Science, 51*(2), 299–318.

Bishop, E. (1967). *Foundations of constructive analysis*. New York: McGraw-Hill.

Bishop, E., & Bridges, D. (1985). *Constructive analysis*. Heidelberg: Springer Verlag.

Bostock, D. (1979). *Logic and arithmetic* (Vol. 2). Oxford: Clarendon Press.

Bostock, D. (1980). A study of type-neutrality. *Journal of Philosophical Logic, 9*, 211–296, 363–414.

Bostock, D. (2009). *Philosophy of mathematics*. Chichester: Wiley-Blackwell.

Bridges, D. S. (1995). Constructive mathematics and unbounded operators: A reply to Hellman. *Journal of Philosophical Logic, 24*(5), 549–561.

Bridges, D. S. (1999). Can constructive mathematics be applied in physics? *Journal of Philosophical Logic, 28*(5), 439–453.

Bridges, D. S. (2016). The continuum hypothesis implies excluded middle. In D. Probst & P. Schuster (Eds.), *Concepts of proof in mathematics, philosophy, and computer science*. Berlin: Walter De Gruyter.

Brouwer, L. E. J. (1907). *On the foundations of mathematics*. PhD thesis, Universiteit van Amsterdam. English translation in Brouwer, L. E. J. (1975). *Collected Works 1. Philosophy and foundations of mathematics* (Heyting, A., Ed.). North-Holland.

Brouwer, L. E. J. (1981). *Brouwer's Cambridge lectures on intuitionism* (D. van Dalen, Ed.). Cambridge: Cambridge University Press.

Bueno, O., & Linnebo, O. (Eds.). (2009). *New waves in philosophy of mathematics*. Basingstoke: Palgrave Macmillan.

Cartwright, N. (1984). Causation in physics: Causal processes and mathematical derivations. *Proceedings of the Biennial Meeting of the Philosophy of Science Association*, Vol. 1984. Volume Two: Symposia and Invited Papers, pp. 391–-404.

Chabert, J.-L. (2008). Algorithms. In T. Gowers (Ed.), *The Princeton companion to mathematics*. Princeton: Princeton University Press.

Cooper, J. M. (Ed.). (1997). *Plato: Complete works*. Indianapolis: Hackett Publishing Company.

Courant, R., & Robbins, H. (1941). *What is mathematics?* Oxford: Oxford University Press.

Davis, P. J., & Hersh, R. (1981). *The mathematical experience*. Boston: Birkhäuser.

Dedekind, R. (1888). *The nature and meaning of numbers*, translated from the German by WW Beman in Dedekind, R. (1963). *Essays on the theory of numbers*. New York: Dover.

Deutsch, D. (2011). *The beginning of infinity: Explanations that transform the world*. London: Allen Lane.

Dummett, D. (1977). *Elements of intuitionism*. Oxford: Clarendon Press.

Extance, A. (2016). How DNA could store all the world's data. *Nature, 537*, 22–24.

Farmelo, G. (2002). *It must be beautiful: Great equations of modern science*. London: Granta Books.

Ferreirós, J. (2008). The crisis in the foundations of mathematics. In T. Gowers (Ed.), *The Princeton companion to mathematics*. Princeton: Princeton University Press.

Field, H. (1980). *Science without numbers*. Princeton: Princeton University Press.

Field, H. (1992). A nominalistic proof of the conservativeness of set theory. *Journal of Philosophical Logic, 21*, 111–123.

Gardner, M. (1970). Mathematical games – The fantastic combinations of John Conway's new solitaire game "life". *Scientific American, 223*, 120–123.
Gödel, K. (1931). Über formal unentscheidbare Sätze der Principia Mathematica und verwandter Systeme I. *Monatshefte für Mathematik Physik, 38*, 173–198. English translation in Gödel, K. (1986). *Collected works I. Publications 1929–1936* (Feferman, S., et al., Eds.). Oxford: Oxford University Press.
Gowers, T. (2008). *The Princeton companion to mathematics*. Princeton: Princeton University Press.
Hardy, G. H. (1940). *A mathematician's apology*. Cambridge: Cambridge University Press. Reprinted in 1992.
Hellman, G. (1993). Constructive mathematics and quantum mechanics: Unbounded operators and the spectral theorem. *Journal of Philosophical Logic, 22*(3), 221–248.
Hellman, G. (1997). Quantum mechanical unbounded operators and constructive mathematics – a rejoinder to Bridges. *Journal of Philosophical Logic, 26*(2), 121–127.
Hellman, G. (1998). Mathematical constructivism in spacetime. *The British Journal for the Philosophy of Science, 49*(3), 425–450.
Higham, N. J. (Ed.). (2015a). *The Princeton companion to applied mathematics*. Princeton: Princeton University Press.
Higham, N. J. (2015b). What is applied mathematics. In N. J. Higham (Ed.), *The Princeton companion to applied mathematics*. Princeton: Princeton University Press.
Hilbert, D. (1899). *The foundations of geometry*. Translated from German by EJ Townsend, 1950. La Salle: Open Court Publishing Company.
Hilbert, D. (1902). Mathematical problems. *Bulletin of the American Mathematical Society, 8*, 437–479, translated from the German by Newson, MW.
Hilbert, D. (1926). On the infinite. *Mathematische Annalen, 95*, 161–190. Translated from German by E Putnam and GJ Massey, in Benacerraf, P., & Putnam, H. (Eds.). (1983). *Philosophy of mathematics: Selected readings* (2nd ed.). Cambridge: Cambridge University Press.
Horsten, L. (2016). Philosophy of mathematics. In E. N. Zalta (Ed.), *The Stanford encyclopedia of philosophy (Winter 2016 edition)*, viewed 12 Feb 2017. https://plato.stanford.edu/archives/win2016/entries/philosophy-mathematics/
Krauss, L. (2012). *A universe from nothing: Why there is something rather than nothing*. New York: Free Press.
Linnebo, Ø. (2013). Platonism in the philosophy of mathematics. In E. N. Zalta (Ed.), *The Stanford encyclopedia of philosophy* (Winter 2013 Edition), viewed 26 Feb 2017. https://plato.stanford.edu/archives/win2013/entries/platonism-mathematics/
Linsky, B., & Zalta, E. N. (1995). Naturalized platonism versus platonized naturalism. *Journal of Philosophy, 92*(10), 525–555.
Linsky, B., & Zalta, E. N. (2006). What is neologicism? *The Bulletin of Symbolic Logic, 12*(1), 60–99.
Maddy, P. (1990). *Realism in mathematics*. Oxford: Clarendon Press.
Mancosu, P. (Ed.). (2008). *The philosophy of mathematical practice*. Oxford: Oxford University Press.
Mohr, H. (1977). *Lectures on structure and significance of science*. New York: Springer-Verlag. Reprinted 2013.
Musser, G. (2013). *Does some deeper level of physics underlie quantum mechanics an interview with Nobelist Gerard 't Hooft*, viewed 1 Feb 2017. https://blogs.scientificamerican.com/critical-opalescence/does-some-deeper-level-of-physics-underlie-quantum-mechanics-an-interview-with-nobelist-gerard-e28099t-hooft/
Parsons, C. (1979/1980). Mathematical intuition. *Proceedings of the Aristotelian Society, 80*, 145–168.
Paseau, A. (2007). Scientific platonism. In M. Leng, A. Paseau, & M. Potter (Eds.), *Mathematical knowledge*. Oxford: Oxford University Press.

Pincock, C. (2009). Towards a philosophy of applied mathematics. In O. Bueno & O. Linnebo (Eds.), *New waves in philosophy of mathematics* (pp. 173–194). Basingstoke: Palgrave Macmillan.

Putnam, H. (1971). *Philosophical logic*. New York: Harper & Row.

Quine, W. V. (1948). On what there is. *Review of Metaphysics, 2*, 21–38.

Reid, C. (1996). *Hilbert*. New York: Springer-Verlag.

Ruelle, D. (2007). *The mathematician's brain*. Princeton: Princeton University Press.

Russell, B., & Whitehead, A. N. (1910). *Principia mathematica* (Vol. 1). Cambridge: Cambridge University Press.

Smolin, L. (2000). *Three roads to quantum gravity*. London: Weidenfeld & Nicolson.

Smoryński, C. (1977). The incompleteness theorems. In J. Barwise (Ed.), *Handbook of mathematical logic* (pp. 821–866). Amsterdam: North-Holland.

Sokal, A. D. (1996). Transgressing the boundaries: Towards a transformative hermeneutics of quantum gravity. *Social Text, 46/47*, 217–252.

Sokal, A. D. (2008). *Beyond the hoax: Science, philosophy and culture*. Oxford: Oxford University Press.

Stewart, I. (2007). *Why beauty is truth*. New York: Basic books.

Tegmark, M. (2008). The mathematical universe. *Foundations of Physics, 38*(2), 101–150.

Tegmark, M. (2014). *Our mathematical universe*. New York: Vintage books.

van Dalen, D. (2008). Luitzen Egbertus Jan Brouwer. In T. Gowers (Ed.), *The Princeton companion to mathematics*. Princeton: Princeton University Press.

Weinberg, S. (1992). *Dreams of a final theory*. New York: Pantheon books.

Wigner, E. (1960). The unreasonable effectiveness of mathematics in the natural sciences. *Communications on Pure and Applied Mathematics, 13*, 1–14.

Wikipedia. (2017). *Mathematical universe hypothesis*, viewed 26 Feb 2017. https://en.wikipedia.org/wiki/Mathematical_universe_hypothesis

Wilson, P. L. (2014). The philosophy of applied mathematics. In S. Parc (Ed.), *50 visions of mathematics* (pp. 176–179). Oxford: Oxford University Press.

WolframAlpha. (2017). *2016 weather in Christchurch, New Zealand*, viewed 21 Feb 2017. http://www.wolframalpha.com/input/?i=2016+weather+in+christchurch,+new+zealand

The manufacturer's authorised representative in the EU is Springer Nature Customer Service Centre GmbH, Europaplatz 3, 69115 Heidelberg, Germany. If you have any concerns regarding our products, please contact ProductSafety@springernature.com

Printed and bound by CPI Group (UK) Ltd, Croydon, CR0 4YY
30/09/2026
02225659-0011